INTRODUCTION TO
APPLIED SOLID
STATE PHYSICS

TOPICS IN THE APPLICATIONS OF
SEMICONDUCTORS, SUPERCONDUCTORS, AND THE
NONLINEAR OPTICAL PROPERTIES OF SOLIDS

INTRODUCTION TO
APPLIED SOLID
STATE PHYSICS

TOPICS IN THE APPLICATIONS OF
SEMICONDUCTORS, SUPERCONDUCTORS, AND THE
NONLINEAR OPTICAL PROPERTIES OF SOLIDS

RICHARD DALVEN
Department of Physics
University of California
Berkeley, California

PLENUM PRESS · NEW YORK AND LONDON

Library of Congress Cataloging in Publication Data

Dalven, Richard.
  Introduction to applied solid state physics.

  Includes bibliographies and index.
  1. Solid state physics. 2. Semiconductors. I. Title.
QC176.D24   621.3'028   79-21902
ISBN 0-306-40385-4

© 1980 Plenum Press, New York
A Division of Plenum Publishing Corporation
227 West 17th Street, New York, N.Y. 10011

Printed in the United States of America

To my father, JOSEPH DALVEN
and to the memory of my mother
RUTH NEWTON DALVEN

Preface

The aim of this book is a discussion, at the introductory level, of some applications of solid state physics.

The book evolved from notes written for a course offered three times in the Department of Physics of the University of California at Berkeley. The objects of the course were (a) to broaden the knowledge of graduate students in physics, especially those in solid state physics; (b) to provide a useful course covering the physics of a variety of solid state devices for students in several areas of physics; (c) to indicate some areas of research in applied solid state physics.

To achieve these ends, this book is designed to be a *survey* of the *physics* of a number of solid state devices. As the italics indicate, the key words in this description are physics and survey. Physics is a key word because the book stresses the basic qualitative physics of the applications, in enough depth to explain the essentials of how a device works but not deeply enough to allow the reader to design one. The question emphasized is how the solid state physics of the application results in the basic useful property of the device. An example is how the physics of the tunnel diode results in a negative dynamic resistance. Specific circuit applications of devices are mentioned, but not emphasized, since expositions are available in the electrical engineering textbooks given as references. To summarize, the aim of the book is the physics underlying the applications, rather than the applications themselves.

The second key word is survey. The book is designed to be broad rather than deep. Although the survey approach is not to everyone's taste, it has proved popular with the approximately 120 Berkeley graduate students (mostly in physics) who took or audited the course in 1973, 1974, and 1977. They seemed to want to learn something, but not everything, about the applications of the solid state physics they already knew. As a survey, the selection of topics is a compromise between recognition of the overwhelming technological importance of semiconductor devices and a desire to have some breadth of coverage. To this end, about 70% of the material covers applications of semiconductors, and the remainder is divided about evenly between nonlinear optical devices and superconductive materials and devices. Since the physics of the applications is the central interest of the book, no special effort was made to select the latest devices or to indicate the present "state of the art."

The book is a textbook ("A textbook explains, a treatise expounds"— J. M. Ziman) in that its aim is frankly tutorial. The book is essentially a collection of material from a number of sources, ranging from introductory textbooks to research journals, organized and presented with the intent of emphasizing the basic physics involved. There is no original work included. More advanced treatments and discussions of fine points are left to the literature. However, the reader is provided with references where fuller and/or more advanced treatments may be found. Further, a special effort has been made to give very specific references, telling where values of parameters, etc., were obtained. A selection of problems can be found at the end of each chapter. These are derivations, illustrative calculations, or invitations to explore the physics of some application. It is believed that these points harmonize with the attempt to provide a broad selection of the applications of solid state physics, while telling the reader where further information may be found.

The order of the first seven chapters is more or less linear. After a first chapter that is partly review and partly new material that will be useful, Chapter 2 treats the semiconductor $p-n$ junction in some detail. The third chapter exploits this treatment in a discussion of several device applications. Chapter 4 treats the physics of metal–semiconductor and metal–insulator–semiconductor junctions, and the results are used in Chapter 5 to explore a few applications. In Chapter 6, a potpourri of "other" devices is discussed; they were chosen principally on the basis of my own interests. The seventh chapter treats a number of detectors and generators (principally semiconductors) of electromagnetic radiation. Chapter 8 is mostly concerned with the physics of Josephson junction devices, but concludes with a short discussion of the transition temperature in superconductors. Finally, Chapter 9 covers the interaction of electromagnetic waves in nonlinear solids, and concludes with a few applications.

In teaching a course on these topics, I have found that this book contains too much material for a one-quarter course. One semester would seem about right, particularly if appropriate review material were included. The notation used is standard, except perhaps that I have used $\mathscr{E}$ for the electric field vector to avoid confusion with energy E, particularly in band diagrams. The chapter on nonlinear optics reverts to the more common **E** for electric field because there seemed little possibility of ambiguity.

The presentation relies on a number of standard sources. Charles Kittel's classic introductory text on solid state physics is constantly quoted and used as a reference. Other books on which I have drawn particularly are *Solid State Electronic Devices* by B. G. Streetman; "Optical Second Harmonic Generation and Parametric Oscillation" by A. Yariv, in *Topics in Solid State and Quantum Electronics*, W. D. Hershberger (editor); *The Feynman Lectures on Physics* by R. P. Feynman, R. B. Leighton, and M. Sands; and *Long-Range Order in Solids* by R. M. White and T. H. Geballe.

This book is at the introductory level in that no particular prior knowledge of solid state device physics is assumed. However, the introductory level is not the same for all topics. For example, the treatment of the applications of nonlinear optical effects in solids is more complex than the treatment of the $p–n$ junction. As for prerequisites, it is assumed that the reader has had an introductory course in solid state physics at the level of Kittel's *Introduction to Solid State Physics*, Fifth Edition. In particular, it is assumed that the reader has a knowledge of energy bands, semiconductors, and superconductivity equivalent to that covered in Chapters 7, 8, and 12 of Kittel's book. In addition, this book assumes a knowledge of electromagnetic theory at the level of Reitz and Milford's *Foundations of Electromagnetic Theory*, of optics at the level of Stone's *Radiation and Optics* or Fowles's *Modern Optics*, and of quantum mechanics at the level of Bohm's *Quantum Theory*.

Many people have shared their expertise with me and have commented on the manuscript at various stages. I would like to thank N. Amer, T. Andrade, B. Black, R. W. Boyd, J. Clarke, M. L. Cohen, L. M. Falicov, L. T. Greenberg, E. L. Hahn, G. I. Hoffer, M. B. Ketchen, A. F. Kip, R. U. Martinelli, R. S. Muller, W. G. Oldham, and P. L. Richards for helping me improve the book. However, the responsibility for errors and misconceptions is mine alone. Special thanks are due M. L. Cohen and C. Kittel for their encouragement during the development of the course. T. H. Geballe kindly provided me with a prepublication copy of his work. I would like to thank M. L. Cohen, D. Long, G. S. Kino, W. F. Oldham, J. Tauc, D. Adler, E. Gutsche, J. Millman, B. G. Streetman, T. C. Harman, H. Y. Fan, and J. Clarke for permission to use figures from their publications. The hospitality extended by John Clarke was invaluable and is sincerely appreciated. Linda Billard typed part of the manuscript with great skill, and Leslie Hausman typed the first draft. Gloria Pelatowski executed the drawings with exceptional skill and enthusiasm.

Last, but also first, I would like to thank D. and G. for making this book a reality.

Berkeley, California   RICHARD DALVEN

Contents

1. Review of Semiconductor Physics

Introduction 1
Metals, Insulators, and Semiconductors 1
Band Structure Diagrams 3
Holes in Semiconductors 6
Effective Mass of Carriers in Semiconductors 7
Conductivity of Semiconductors 10
Carrier Density in an Intrinsic Semiconductor 12
Impurity Conductivity (Extrinsic Conductivity). 14
Fermi Level Position in Extrinsic Semiconductors 17
Carrier Lifetime in Semiconductors 21
Problems 22
References and Comments 23
Suggested Reading 24

2. The Semiconductor p–n Junction

Introduction 25
Qualitative Discussion of the p–n Junction in Equilibrium 25
Quantitative Treatment of the p–n Junction in Equilibrium 33
Effect of an Applied Potential on Electron Energy Bands 48
Diffusion and Recombination of Excess Carriers 49
Qualitative Discussion of a Junction under an Applied Potential . . 53
Qualitative Discussion of Current Flow in the Biased Junction . . . 57
Quantitative Treatment of Carrier Injection in the Junction 59
Calculation of the Current through the Junction 62
Majority and Minority Carrier Components of the Junction Current. 69
Summary of the Basic Physics of the p–n Junction 72
Reverse Breakdown in p–n Junctions 73
Other Topics on p–n Junctions 75
Problems 75
References and Comments 76
Suggested Reading 77

3. Semiconductor *p–n* Junction Devices

Introduction   79
Semiconductor *p–n* Junction Diodes   79
The Bipolar Junction Transistor   81
Amplification in the Bipolar Transistor   85
Current Gain in the Bipolar Transistor   86
Circuit Configurations for Amplification with the Bipolar Transistor.   88
Quantitative Discussion of the Bipolar Transistor   91
Summary of the Physics of Amplification in the Bipolar Transistor .   96
Tunnel Diodes   96
The Junction Field Effect Transistor (JFET) 101
Physical Basis of the Current–Voltage Characteristic of the JFET. . . 102
Problems 106
References and Comments 106
Suggested Reading 108

4. Physics of Metal–Semiconductor and Metal–Insulator–Semiconductor Junctions

Introduction 109
The Metal–Semiconductor Junction at Equilibrium 109
Effect of an Applied Potential on the Metal–Semiconductor Junction 115
Physics of the Metal–Insulator–Semiconductor Structure 119
Problems 123
References and Comments 124
Suggested Reading 125

5. Metal–Semiconductor and Metal–Insulator–Semiconductor Devices

Introduction 127
Metal–Semiconductor (Schottky) Diodes 127
The Insulated-Gate Field-Effect Transistor (IGFET) 128
The Induced-Channel MOSFET 131
Summary of the Physics of Field-Effect Transistors 132
Applications of the MOSFET 133
Charge-Coupled Devices 134
Problems 136
References and Comments 137
Suggested Reading 138

6. Other Semiconductor Devices

Introduction 139
Semiconductor Surface States 139
Band Structure at the Semiconductor Surface 141

Calculation of the Amount $e\Phi_s$ of Band Bending 145
Effect of Surface States on Metal–Semiconductor Contacts 147
Photoemission from Semiconductors 149
Effect of the Surface on Photoemission 151
Negative Electron Affinity in Semiconductors 154
Physics of the Transferred Electron (Gunn) Effect 157
Physics of Amorphous Semiconductors 164
Amorphous Semiconductor Devices 168
Problems 170
References and Comments 170
Suggested Reading 173

7. Detectors and Generators of Electromagnetic Radiation

Introduction 175
Intrinsic Photon Absorption in Semiconductors 175
Photon Absorption by Bound States of Impurities in Semiconductors   180
Threshold Energies for Photon Absorption 181
Photoconductivity in Semiconductors 185
Photodiodes 188
Photovoltaic Devices 189
Other Applications of Intrinsic Photoconductivity 191
Summary on Semiconductor Photon Detectors 195
Emission of Photons in Semiconductors 195
p–n Junction Luminescence 199
Light Amplification by Stimulated Emission of Radiation. 200
Solid State Lasers. 204
Semiconductor Injection Lasers 206
Summary on Solid State Lasers 209
Problems 210
References and Comments 210
Suggested Reading 213

8. Superconductive Devices and Materials

Introduction 215
Review of Some Aspects of Superconductivity 215
Wave Function of the Condensed Phase of Pairs 216
The Josephson Effects 220
Physics of the DC Josephson Effect 220
Physics of the AC Josephson Effect 226
Voltage–Current Curves for Josephson Junctions 228
Effect of Electromagnetic Radiation on the Junction 229
Quantization of Magnetic Flux in a Superconducting Ring 233
Superconducting Quantum Interference 236
The Superconducting Quantum Interference Device (Squid) 239
Superconducting Materials: The Transition Temperature 242

Problems   251
References and Comments   252
Suggested Reading   255

9. Physics and Applications of the Nonlinear Optical Properties of Solids

Introduction   257
Review of Electromagnetic Wave Propagation in Solids   257
Electric Polarization in an Isotropic Linear Dielectric Solid   260
Nonlinear Polarization and Nonlinear Susceptibility   264
Anharmonic Oscillator Model of a Nonlinear Solid   265
Summary of the Physical Picture of Nonlinear Polarization   270
Tensor Nature of the Nonlinear Susceptibility   271
Solid State Physics Factors Affecting the Nonlinear Susceptibility . .   271
Magnitude of the Nonlinear Susceptibility   275
Wave Equation for the Nonlinear Crystal   276
Wave Propagation and Interaction in a Nonlinear Crystal   278
Optical Second Harmonic Generation   284
Phase Matching (Index Matching) in Second Harmonic Generation .   292
Frequency Mixing and Up-Conversion   293
Parametric Amplification   298
Summary   302
Problems   302
References and Comments   303
Suggested Reading   306

Appendix: References on Some Other Topics   309
Author Index   311
Subject Index   315

1

Review of Semiconductor Physics

Introduction

The aim of this chapter is a brief discussion of some topics in semiconductor physics that will be useful in our discussion of applications. Since it is assumed that the reader has had an introductory course in solid state physics at the level of the book by Kittel,[1] some of the chapter will be review material. However, since some of the topics may be new to some readers, references to more complete and/or advanced treatments are given.

Metals, Insulators, and Semiconductors

We recall[2] that there are $2N$ independent states in each energy band of the band structure of a crystal containing N primitive unit cells. Each energy band can therefore hold $2N$ electrons. If the atomic arrangement in the unit cell is such that each primitive cell contains one valence electron (e.g., the $3s$ electron in sodium), there will be a total of N electrons occupying the $2N$ states in the band formed from the $3s$ atomic level. This $3s$ band is therefore only half-full and we expect sodium to be a metal. This situation is illustrated schematically in Figure 1.1, in which the vertical axis is electron energy and the horizontal axis[†] is distance (in real space) within the crystal.

[†] This type of drawing is also used without ascribing any particular meaning to the horizontal axis. We, however, will usually assign the meaning of distance to the horizontal axis.

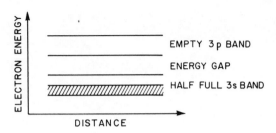

Figure 1.1. Energy bands (schematic) in sodium, showing the half-filled 3s band.

In this figure, we see a half-full 3s valence band separated by an energy gap from the next higher band, the completely empty 3p band.

If the number of valence electrons per primitive unit cell is an even integer, then we will have, if all other things are equal,[†] a filled valence band and the crystal will be an insulator. An example is diamond, which has eight electrons per primitive unit cell. This situation is indicated schematically in Figure 1.2, which shows the energy gap E_g between the highest of the four filled valence bands in diamond and the lowest empty band. We expect diamond to be an insulator, at least at 0 K. If the magnitude of the energy gap E_g in Figure 1.2 is not too large, then, at temperatures greater than 0 K, we would expect a few electrons in the highest filled valence band to be thermally excited into the lowest empty band, as indicated schematically in Figure 1.3. The presence of a few mobile electrons in the lowest empty band gives the crystal a weak electrical conductivity at temperatures above 0 K. For this reason, the lowest empty band, above the highest valence band, is called the conduction band. In such a case,

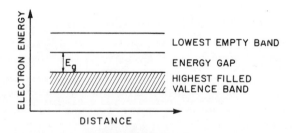

Figure 1.2. Diagram (schematic) of the highest filled valence band and lowest empty band separated by an energy gap E_g in an insulator (e.g., diamond).

[†] In this case, "all other things equal" means that there are no bands that overlap in energy making the crystal a metal[2] rather than an insulator.

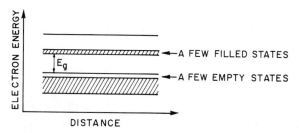

Figure 1.3. Energy bands (schematic) in a crystal in which a few electrons have been thermally excited from the filled band to the empty band.

the crystal is a semiconductor. A semiconductor differs only quantitatively from an insulator. The magnitude of the energy gap is smaller (about 1 eV) in a semiconductor than in an insulator (several eV). At temperatures above 0 K, a semiconductor will, because of its smaller energy gap, have more electrons thermally excited into its conduction band than will an insulator. The semiconductor will thus have a greater electrical conductivity than does the insulator.

Band Structure Diagrams

Consider a perfect crystal of diamond. The carbon atom has the electronic configuration $(1s)^2(2s)^2(2p)^2$, in which the $2s$ and $2p$ valence electrons are four in number. There are two atoms, and hence eight electrons, per primitive unit cell in the diamond crystal. We therefore expect diamond to have completely filled bands and to be an insulator. In an approximate way, we may think of the one $2s$ level and the three $2p$ levels in the carbon atom as each giving rise to a band in the crystal.[3,4] This gives a total of four bands arising from the atomic energy levels containing the valence electrons of carbon. These four bands are called the valence bands of the band structure of diamond. The valence bands of diamond (including spin) are shown[5] in the reduced zone scheme in Figure 1.4. The valence band structure in Figure 1.4 is shown for two directions, [111] and [100], in k space. The right-hand half of the drawing shows the valence bands for the [100] direction; the left-hand half, for the [111] direction. The symbols Γ, X, etc., in Figure 1.4 label certain special points in the Brillouin zone. While they are connected with the group-theoretical description of the symmetry properties of the crystal lattice, we will use them simply as "band labels."

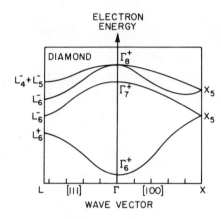

Figure 1.4. Valence band structure of diamond (reduced zone scheme) in the [111] and [100] directions in k-space. (Adapted from Long.[5])

We now consider the allocation of the eight valence electrons of carbon among the four valence bands. Since we consider $2N$ atoms in N primitive unit cells of the diamond lattice, we have a total of $8N$ electrons to put into four valence bands. Considering, for example, the [100] direction, the lowest valence band is the $\Gamma_6^+ X_5$ band; it has N energy levels, which can accept $2N$ electrons (including spin). The same is true of the $\Gamma_7^+ X_5$ band and the two $\Gamma_8^+ X_5$ valence bands. We find, as expected, a grand total of $8N$ electrons in the four valence bands ($\Gamma_6^+ X_5$, $\Gamma_7^+ X_5$, $\Gamma_8^+ X_5$, $\Gamma_8^+ X_5$) of diamond.

We see from Figure 1.4 that the point in k space at which a valence electron has the highest energy is the Γ point at the center of the Brillouin zone. We therefore say that the valence band maximum occurs at the center of the Brillouin zone. At 0 K, all of the valence band states shown in Figure 1.4 are occupied by electrons, and all of the higher bands of the diamond band structure (which are not shown in Figure 1.4) are empty. At temperatures above 0 K, a few electrons will be thermally excited from the highest valence band into the vacant next higher band. The next empty band above the highest valence band is called the conduction band, because electrons thermally excited into it will find empty states available for the electrical conduction process.

The element silicon, which is of great technological importance as a semiconductor, has the same crystal structure as diamond. Figure 1.5 shows the band structure[6] (excluding spin) of silicon in the [111] and [100] directions. The upper valence band is doubly degenerate in the absence of spin–orbit coupling. The valence band (abbreviated VB) maximum energy is at the Γ point at the center of the Brillouin zone (abbreviated BZ). The conduction band (abbreviated CB) minimum energy is at a point in the

[100] direction, 0.86 of the distance between the zone center at Γ and the edge of the zone at the point X. The minimum energy gap E_g is the energy difference between the conduction band minimum and the valence band maximum, so the magnitude E_g of the energy gap in silicon is given by the relation

$$E_g = E(\text{CB minimum}) - E(\text{VB maximum}) \qquad (1.1)$$

Since the conduction band minimum and the valence band maximum occur at different points of the zone, silicon is said to have an indirect minimum energy gap. If the conduction band minimum and the valence band maximum occur at the same point of the zone, the semiconductor is said to have a direct minimum energy gap. Examples of semiconductors with direct energy gaps are indium antimonide (InSb) and lead sulfide (PbS). Figure 1.6 shows the band structure[7] of InSb, in the [100] and [111] directions, near the center of the Brillouin zone. The zone edge is not shown. The conduction band minimum is at the Γ point at the zone center. The valence band maximum is (almost exactly[7]) at the zone center also, so the energy gap in InSb is a direct gap. (As a point of terminology, the conduction band minimum is often called the conduction band edge, and the valence band maximum is called the valence band edge.) In InSb, there are four valence bands V_1, V_2, V_3 (the V_3 band is twofold degenerate) which contain the eight valence electrons in this III–V semiconductor.

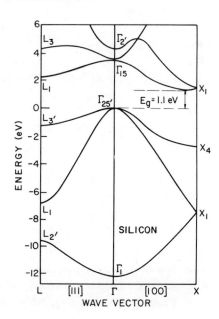

Figure 1.5. Band structure of silicon in the [111] and [100] directions in k-space. The minimum energy gap $E_g = 1.1$ eV at room temperature. (Adapted from Chelikowsky and Cohen.[6])

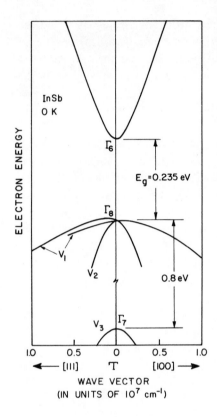

Figure 1.6. Band structure of InSb in the [100] and [111] directions in k-space. The energy separations, including the energy gap E_g, are at 0 K. (Adapted from Long.[7])

Holes in Semiconductors

We recall[8] the picture of a single electron missing from a state at the top ($k = 0$) of an otherwise filled band. If an electric field is applied, then the collective motion of the electrons in the band is equivalent to the motion of the empty state in the direction of decreasing (i.e., more negative) k. If an electron is missing from a state of wave vector k_e in a band, we say that there is a hole in that state. To quote Kittel,[9] "The physical properties of the hole follow from those of the totality of electrons in the band." We recall that, if an electron of effective mass m_e (effective mass is reviewed below) is missing from a state of wave vector k_e and energy $E(k_e)$, then the hole has the following properties:

1. Wave vector $k_h = -k_e$
2. Effective mass $m_h = -m_e$; the hole effective mass is positive
3. Energy $E(k_h) = -E(k_e)$
4. Positive charge ($+e$), where ($-e$) is the electron charge

Note that relation (3) means that, if we plot electron energy increasing upward, hole energy increases downward. In other words, electrons tend to "sink"; holes tend to "float," on the usual plots of *electron* energy.

Effective Mass of Carriers in Semiconductors

Consider an electron in the periodic potential of the lattice. If an electric or magnetic field is applied, the electron is accelerated relative to the lattice *as if* it had a mass m^* whose value is different from the free electron mass m_0. The effective mass m^* of an electron in a state in an energy band whose energy–wave-vector relation is $E = E(k)$ is given by the relation

$$m^* = \frac{\hbar^2}{d^2E/dk^2} \tag{1.2}$$

For an electron in such a band, we see from (1.2) that the effective mass m^* is related to the curvature of the band $E = E(k)$. Consider two cases in which the $E(k)$ relation is quadratic in k and given by

$$E = \frac{\hbar^2}{2m^*} k^2 \tag{1.3}$$

but for which the effective masses are different. Figures 1.7a and 1.7b show the two cases; in Figure 1.7a, the electron effective mass is m_1^* and in Figure 1.7b it is m_2^*. From the figure, we see that the $E(k)$ function is, qualitatively speaking, more "curved" in Figure 1.7a than it is in Figure 1.7b, so d^2E_1/dk^2 is larger than d^2E_2/dk^2, using $E = E_1(k)$ and $E = E_2(k)$ for the bands in Figures 1.7a and 1.7b, respectively. From equation (1.2), we have the result that the effective mass m_1^* is smaller than the effective mass m_2^*.

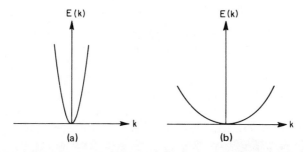

Figure 1.7. (a) Band with a small effective mass $[E_1 = (\hbar^2/2m_1^*)k^2]$. (b) Band with a larger effective mass $[E_2 = (\hbar^2/2m_2^*)k^2]$.

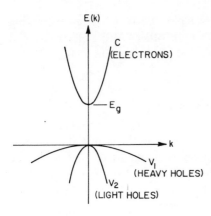

Figure 1.8. Schematic band structure with a direct energy gap E_g and two valence bands V_1 and V_2. The conduction band is denoted by C.

Note that, if the $E(k)$ function is proportional only to k^2 (a parabolic band), then (1.2) shows that the effective mass m^* is independent of k and is a constant over the whole band in question.

As an example of these definitions, consider the semiconductor band structure shown in Figure 1.8. This band structure shows a conduction band (C) and two valence bands $(V_1$ and $V_2)$. The energy gap is direct at the center of the zone and is of magnitude E_g, since the zero of energy has been taken at the valence band maximum. In the conduction band, the dispersion relation (i.e., the energy–wave-vector relation) is

$$E(k) = E_g + \frac{\hbar^2}{2m_c} k^2 \tag{1.4}$$

where the quantity m_c in (1.4) is seen to be positive. The dispersion relations for the valence bands V_1 and V_2 are

$$E(k) = -\frac{\hbar^2}{2m_{V_1}} k^2 \tag{1.5}$$

and

$$E(k) = -\frac{\hbar^2}{2m_{V_2}} k^2 \tag{1.6}$$

where m_{V_1} and m_{V_2} are positive quantities. We are thus considering a semiconductor in which all of the bands under consideration are parabolic.

To find the effective mass m_e^* of an electron in the conduction band, we apply equation (1.2) to the conduction band dispersion relation (1.4) and find, since $d^2E/dk^2 = \hbar^2/m_c$,

$$m_e^* = m_c \tag{1.7}$$

Equation (1.7) tells us that the effective mass m_e^* of an electron in this conduction band is equal to m_c and is therefore positive. Note also that, as expected, the electron effective mass m_e^* is constant because the dispersion relation (1.4) for the conduction band is quadratic in k.

We now want to find the effective mass $m_{h_1}^*$ of a hole in the valence band V_1. In this band, the dispersion relation is given by (1.5), so the definition (1.2) of the effective mass gives, since $d^2E/dk^2 = -\hbar^2/m_{V_1}$, the result

$$m^* = -m_{V_1} \tag{1.8}$$

However, m^* in equation (1.8) is the mass of an *electron* in the valence band V_1. From relation (2) concerning holes given earlier, we see that the effective mass of a *hole* in a state in the valence band is the *negative* of the effective mass of an electron in the same state. Applying this result to equation (1.8) gives

$$m_{h_1}^* = -m^* = -(-m_{V_1}) = +m_{V_1} \tag{1.9}$$

We note that the hole effective mass $m_{h_1}^*$ is positive because the quantity m_{V_1} is positive and that $m_{h_1}^*$ is independent of k because equation (1.5) tells us that the valence band V_1 is parabolic.

We note also from Figure 1.8 the difference in curvature between the conduction band C and the valence band V_1. For this band structure, then, we have the result that the hole effective mass $m_{h_1}^*$ is larger than the electron effective mass m_e^*. This is usually the case for real semiconductors: the hole effective mass is larger than the electron effective mass.

Note also that we have tacitly been discussing electrons in states near the conduction band minimum and holes in states near the valence band maximum. These states[10] are important because they are where the carriers are located in most semiconductor devices. The results we have obtained are, of course, for parabolic bands. This assumption is at least approximately valid for the band edges of most of the semiconductors of device interest that we will discuss.

Some semiconductors have more than one valence band (like the band structure in Figure 1.8) which may (possibly) contribute holes to the process of electrical conduction. From the curvatures of the valence bands in Figure 1.8, we see that the hole effective mass $m_{h_1}^* = m_{V_1}$ in valence band V_1 will be larger than the hole effective mass $m_{h_2}^* = m_{V_2}$ in band V_2, where m_{V_2} is the quantity in the dispersion relation (1.6) for valence band V_2. For this reason, the holes in band V_1 are called "heavy holes" and those in band V_2 are called "light holes." An example[11] is GaAs, where (near $k = 0$), $m_{h_1}^* \cong 0.7m_0$ and $m_{h_2}^* \cong 0.1m_0$ are the hole masses.

Conductivity of Semiconductors

Consider a pure semiconductor crystal at 0 K. None of the electrons in the conduction band are thermally excited into the valence band. Suppose one electron is excited from the valence band into the conduction band. The result is that there is a hole (i.e., a missing electron) in the valence band and an electron in the conduction band. The situation is shown schematically in Figure 1.9, which is of the same type as Figure 1.2. If an electric field $\mathscr{E}$ is now applied to the crystal, the electron and hole will have opposite velocities. This is shown in Figure 1.10, in which q is the electric charge of a carrier, $\mathbf{v}$ is its velocity, and $\mathbf{J}$ is the current density. Since

$$\mathbf{J} = nq\mathbf{v} \tag{1.10}$$

where n (or p) is the number density of electrons (or of holes), we have that the current densities $\mathbf{J}_n$ and $\mathbf{J}_p$ of electrons and holes, respectively, are given by

$$\mathbf{J}_n = -ne\mathbf{v}_n, \qquad \mathbf{J}_p = pe\mathbf{v}_p \tag{1.11}$$

This leads to the result that $\mathbf{J}_n$ and $\mathbf{J}_p$ are in the same direction, as shown in Figure 1.10, because the electron velocity $\mathbf{v}_n$ and the hole velocity $\mathbf{v}_p$ are in opposite directions. The electrons and holes in a semiconductor therefore both contribute in the same way to the total current density.

From Ohm's law, we have, for electrons,

$$\mathbf{J}_n = \sigma_n \mathscr{E} = -ne\mathbf{v}_n \tag{1.12}$$

where σ_n is the electrical conductivity due to electrons. For holes,

$$\mathbf{J}_p = \sigma_p \mathscr{E} = pe\mathbf{v}_p \tag{1.13}$$

We define the electron and hole mobilities μ_n and μ_p by the vector equations

$$\mathbf{v}_n \equiv -\mu_n \mathscr{E} \tag{1.14}$$

$$\mathbf{v}_p \equiv \mu_p \mathscr{E} \tag{1.15}$$

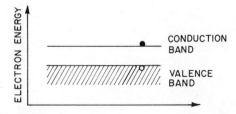

Figure 1.9. Excitation of an electron from the valence band to the conduction band, producing a free electron (●) and a free hole (○).

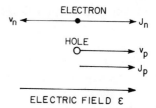

Figure 1.10. Velocity vectors $\mathbf{v}_n$ and $\mathbf{v}_p$, and current density vectors $\mathbf{J}_n$ and $\mathbf{J}_p$ for an electron ($\bullet$) and a hole ($\bigcirc$) in an electric field $\mathscr{E}$. The current density vectors $\mathbf{J}_n$ and $\mathbf{J}_p$ are in the same direction.

where the negative sign in (1.14) means the electron velocity $\mathbf{v}_n$ is opposite to the direction of the electric field $\mathscr{E}$. The mobilities μ_n and μ_p are both positive quantities. Combining (1.12)–(1.15) gives

$$\mathbf{J}_n = ne\mu_n\mathscr{E} = \sigma_n\mathscr{E} \tag{1.16}$$

$$\mathbf{J}_p = pe\mu_p\mathscr{E} = \sigma_p\mathscr{E} \tag{1.17}$$

and the total current density $\mathbf{J} = \mathbf{J}_n + \mathbf{J}_p$ is given by

$$\mathbf{J} = (ne\mu_n + pe\mu_p)\mathscr{E} = \sigma\mathscr{E} \tag{1.18}$$

where the total conductivity σ due to both electrons and holes is

$$\sigma = e(n\mu_n + p\mu_p) = \sigma_n + \sigma_p \tag{1.19}$$

The carrier mobilities are determined by the collisions of the carriers with various imperfections (e.g., phonons) in the crystal lattice. We recall[12] the expression

$$\sigma = ne^2\tau_n/m_0 \tag{1.20}$$

for the DC conductivity σ in the free-electron model of a solid, where n is the free-electron concentration, m_0 is the free-electron mass, and τ_n is the collision time, or relaxation time, for electrons. Comparing (1.20) and (1.19), we have

$$\mu_n = e\tau_n/m_0 \tag{1.21}$$

If the electrons have an effective mass m_e^*, (1.21) becomes

$$\mu_n = e\tau_n/m_e^* \tag{1.22}$$

for electrons, while for holes the analogous relation is

$$\mu_p = e\tau_p/m_h^* \tag{1.23}$$

where m_h^* is the hole effective mass and τ_p is the collision time for holes.

In a semiconductor at 300 K, the dominant carrier scattering process is usually scattering by phonons. It is thus this type of scattering[13] that determines the collision time τ in this case. In some cases, however, scattering by ionized impurities may dominate.

We note from equations (1.22) and (1.23) that the carrier mobility μ is, for a given scattering time τ, inversely proportional to the carrier effective mass m^*. Since the hole effective mass in a given semiconductor is generally larger than the electron effective mass, it is usually true that the electron mobility μ_n is larger than the hole mobility μ_p in a semiconductor. Further, it is known[14] that the electron effective mass varies directly with the magnitude of the energy gap E_g, semiconductors with small energy gaps will have small values of effective mass, and hence large values of the electron mobility.[15] (This conclusion is often true for holes also, but is not as clear-cut because of the relatively complex structures of semiconductor valence bands leading to light and heavy holes and other complications.)

Carrier Density in an Intrinsic Semiconductor

In a pure semiconductor crystal, each electron thermally excited into the conduction band creates a hole in the valence band. One speaks of the creation of electron–hole pairs. In this case the number of electrons is necessarily equal to the number of holes. A semiconductor for which this is true is called an intrinsic semiconductor, defined by the condition

$$n = p \tag{1.24}$$

where n is the electron density and p is the hole density. It is important to note that electrons and holes can, and do, simultaneously coexist in semiconductors. This fact is the physical basis for the action of the bipolar transistor.

We now recall[16] a few standard results concerning the intrinsic carrier density in a pure semiconductor. Consider the Fermi–Dirac distribution function $f(E)$, where

$$f(E) = \{\exp[(E - E_F)/k_B T] + 1\}^{-1} \tag{1.25}$$

and where E is the energy, E_F is the Fermi energy, and k_B is Boltzmann's constant; the function $f(E)$ gives the probability that the state of energy E is occupied by a particle. If "not too many" electrons[17] have been excited into the conduction band, then it will be true that $(E - E_F) \gg k_B T$, the

Fermi function $f(E) \ll 1$, and the semiconductor is referred to as non-degenerate. In such a case, the Fermi function may be approximated by

$$f(E) \cong \exp[-(E - E_F)/k_B T] \qquad (1.26)$$

meaning that classical statistics furnish a valid description of the carrier densities involved. If the carrier densities are large enough that it is not true that $f(E) \ll 1$, then the approximation (1.26) is not valid, the exact Fermi function (1.25) must be used, and the semiconductor is referred to as degenerate.

Using equation (1.26) for $f(E)$, and a parabolic energy dependence of the densities of states in the valence and conduction bands of the semiconductor, one may calculate[16] the intrinsic electron and hole densities n and p. The results are

$$n = N_C \exp[(E_F - E_g)/k_B T] \qquad (1.27)$$

$$p = N_V \exp[-E_F/k_B T] \qquad (1.28)$$

where E_g is the energy gap and

$$N_C \equiv 2(m_e^* k_B T/2\pi\hbar^2)^{3/2} \qquad (1.29)$$

$$N_V \equiv 2(m_h^* k_B T/2\pi\hbar^2)^{3/2} \qquad (1.30)$$

If we multiply the values of n and p given by (1.27) and (1.28), we obtain

$$np = 4(k_B T/2\pi\hbar^2)^3 (m_e^* m_h^*)^{3/2} \exp(-E_g/k_B T) \qquad (1.31)$$

showing that the np product is independent of the Fermi energy in a non-degenerate intrinsic semiconductor. The np product is thus a constant at a given temperature; this constant has the symbol n_i^2, where the subscript i stands for intrinsic. At 300 K the value of n_i^2 for silicon is 2.2×10^{18} cm^{-6}, calculated using the values[16] $m_e^* = 0.26m_0$ and $m_h^* = 0.49m_0$ (where m_0 is the free-electron mass) for the electron and hole effective masses.

We next invoke the condition (1.24) that the semiconductor is intrinsic, and equate equations (1.27) and (1.28) for n and p, solving for the Fermi energy E_F. The result is

$$E_F = \tfrac{1}{2}E_g + \tfrac{3}{4}k_B T \ln(m_h^*/m_e^*) \qquad (1.32)$$

giving the position of the Fermi energy (relative to the energy zero chosen at the valence band maximum) in an intrinsic semiconductor. The Fermi energy is seen to be a function of the carrier effective masses m_e^* and m_h^*. If $m_e^* = m_h^*$, then $E_F = E_g/2$ and the Fermi energy is exactly in the center of the energy gap. If, as is usual in semiconductors, m_h^* is larger than m_e^*, the second term in (1.32) is positive, and the Fermi energy lies somewhat above the center of the energy gap. It is conventional in semiconductor physics to refer to E_F as the Fermi level.

Impurity Conductivity (Extrinsic Conductivity)

In general, for device applications, we will be interested in semiconductors that are not intrinsic, but in which the concentration of electrons or of holes is controlled by deliberately added impurities. Such semiconductors are called extrinsic semiconductors or impurity semiconductors.[18] Certain impurities (e.g., arsenic) added to silicon enter the lattice substitutionally and can give up an electron to the conduction band. Such impurities, called donors, add free electrons to the crystal. The amount of energy E_d necessary to free the electron from the donor atom and liberate it into the conduction band is called the donor ionization energy. Values of E_d are about 0.01 eV and 0.05 eV for various pentavalent impurities in germanium and silicon, respectively. Figure 1.11 shows schematically the energies involved in donor ionization. Figure 1.11a shows the electron (–●–) bound to a donor atom; this energy level is an energy E_d below the energy of the conduction band minimum, so the addition of the donor ionization energy E_d will promote the electron into a conduction band (CB) state, as shown in Figure 1.11b. An empty donor state (—) remains in the energy gap. The process may be summarized by the equation

$$D^0 + E_d \rightarrow D^+ + e^- \tag{1.33}$$

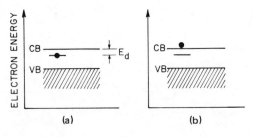

Figure 1.11. Ionization (schematic) of electron bound to a donor atom. In (a), the electron (–●–) is in a bound state an energy E_d below the energy of the conduction band (CB) minimum. In (b), the electron (●) has been promoted into a CB state by adding the ionization energy E_d, leaving behind an empty donor state (—) in the energy gap. (The energy spacings are not to scale.)

Figure 1.12. Ionization (schematic) of hole bound to an acceptor atom. In (a), the hole $(-\!\bigcirc\!-)$ is in a bound state at an energy E_a above the energy of the valence band (VB) maximum. In (b), the hole $(\bigcirc)$ has been raised to a valence band state by adding the ionization energy E_a,

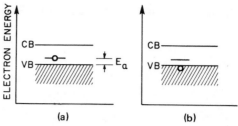

leaving behind an empty acceptor state $(-)$ in the energy gap. (The energy spacings are not to scale.)

where D^0 is the neutral donor impurity, E_d is the ionization energy of the electron e^-, and D^+ is the positively charged ionized donor.

In the same way, a trivalent impurity (e.g., boron) in silicon can accept an electron from the valence band, thereby generating a free hole. In this process the impurity atom is called an acceptor, and adding an energy E_a promotes an electron from the valence band (VB) maximum to the energy level of the neutral acceptor, leaving behind a free hole and creating a negatively charged acceptor center. One can write the equation

$$VB + E_a + A^0 \rightarrow A^- + (VB - e^-) \tag{1.34}$$

where A^0 is the neutral acceptor, A^- is the negatively charged acceptor, VB represents the valence band, and $VB - e^-$ represents the valence band minus one electron, or, in other words, a hole. It is more usual to write (1.34) as

$$E_a + A^0 \rightarrow A^- + h^+ \tag{1.35}$$

where h^+ is a hole, and the energy E_a is called the ionization energy of the acceptor. Values of E_a for trivalent impurities in silicon range from 0.016 to 0.065 eV; the value of E_a for germanium is about 0.01 eV.

It is also common to describe the process in terms of holes rather than electrons. One says that the hole is bound to the acceptor forming the neutral acceptor A^0. Adding the ionization energy E_a frees the hole into the valence band, leaving a negatively charged acceptor A^-. This description is shown in Figure 1.12, which is the analog for holes of Figure 1.11. Figure 1.12a shows the hole $(-\!\bigcirc\!-)$, bound to the acceptor atom, at an energy E_a above the valence band maximum, and where we recall that hole energy increases in the downward direction on a plot of electron energy like Figure 1.12. The addition of the acceptor ionization energy E_a raises the hole $(\bigcirc)$ into the valence band, as shown in Figure 1.12b. An empty hole state $(-)$ remains behind in the energy gap.

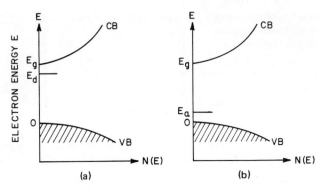

Figure 1.13. Density of states $N(E)$ as a function of electron energy E for (a) an n-type, and (b) a p-type semiconductor. In (a), the donor levels are an energy E_d below the conduction band edge, where E_d is the donor ionization energy. In (b), the acceptor levels are an energy E_a above the valence band edge, where E_a is the acceptor ionization energy. (In both drawings, the energy separations are exaggerated for clarity.)

For future reference, we may also display the ideas of Figures 1.11 and 1.12 on a plot of the density of electron states $N(E)$ as a function of energy E. This is shown in Figure 1.13, in which $N(E)$ is plotted horizontally and E is plotted vertically. Figure 1.13a shows an n-type semiconductor, in which the donor states are an energy E_d below the conduction band, and Figure 1.13b shows a p-type semiconductor with acceptor levels an energy E_a above the valence band. (The energy spacings of the impurity levels from the bands are greatly exaggerated for clarity.)

The addition of donors to a pure semiconductor adds electrons to the conduction band. Such a semiconductor is called n type, meaning that its electrical conduction is due primarily to electrons. Similarly, the addition of acceptors adds holes to the valence band, producing a semiconductor, called p type, in which electrical conduction is mainly due to holes. Equation (1.31) tells us that the np product n_i^2 is constant at a given temperature. This means that an increase in the density of electrons in a semiconductor results in a decrease in the density of holes, and vice versa.

Suppose that the donor and acceptor ionization energies in a semiconductor are E_d and E_a, respectively, and that only one or the other type of impurity is present. If, in the n-type semiconductor, the donor atom density is N_d, and if the temperature is low enough that the condition $E_d \gg k_B T$ is satisfied, then the electron density n due to the ionization of donors is given by[18]

$$n \cong (n_0 N_d)^{1/2} \exp[-E_d/2k_B T] \qquad (1.36)$$

where the quantity n_0 is equal to $[2(m_e^* k_B T / 2\pi \hbar^2)^{3/2}]$. A similar equation holds for the hole density p in a p-type semiconductor in which the acceptor atom density is N_a. (The case in which both donors and acceptors are present in comparable concentrations is more complicated.[19])

Most semiconductor devices are designed to operate at 300 K, at which temperature the usual donor or acceptor impurities in germanium or silicon are essentially completely ionized. We will therefore usually be interested in intrinsic n-type or p-type semiconductors in which $n = N_d$ or $p = N_a$, i.e., the electron concentration n will be taken as equal to the donor atom concentration N_d, and similarly for holes.

Suppose we have a semiconductor containing N_d ionized donors and N_a ionized acceptors per unit volume, where N_d is larger than N_a. Then this semiconductor would have a net extrinsic electron concentration equal to $N_d - N_a$. We say that the N_a acceptors have neutralized or compensated N_a of the donors, leaving $N_d - N_a$ donors uncompensated and able to supply free electrons. This semiconductor would be referred to as a compensated n-type semiconductor.

Fermi Level Position in Extrinsic Semiconductors

We are next interested in considering the Fermi level in an impurity or extrinsic semiconductor[19-23] since these are the semiconductors of greatest interest in applications. Equations (1.27) and (1.28) give the electron and hole concentrations n and p in terms of the Fermi level E_F and the various semiconductor parameters. These equations apply to any nondegenerate semiconductor with parabolic conduction and valence bands, and hence hold for an extrinsic semiconductor with those properties. We may therefore use these equations to find out how the position of the Fermi level will vary with the carrier concentration in an extrinsic semiconductor.

We repeat the equations for convenience:

$$n = N_C \exp[(E_F - E_g)/k_B T] \tag{1.27}$$

$$p = N_V \exp[-E_F/k_B T] \tag{1.28}$$

and consider an n-type semiconductor containing N_d ionized donors per unit volume, where the electron density n is equal to N_d. Consider two semiconductor samples with ionized donor densities N_d and N_d', where N_d' is larger than N_d; the electron densities are $n = N_d$ and $n' = N_d'$, where

n' is larger than n. Equation (1.27) must hold, so it must be true that

$$\exp[(E_{F}' - E_{g})/k_{B}T] > \exp[(E_{F} - E_{g})/k_{B}T] \tag{1.37}$$

where E_{F}' and E_{F} are, respectively, the Fermi level position in the samples of electron density n' and n. Equation (1.37) requires that

$$E_{F}' - E_{g} > E_{F} - E_{g} \tag{1.38}$$

so that it is also true that

$$E_{g} - E_{F}' < E_{g} - E_{F} \tag{1.39}$$

Equation (1.39) tells us that the energy separation between the energy E_{g} of the conduction band and the energy E_{F} of the Fermi level is smaller in the sample with the larger electron density n'. Put another way, the Fermi level will be closer to the conduction band in the sample with the larger electron density. For example, if we compare an n-type extrinsic semiconductor with an intrinsic semiconductor, the extrinsic sample will have the larger electron density. We expect, therefore, that the Fermi level will be closer to the conduction band in the n-type extrinsic sample, as shown schematically in Figure 1.14. In that figure, the electron energy zero is taken at the valence band maximum, the conduction band minimum is at the forbidden gap energy E_{g}, and the Fermi level energy E_{F} is close to that of the conduction band. If we compare Fermi level positions in two n-type samples, we expect the Fermi level to be closer to the conduction band for the sample with the larger donor density and thus with the larger electron concentration.

We may discuss a p-type semiconductor in the same way. From equation (1.28), as the hole density p becomes larger, $\exp[-E_{F}/k_{B}T]$ becomes larger, and thus $\exp[E_{F}/k_{B}T]$ becomes smaller. This results in the Fermi

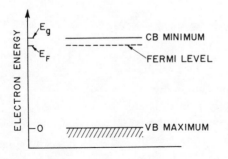

Figure 1.14. Position (schematic) of the Fermi level in an n-type semiconductor of energy gap E_{g}. The Fermi level is the dashed line.

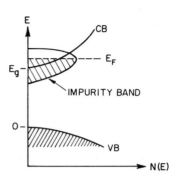

Figure 1.15. Plot of density of electron states $N(E)$ as a function of electron energy E for a high donor density and impurity band formation. States occupied by electrons are shown shaded. The impurity band of donor states overlaps the conduction band, and the position of the Fermi level E_F is above the conduction band minimum energy E_g. (The width of the impurity band is exaggerated for clarity.)

energy E_F becoming smaller as p increases, meaning that the position E_F of the Fermi level (relative to the energy zero at the valence band maximum) becomes closer to the valence band. If we compare the Fermi level positions in two p-type samples, we expect the Fermi level to be closer to the valence band in the sample with the larger acceptor density and thus with the larger hole concentration.

As the carrier density is increased further, the position of the Fermi level moves closer to the relevant band edge. Eventually the assumption that $E - E_F \gg k_B T$, made in the derivation of equations (1.27) and (1.28), fails and the classical approximation (1.26) is no longer valid. In that case, use of the exact Fermi function (1.25) is required, and the semiconductor is referred to as degenerate. Such semiconductors with very high carrier densities are of considerable interest for applications, and we will discuss, qualitatively, their physical situation, considering donors in an n-type semiconductor for concreteness. As the donor and electron densities are increased, the donor centers are closer together, and their wave functions overlap, forming a band called an impurity band.[24-26] The donor states are no longer discrete, with a well-defined ionization energy; the ionization energy approaches zero[24] with increasing donor, and electron, density. The impurity band may, at high donor concentrations, be wide enough to overlap the conduction band, as shown schematically in Figure 1.15. This figure plots density of electron states $N(E)$ as a function of electron energy E. The donor electrons are free carriers in conduction band states, and there are electrons in the conduction band up to the energy E_F of the Fermi level. From the point of view of applications, the important results are that, in a degenerate heavily doped n-type semiconductor, the Fermi level is above the conduction band minimum energy, as shown in Figure 1.16, and that there is a density of free electrons in the conduction band.

Similar statements may be made about impurity banding in heavily doped p-type semiconductors. The analogous results are that, for such a

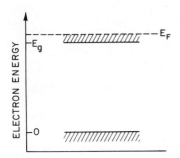

Figure 1.16. Position (schematic) of the Fermi level E_F in a degenerate n-type semiconductor. The states occupied by electrons are shown shaded, and the impurity states are not shown.

degenerate p-type semiconductor, the Fermi level is below the valence band maximum, and there is a density of free holes in the valence band.

We may also discuss the temperature dependence of the Fermi level in a nondegenerate extrinsic semiconductor,[21,23] using an n-type semiconductor as an example. We consider first such a semiconductor at 0 K. The situation is shown in Figure 1.17. At 0 K, all of the donors are un-ionized so the probability that a donor level *is* occupied is unity. Since there are no electrons in the conduction band at 0 K, the probability that a conduction band state is occupied is zero. Since, at 0 K, the Fermi function $f(E)$ has the shape[27] shown in Figure 1.17b, and since $f(E_F) = \frac{1}{2}$, the Fermi energy E_F must lie between the donor level energy and the conduction band edge at 0 K. It can be shown[28] that, at 0 K, the Fermi level E_F lies exactly half-way between the donor level energy and the conduction band edge in an uncompensated n-type semiconductor.

We now consider the same semiconductor at a temperature greater than 0 K, so some of the donor are ionized, as shown in Figure 1.18. Since the donors are no longer all un-ionized, the probability that a donor energy level is occupied is no longer exactly unity. Similarly, the probability that a conduction band state is occupied is no longer zero. This results in the energy E_F (at which the probability of occupancy of a state is exactly $\frac{1}{2}$) being lower (i.e., further from the conduction band edge) at a temperature above 0 K

Figure 1.17. (a) Band diagram for an n-type nondegenerate semiconductor at 0 K. The energy separations are exaggerated for clarity. The symbol (–●–) represents an un-ionized donor. (b) Plot of the Fermi function $f(E)$, at 0 K, as a function of electron energy E.

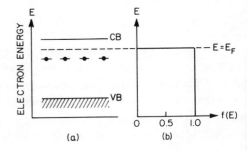

Figure 1.18. (a) Band diagram for an *n*-type nondegenerate semiconductor at a temperature above 0 K. The symbols (–●–), (——), and (●) refer, respectively, to an un-ionized donor, an ionized donor, and a free electron. The energy separations are exaggerated for clarity. (b) Plot of the Fermi function $f(E)$, at a temperature above 0 K, as a function of electron energy E.

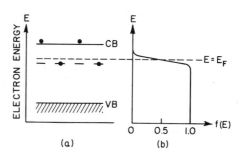

than it is at 0 K. This result is exhibited in Figure 1.18b. As the temperature is increased further, the Fermi level E_F continues to "drop" toward the position it would occupy if the semiconductor were intrinsic.[21,23]

Analogous results may be obtained for the temperature dependence of the Fermi level in a *p*-type extrinsic semiconductor.[23]

Carrier Lifetime in Semiconductors

If a carrier is excited into the conduction band, it will eventually recombine and disappear after a time called its lifetime.[29,30] The lifetime of a carrier in a given semiconductor will depend on the kinetics of its mode of recombination, either by direct recombination, or by recombination via an intermediate state or "trap." While the subject of carrier lifetimes in semiconductors is a complex one, a few qualitative observations which will be useful in discussing applications may be made here. In semiconductors with direct energy gaps (e.g., InSb, GaAs) direct recombination of an electron and a hole, both with the same value of wave vector, to produce a photon, is a process of high probability and is often the dominant process determining carrier lifetime. However, in a semiconductor with an indirect energy gap (e.g., germanium or silicon), direct recombination of an electron and hole, with different values of wave vector, must involve a phonon in the process. Because this is essentially a three-body process, direct recombination of an electron and hole, in an indirect-gap semiconductor, with the emission of a photon and the participation of a phonon to conserve wave vector, is a process of low probability. For this reason, the dominant recombination process in indirect-gap semiconductors usually proceeds through an intermediate "trap" state. The kinetics of these recombination processes taking place via traps can be very complex.[29,30]

However, for our purposes, an important qualitative conclusion concerning carrier lifetime may be reached. We expect carrier lifetimes to be longer in semiconductors with indirect energy gaps. Further, since electron–hole recombination in indirect-gap semiconductors generally proceeds through trap states which may be impurity levels, one would expect a decreased trap density to favor longer lifetimes. For this reason, a smaller doping impurity concentration in a semiconductor with an indirect energy gap will favor a longer carrier lifetime. To summarize, we expect that an indirect band gap and a low impurity density in a semiconductor will favor longer carrier lifetimes. We will use these ideas in later chapters on semiconductor p–n junctions.

While the complexity of the recombination kinetics involved makes it difficult to state typical values for carrier lifetimes, an example may be given. In silicon, the electron lifetime may range from milliseconds[31] for pure material to nanoseconds[32] for silicon doped with gold, which acts as a recombination center. Tables of values of carrier lifetimes in silicon and in GaAs are given by Hovel.[33]

Problems

1.1. *Band Structure of GaAs.* (a) Using any convenient reference, sketch the band structure of GaAs in the reduced zone scheme, indicating the highest valence band and the lowest conduction band. (b) Allocate the valence electrons in GaAs to the valence bands. Is GaAs a metal or a semiconductor? (c) Where in the Brillouin zone is the minimum energy gap located? Is the gap direct or indirect? (d) Is there a light hole band in GaAs? (e) In which valence band would you expect to find holes in intrinsic GaAs at room temperature?

1.2. *Intrinsic Carriers in Silicon.* (a) Calculate the concentration of intrinsic electrons in pure silicon at 300 K. Use a value of 1.14 eV for the energy gap, and $0.26m_0$ and $0.49m_0$ for the electron and hole effective masses, respectively, where m_0 is the free electron mass. (b) What is the hole concentration in pure silicon at 300 K? Why?

1.3. *Fermi Level Position in Extrinsic Silicon.* (a) Consider a sample of silicon in which the density of ionized arsenic atoms in substitutional lattice sites is 10^{17} cm^{-3}. Using the carrier effective mass values given in Problem 1.2 above, find the position of the Fermi level (relative to the appropriate band edge) at 300 K. (b) Sketch the band diagram, taking the impurity ionization energy of arsenic in silicon as 0.049 eV.

1.4. *Band Diagrams for Degenerate p-Type Semiconductor.* For a heavily doped, p-type semiconductor, make drawings that are analogous to Figures 1.15 and 1.16 for the n-type case.

References and Comments

1. C. Kittel, *Introduction to Solid State Physics*, Fifth Edition, John Wiley, New York (1976).
2. C. Kittel, Reference 1, pages 201–203.
3. See, for example, L. Pincherle, *Electronic Energy Bands in Solids*, Macdonald, London (1971), Section 6.3, page 172.
4. M. Tinkham, *Group Theory and Quantum Mechanics*, McGraw-Hill, New York (1964), Section 8.3, page 277.
5. Adapted from D. Long, *Energy Bands in Semiconductors*, John Wiley, New York (1968), Figure 2.7(a), page 39.
6. Adapted from J. R. Chelikowsky and M. L. Cohen, *Physical Review B*, **10**, 5095 (1974), Figure 2 (using the nonlocal pseudopotential calculation).
7. Adapted from D. Long, Reference 5, Figure 6.1, page 101.
8. C. Kittel, Reference 1, pages 214–217.
9. C. Kittel, *Introduction to Solid State Physics*, Fourth Edition, John Wiley, New York (1971), page 328.
10. A. J. Dekker, *Solid State Physics*, Prentice-Hall, New York (1957), Section 10.4, gives a brief discussion of electron motion at higher energies in the bands.
11. D. Long, Reference 5, page 105.
12. C. Kittel, Reference 1, page 169.
13. See, for example, A. J. Dekker, Reference 10, pages 329–331.
14. See, for example, C. Kittel, *Quantum Theory of Solids*, John Wiley, New York (1963), page 187.
15. C. Kittel, Reference 1, Chapter 8, Tables 1 and 3, pages 210 and 231, gives values of energy gaps and carrier mobilities for a number of semiconductors.
16. M. L. Schultz, *Infrared Physics*, **4**, 93–112 (1964), page 96.
17. See, for example, R. A. Smith, *Semiconductors*, Cambridge University Press, Cambridge (1961), page 79.
18. C. Kittel, Reference 1, pages 231–237.
19. R. A. Smith, Reference 17, Section 4.3, pages 82–92.
20. A. S. Grove, *Physics and Technology of Semiconductor Devices*, John Wiley, New York (1967), Section 4.4, pages 100–106, especially Figure 4.7, page 104.
21. S. Wang, *Solid State Electronics*, McGraw-Hill, New York (1966), Section 3.5, pages 146–152.
22. S. M. Sze, *Physics of Semiconductor Devices*, John Wiley, New York (1969), pages 32–38.
23. A. J. Dekker, Reference 10, Section 12.4, pages 310–314.
24. J. S. Blakemore, *Semiconductor Statistics*, Pergamon Press, New York (1962), Section 3.5, pages 166–169.
25. N. W. Ashcroft and N. D. Mermin, *Solid State Physics*, Holt, Rinehart and Winston, New York (1976), page 584.
26. H. M. Rosenberg, *Low Temperature Solid State Physics*, Oxford University Press, Oxford (1963), pages 237–240.
27. See, for example, C. Kittel, *Thermal Physics*, John Wiley, New York (1968), Chapters 9 and 14.
28. S. Wang, Reference 21, page 149, equation (3.57).

29. See, for example, W. R. Beam, *Electronics of Solids*, McGraw-Hill, New York (1965), Section 4.6, pages 190–200.
30. S. Wang, Reference 21, Section 5.5, pages 275–282.
31. W. R. Beam, Reference 29, page 165.
32. A. S. Grove, Reference 20, page 142.
33. H. J. Hovel, *Semiconductors and Semimetals*, R. K. Willardson and A. C. Beer (editors), Academic Press, New York (1975), Volume 11, pages 11, 12, and 14.

Suggested Reading

C. KITTEL, *Introduction to Solid State Physics*, Fifth Edition, John Wiley, New York (1976). The current edition of this modern classic is our basic background reference on solid state physics at the introductory level. Chapter 8 provides an introduction to semiconductor physics. We will sometimes refer to earlier editions of this book.

A. J. DEKKER, *Solid State Physics*, Prentice-Hall, New York (1957). This text is now somewhat out of date but is clearly written. It also contains some interesting material not readily found elsewhere at the introductory level. Chapters 12 and 13 discuss semiconductors.

R. A. SMITH, *Semiconductors*, Cambridge University Press (1959). This book provides detailed discussions of many topics (especially transport properties) in semiconductor physics. Many of the data provided on specific semiconductors are now out of date; beware of typographical errors in early printings. (A second edition of this book was published in 1978.)

D. LONG, *Energy Bands in Semiconductors* John Wiley, New York (1968). This short book provides, among other things, a compendium of band structures and other data on semiconductors.

B. G. STREETMAN, *Solid State Electronic Devices*, Prentice-Hall, New York (1972). Chapter 3 of this fine textbook provides an introduction to semiconductor physics, written with applications in mind.

N. W. ASHCROFT and N. D. MERMIN, *Solid State Physics*, Holt, Rinehart and Winston, New York (1976). Chapter 28 of this advanced-level textbook discusses the physics of semiconductors.

W. R. BEAM, *Electronics of Solids*, McGraw-Hill, New York (1965). Chapter 4 of this electrical engineering text discusses semiconductor physics, again with an eye toward applications.

J. C. PHILLIPS, *Bonds and Bands in Semiconductors*, Academic Press, New York (1973). An interdisciplinary discussion of many topics in semiconductor physics, emphasizing relationships between structure and properties.

C. A. HOGARTH (editor), *Materials Used in Semiconductor Devices*, Interscience Publishers, New York (1965). While now somewhat old, this book summarizes some of the physics and properties of a number of useful semiconductors, including silicon, InSb, and the lead salt semiconductors PbS, PbSe, and PbTe.

2

The Semiconductor *p–n* Junction

Introduction

The aim of this chapter is a discussion of the physics of a semiconductor *p–n* junction, i.e., a semiconductor structure in which there is a change from *n* type to *p* type over some region of space. A simple qualitative picture is used first to obtain the energy band diagram of a *p–n* junction; a quantitative treatment follows. The important ideas underlying the effect of an applied potential on the junction are then discussed, both qualitatively and quantitatively. These are followed by a calculation of the current through the junction, culminating in the celebrated Shockley equation. Finally, the majority and minority carrier components of the current are discussed.

Qualitative Discussion of the *p–n* Junction in Equilibrium

The main applications of solid state physics are in the area of solid state electronics. The semiconductor *p–n* junction is not only of interest technologically but, more important for our purposes, illustrates a wide range of interesting phenomena of importance in applied solid state physics.

We begin with a qualitative physical discussion of the electronic processes that take place during the formation of a semiconductor *p–n* junction. Consider a semiconductor in which there is a change from *n* type to *p* type over a very small distance, as shown schematically in Figure 2.1. Let there be N_a ionized acceptors per unit volume in the *p*-type region and N_d ionized

p - TYPE	n - TYPE
$p = N_a$	$n = N_d$

Figure 2.1. Schematic representation of a semiconductor with an abrupt change from *p* type (with N_a ionized acceptors per unit volume) to *n* type (with N_d ionized donors per unit volume).

donors per unit volume in the *n*-type region, so the hole concentration *p* equals N_a in the *p*-type region and the electron concentration *n* equals N_d in the *n*-type region. The structure shown in Figure 2.1 is thus an idealized semiconductor *p–n* junction in which the change of conductivity type takes place abruptly at a certain point in space. This structure is called an ideal abrupt *p–n* junction.

We will discuss the ideal abrupt junction by considering the process of bringing together a piece of *n*-type semiconductor and a piece of *p*-type semiconductor to form the junction. The band diagrams of the *n*- and *p*-type semiconductors, when still separated in space, are shown in Figure 2.2. In Figure 2.2, the vacuum level is the energy required to remove an electron from the bottom of the conduction band to the vacuum and is the same for the *n*- and *p*-type semiconductors.

If the two semiconductors are brought together in space, the band diagram in Figure 2.3 results because the two semiconductors have the same vacuum level. The junction between the *n*- and *p*-type regions is the vertical line between the two regions. From Figure 2.3 we see that, just after contact, the *n*-type side has an excess of electrons relative to the *p*-type side. Similarly, the *p*-type side has an excess of holes relative to the *n*-type side. There thus exists a concentration gradient of both electrons and holes at the junction between the *n*- and *p*-type sides.

In order to attain equilibrium, a diffusion of electrons and holes,

Figure 2.2. Band diagrams of *p*- and *n*-type semiconductors when separated in space. Shaded regions indicate filled electron states; dots (●) indicate free electrons, and circles (○) indicate free holes; CB and VB mean conduction band and valence band, respectively. (Fermi levels are not shown.)

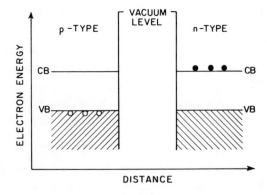

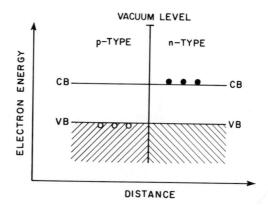

Figure 2.3. Band diagram of *n*- and *p*-type semiconductors just after contact and before any flow of holes or electrons. The junction is the vertical line separating the *n* and *p* regions.

driven by their respective concentration gradients, takes place. Holes diffuse across the junction and into the *n*-type side, and electrons diffuse into the *p*-type side. Considering the diffusion of holes, the diffusing holes enter the *n*-type side, where they combine with the free electrons present there. Recalling that the recombination of a free electron and a free hole "annihilates" both carriers, the result of the diffusion of holes across the junction is to remove free electrons from the *n*-type side. A region in the *n*-type side, near the junction, thus becomes deficient in free electrons. This region in the *n*-type side therefore has an excess of positively charged ionized donor centers. Since these ionized donors are at fixed positions in the semiconductor crystal lattice, a region of positive space charge (with a concentration of N_d positive charges per unit volume) is created in the *n*-type region near the junction.

In a similar fashion, electrons diffusing from the *n*-type region into the *p*-type region create a region of fixed negative charge (of concentration N_a negative charges per unit volume) in the *p*-type region in the vicinity of the junction. The sum of these two regions, near the junction, containing fixed space charges is called the space charge region. The space charge region is also called the depletion layer because it is depleted of mobile charges due to electron–hole recombination.

Figure 2.4 shows a band diagram of the situation at this stage, i.e., after carrier diffusion is complete and equilibrium has been attained. In Figure 2.4, the plus and minus signs represent fixed ionized donors and acceptors, respectively, and the space charge region or depletion layer is the region between the vertical dashed lines on both sides of the junction. Outside the space charge region, further from the junction, there has been no recombination of electrons on the *n*-type side with holes diffusing from the *p* side, or of holes on the *p*-type side with electrons diffusing from the

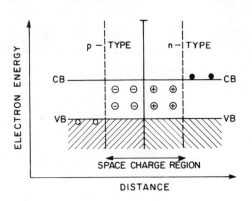

Figure 2.4. Band diagram (electron energy plotted as a function of distance) of an ideal abrupt *p–n* junction after equilibrium is attained and the space charge region has formed. The plus and minus signs represent fixed ionized donors and acceptors, respectively.

n side. We see that, outside the space charge region, electrical neutrality continues to be maintained between electrons and positively charged ionized donors (on the *n*-type side), and between holes and negatively charged ionized acceptors (on the *p*-type side). The regions outside the space-charge layer are called the *n*- and *p*-type neutral regions; these regions are the parts of Figure 2.4 to the right and left, respectively, of the vertical dashed lines which delineate the space charge region. From Figure 2.4 we see also that the space charge region extends a certain distance into the *n*-and *p*-type material on either side of the junction between the *n*- and *p*-type sides. (The *n* and *p* sides of the space charge layer are drawn as equal in extent in Figure 2.4; this need not be the case, as we shall see later.) Finally, we note that the entire semiconductor structure remains electrically neutral *overall* since no net charge has been created or destroyed.

We note also that the creation of these two regions of space charge of opposite sign sets up an electric field $\mathscr{E}$ which is known as the "built-in" electric field. The situation is shown schematically in Figure 2.5, which includes the built-in electric field extending over the space charge region between the vertical dashed lines on either side of the junction. The field $\mathscr{E}$ is directed from the *n*-type space charge region toward the *p*-type space-charge region. Figure 2.5 also shows the forces $\mathbf{F}_e$ and $\mathbf{F}_h$ which the electric field $\mathscr{E}$ exerts on electrons and holes, respectively. It is seen that the force exerted by the built-in electric field opposes the diffusion of electrons out

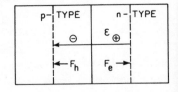

Figure 2.5. Built-in electric field $\mathscr{E}$ in the space charge region.

of the *n*-type side and of holes out of the *p*-type side. In this way, the development of the built-in field brings about a condition of equilibrium in the *p–n* junction.

The built-in electric field $\mathscr{E}$ corresponds to a gradient $-dV/dx$ of the electrostatic potential $V(x)$, where $V(x)$ is a function of the distance x in the semiconductor. For simplicity, we will make the assumption[1] that the built-in electric field is confined to the space charge region of the ideal abrupt junction we are considering. This in turn means that we are assuming that the electrostatic potential $V(x)$ is constant in the neutral regions outside the space charge layer. If we denote the electrostatic potential in the neutral *n*- and *p*-type regions by V_n and V_p, respectively, then Figure 2.6 shows a plot of $V(x)$ as a function of distance x in the semiconductor. Since the vector $\mathscr{E} = -\boldsymbol{\nabla}V$, and $\mathscr{E} = -\mathscr{E}\hat{\mathbf{i}}$ (where $\mathscr{E}$ is positive and $\hat{\mathbf{i}}$ is a unit vector in the x direction), we have, as shown in Figure 2.6, the result that

$$dV/dx > 0 \qquad (2.1)$$

meaning that the electrostatic potential V of an electron in a *p–n* junction increases on going from the *p* region to the *n* region. The existence of the built-in electric field therefore increases the electrostatic potential of an electron in the neutral *n* side of the junction by an amount

$$V_n - V_p \equiv V_0 \qquad (2.2)$$

where V_n and V_v are the constant electrostatic potentials in the neutral

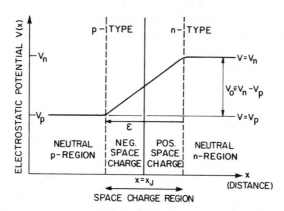

Figure 2.6. Electrostatic potential $V(x)$ shown schematically as a function of distance x in an ideal abrupt *p–n* junction. The built-in electric field $\mathscr{E}$ is confined to the space charge region and the point $x = x_J$ is the location of the junction between the *p*- and *n*-type sides denoted by the vertical solid line.

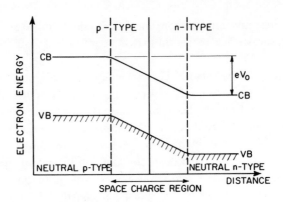

Figure 2.7. Electron energy shown schematically as a function of distance in the ideal abrupt *p–n* junction. The shading indicates filled valence band states.

n and *p* regions, respectively. The quantity V_0 is called the contact potential or the diffusion potential.

We can now interpret these results in terms of electron energy. The potential energy of an electron of electrostatic potential V is $-eV$. We see that the result of the built-in electric field is to change the potential energy of the electrons in the neutral *n* region by an amount

$$-e(V_n - V_p) \equiv -eV_0 \qquad (2.3)$$

relative to the electrons in the neutral *p* region. Since the band diagrams we use are plots of electron energy, the electron energy levels in the neutral *n* region are lowered by an amount eV_0. Figure 2.7 shows electron energy plotted as a function of distance in the junction after equilibrium has been attained. This figure shows the valence band levels and conduction band levels as a function of distance in the junction, and we see that electron energies, in both valence and conduction bands, are lower by an amount eV_0 in the neutral *n* region than they are in the neutral *p* region. There is thus an energy difference eV_0 between valence band states in the neutral *n* region and valence band states in the neutral *p* region, and an energy difference eV_0 between conduction band states in the neutral *n* region and conduction band states in the neutral *p* region. The situation may be described physically by saying that the flow of electrons due to diffusion produces an electric field that opposes that flow by lowering the electron energy in the *n*-type side of the junction. There is thus created an energy barrier of height eV_0 to electron flow from the *n* region to the *p* region.

At equilibrium, there is also a barrier of height of eV_0 to the flow of holes from the *p* region to the *n* region. This can be seen from Figure 2.7

if we recall that, on an electron energy diagram, holes tend to "float," so holes in the valence band in the *p* side "see" an energy barrier of height eV_0 between them and the valence band in the neutral *n*-type region. (Another way of seeing the energy barrier to the flow of holes is to turn Figure 2.7 upside down, which effectively converts it into a plot of hole energy as a function of distance.)

A very useful way of discussing the attainment of equilibrium between *n*- and *p*-type semiconductors forming a *p–n* junction is to consider the Fermi level or chemical potential. An important result[2] of statistical mechanics is that "two systems that can exchange energy and particles are in equilibrium when the temperatures and the chemical potentials are equal." We consider again *n*- and *p*-type semiconductors which are in contact, before the attainment of equilibrium, by redrawing Figure 2.3 to show the Fermi levels on both sides. The result is Figure 2.8, which shows the Fermi level positions in the *n*- and *p*-type semiconductors. In the *p*-type semiconductor, the Fermi level is close to the valence band, while in the *n*-type semiconductor, it is close to the conduction band.

The energy band diagrams for a *p–n* junction at equilibrium may be obtained from Figure 2.8 by making the Fermi energies on the two sides equal. This result is shown in Figure 2.9. This figure shows the change in energy between the *n*- and *p*-type sides taking place discontinuously over a zero length. However, we know physically that the change takes place over a region of nonzero length, so the drawing in Figure 2.10 is more realistic. Figure 2.10 shows the result of equalizing the Fermi energies on the two sides of the junction when equilibrium is attained. This process is equivalent to decreasing the electron energies on the *n* side of the junction. This is, of course, just what we described earlier as the change in the electrostatic potential of the electrons brought about by the built-in electric field.

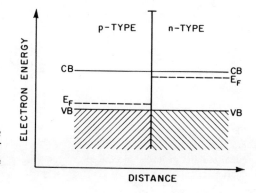

Figure 2.8. Band diagram of a *p–n* junction before the attainment of equilibrium. The Fermi levels E_F are shown as dashed lines.

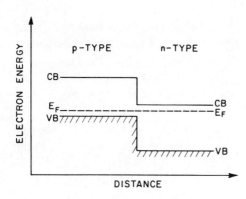

Figure 2.9. Band diagram of a *p–n* junction showing equalization of the Fermi levels.

From Figures 2.8 and 2.9, we see that the shift in energy between *n*- and *p* sides, when their Fermi levels are equalized, is just the difference in the Fermi energies on the two sides. If we call $E_F(n)$ the Fermi energy on the *n* side, and $E_F(p)$ the Fermi energy on the *p* side, then this energy shift is just $E_F(n) - E_F(p)$. However, on comparing Figure 2.10 with Figure 2.7, we see graphically that the energy shift between the two sides is just the height eV_0 of the energy barrier between the two sides. We therefore have the relation that

$$eV_0 = E_F(n) - E_F(p) \tag{2.4}$$

where V_0 is the contact potential defined by equation (2.2). Equation (2.4) shows that the contact potential V_0 is the difference $E_F(n) - E_F(p)$ between the Fermi levels on the two sides, divided by the charge e on the proton.

Finally, we can calculate the width of the space charge region of the *p–n* junction. If Q_+ is the number of positive charges on the *n* side of the space charge region, and Q_- is the number of negative charges on the *p*

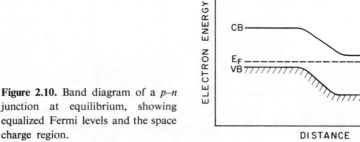

Figure 2.10. Band diagram of a *p–n* junction at equilibrium, showing equalized Fermi levels and the space charge region.

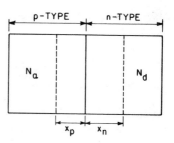

Figure 2.11. Widths x_p and x_n of the space charge regions in the *p*- and *n*-type sides of an ideal abrupt *p–n* junction (shown schematically). The donor and acceptor densities are N_d and N_a, respectively, and the cross-section area of the sample is assumed to be a constant.

side, then electrical neutrality requires that

$$Q_+ = Q_- \tag{2.5}$$

If there are N_d ionized donors per unit volume on the *n* side, and N_a ionized acceptors on the *p* side, then (2.5) becomes

$$N_a x_p = N_d x_n \tag{2.6}$$

where x_p is the distance the space-charge region extends into the *p*-type side of the junction, and x_n is the distance the space charge region extends into the *n*-type side of the junction, and assuming that the cross-section area is the same throughout the sample. This result, and its consequence that x_p and x_n are unequal if N_a and N_d are unequal, are shown in Figure 2.11.

Quantitative Treatment of the *p–n* Junction in Equilibrium

We now have a physical picture of the semiconductor *p–n* junction in equilibrium and have obtained, in Figure 2.7, a qualitative idea of the electron energy level diagram of the junction. We now consider a quantitative discussion[3,4] of the *p–n* junction in equilibrium.

The equilibrium described in the previous section is a dynamic state, in which there are equal and opposite fluxes of both carrier types. For electrons, we have

$\mathbf{J}_{n1}$ = current density of electrons from *n* to *p* due to diffusion

$\mathbf{J}_{n2}$ = current density of electrons from *p* to *n* due to the built-in electric field

The current $\mathbf{J}_{n1}$ is called the diffusion current of electrons; $\mathbf{J}_{n2}$ is called the generation or drift current of electrons. At equilibrium, these two current densities are equal in magnitude and opposite in direction, so the net electron current density is equal to zero. Similarly, we have the hole current densities

J_{p1} = current density of holes from p to n due to diffusion

J_{p2} = current density of holes from n to p due to the built-in electric field

At equilibrium, the diffusion current density of holes J_{p1} is equal in magnitude and opposite in direction to the generation current of holes J_{p2}, so the net hole current density vanishes. The conditions for equilibrium are expressed by the equations

$$\mathbf{J}_{n1} + \mathbf{J}_{n2} = 0 \qquad (2.7)$$

and

$$\mathbf{J}_{p1} + \mathbf{J}_{p2} = 0 \qquad (2.8)$$

The diffusion of carriers across the junction is due to the concentration gradient between the n and p sides of the junction. We recall[5] that the flux $\mathbf{F}$ (the number of particles crossing unit area in unit time) is given by

$$\mathbf{F} = -D\nabla C$$

where C is the concentration of particles and D is the diffusion coefficient or diffusivity. The current density $\mathbf{J}$ of particles of charge q due to diffusion is then

$$\mathbf{J} = q\mathbf{F} = -qD\nabla C \qquad (2.9)$$

We first apply equation (2.9) to the diffusion of electrons of charge $-e$, yielding

$$\mathbf{J}_{n1} = eD_n\nabla n \qquad (2.10)$$

where D_n is the diffusion coefficient for electrons and n is the spatially varying electron density.

To obtain the current density $\mathbf{J}_{n2}$ of electrons due to the built-in electric field $\mathscr{E}$, we recall that the current density of charges q of concentration n and drift velocity $\mathbf{v}$ is equal to $nq\mathbf{v}$. Further, we have

$$\mathbf{v}_p = \mu_p\mathscr{E} \qquad (2.11)$$

for holes and

$$\mathbf{v}_n = -\mu_n\mathscr{E} \qquad (2.12)$$

for electrons, where μ_p and μ_n are the hole and electron mobilities, both of which are defined as positive quantities. The minus sign in equation (2.13) shows that the electron drift velocity is in the direction opposite to that of the electric field. Then the current density $\mathbf{J}_{n2}$ is given by

$$\mathbf{J}_{n2} = nq\mathbf{v}_n = n(-e)(-\mu_n\mathscr{E}) = en\mu_n\mathscr{E} \tag{2.13}$$

showing that the drift (or generation) current density $\mathbf{J}_{n2}$ is parallel to the built-in electric field $\mathscr{E}$.

In a similar manner, we may obtain the current density

$$\mathbf{J}_{p1} = -eD_p\nabla p \tag{2.14}$$

of holes due to diffusion, and the current density

$$\mathbf{J}_{p2} = pq\mathbf{v}_p = ep\mu_p\mathscr{E} \tag{2.15}$$

of holes due to the built-in electric field. The conditions (2.7) and (2.8) then become

$$eD_n\nabla n + en\mu_n\mathscr{E} = 0 \tag{2.16}$$

$$-eD_p\nabla p + ep\mu_p\mathscr{E} = 0 \tag{2.17}$$

as the equations that must hold at equilibrium. From equation (2.16), we can see that the vector ∇n is opposite to the direction of the built-in electric field $\mathscr{E}$ because D_n, μ_n, and n are all positive quantities. This is as it should be because the vector $\mathscr{E}$ points from the n side to the p side of the junction and the vector ∇n points in the spatial direction of increasing electron concentration n, i.e., from the p side to the n side.

If we now specialize the equilibrium conditions (2.16) and (2.17) to our one-dimensional case, they become

$$eD_n\frac{dn(x)}{dx} + en(x)\mu_n\mathscr{E}(x) = 0 \tag{2.18a}$$

$$-eD_p\frac{dp(x)}{dx} + ep(x)\mu_p\mathscr{E}(x) = 0 \tag{2.18b}$$

where $n(x)$, $p(x)$, and $\mathscr{E}(x)$ are the spatially varying electron concentration, hole concentration and built-in electric field, respectively. Physically, these last two equations are the conditions for equilibrium in our one-dimensional *p–n* junction, and state the equality of two opposing current densities for both electrons and holes. The first term in each equation represents the diffusion current density and the second term is the current density due to the action of the built-in electric field. This latter current is often called the drift current density. We may use the Einstein relation[6]

$$eD = \mu k_B T$$

where k_B is Boltzmann's constant, relating the mobility μ of a carrier at temperature T to its diffusion coefficient D, to eliminate the mobility from equations (2.18). The results[7] are

$$\frac{dn(x)}{dx} + \left(\frac{e}{k_B T}\right) n(x) \mathscr{E}(x) = 0 \qquad (2.19a)$$

and

$$\frac{dp(x)}{dx} - \left(\frac{e}{k_B T}\right) p(x) \mathscr{E}(x) = 0 \qquad (2.19b)$$

as the differential equations relating the electron and hole concentrations to the built-in electric field $\mathscr{E}(x)$ of the junction. Since $\mathscr{E}(x)$ is equal to the gradient $-dV/dx$ of the electrostatic potential $V(x)$ in the junction, we obtain, on rearranging terms,

$$\frac{1}{n} \frac{dn}{dx} - \left(\frac{e}{k_B T}\right) \frac{dV}{dx} = 0 \qquad (2.20)$$

and

$$\frac{1}{p} \frac{dp}{dx} + \left(\frac{e}{k_B T}\right) \frac{dV}{dx} = 0 \qquad (2.21)$$

as the differential equations relating the spatially varying electrostatic potential $V(x)$ to the spatially varying hole and electron concentrations $p(x)$ and $n(x)$ in the junction.

We next consider the boundary conditions subject to which the differential equations (2.20) and (2.21) will be solved. To do so, we must recall that we have been discussing the idealized case[1,8] of an abrupt junction in which the semiconductor changes from n type to p type at a given point in space. This means that we are assuming that the space charge region has abrupt boundaries and that the built-in electric field is confined to the space charge region. Further, this also means that the semiconductor is assumed to be neutral outside the boundaries of the space charge region. The model of an ideal abrupt junction is clearly an approximation but is amenable to calculation, and more realistic models[8,9] (e.g., graded junctions) do not further illuminate the physics significantly.

We set up the geometry of the junction as shown in Figure 2.12, in which the junction is located at $x = 0$, the region of positive x is n type

Figure 2.12. Idealized abrupt *p–n* junction located at $x = 0$, with $-x_p \leq x \leq 0$ as the negative space charge layer on the *p*-type side, and $0 \leq x \leq x_n$ as the positive space charge layer on the *n*-type side. The ionized donor and acceptor concentrations are, respectively, N_d and N_a.

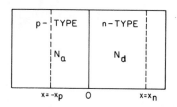

with donor concentration N_d, and the region of negative x is p type with acceptor concentration N_a. The space charge layer in the *n*-type side extends from $x = 0$ to $x = x_n$, and the space charge layer on the *p*-type side extends from $x = 0$ to $x = -x_p$. We note that equation (2.6) implies that the magnitude $|x_n|$ will be unequal to the magnitude $|-x_p|$ if the donor concentration N_d is unequal to the acceptor concentration N_a, so the space charge region may extend different distances into the n and p sides of the junction.

We consider next the differential equation (2.21) for the hole concentration $p(x)$ and set the following boundary conditions on the variables $p(x)$ and $V(x)$. First, that the hole concentration p has its equilibrium values for *n*- and *p*-type material in the neutral regions outside the space charge region. If we use the symbols p_p and p_n for the equilibrium hole concentrations in *p*- and *n*-type material respectively, this boundary condition is

$$p = p_p \qquad \text{for } -x_p \geq x \tag{2.22}$$

and

$$p = p_n \qquad \text{for } x \geq x_n \tag{2.23}$$

Our second boundary condition is that the electrostatic potential $V(x)$ has particular values in the neutral regions outside the space charge region. This condition is expressed as

$$V = V_p \qquad \text{for } -x_p \geq x \tag{2.24}$$

and

$$V = V_n \qquad \text{for } x \geq x_n \tag{2.25}$$

Equations (2.22)–(2.25) tell us that the hole concentration varies from $p = p_p$ to $p = p_n$, and the electrostatic potential varies from $V = V_p$ to $V = V_n$, across the space charge region which extends from $x = -x_p$ to $x = x_n$. Rearranging (2.21) gives

$$\frac{1}{p} dp = -\left(\frac{e}{k_B T}\right) dV \tag{2.26}$$

which can be integrated over the space charge region as

$$-\frac{e}{k_BT}\int_{V_p}^{V_n} dV = \int_{p_p}^{p_n} \frac{1}{p}\, dp \tag{2.27}$$

to give

$$\ln\left(\frac{p_n}{p_p}\right) = \frac{-e}{k_BT}\,(V_n - V_p) = \frac{-eV_0}{k_BT} \tag{2.28}$$

where the definition (2.2) of the contact potential V_0 has been used. Rewriting (2.28) gives

$$p_n/p_p = \exp(-eV_0/k_BT) \tag{2.29}$$

We note in passing that equation (2.4) gives V_0 in terms of the Fermi levels in the neutral n and p regions; the latter are (as discussed in Chapter 1) determined by the carrier concentrations N_d and N_a. The contact potential is thus determined by the impurity densities on the two sides of the junction.

Equation (2.29) is important because it gives the ratio of the hole concentrations in the neutral n and p regions. Since $V_0 \equiv V_n - V_p$ is a positive quantity, (2.29) tells us that p_n is smaller than p_p. This is as it should be; there are fewer holes in n-type material than there are in p-type material. Further, we recall that, at equilibrium at a temperature T, the product

$$np \equiv n_i^2 \tag{2.30}$$

is a constant, so the np product is a constant in both the neutral n and p regions of the junction. Since this is the case, it is true that

$$n_np_n = n_pp_p = n_i^2 \tag{2.31}$$

where n_n is the electron concentration in the neutral n region and n_p is the electron concentration in the neutral p region. Combining (2.31) with (2.29), we get

$$\frac{p_n}{p_p} = \frac{n_p}{n_n} = \exp\left(\frac{-eV_0}{k_BT}\right) \tag{2.32}$$

for the two ratios p_n/p_p and n_p/n_n. We also know that, since $p_p = N_a$, we have from equation (2.30) that

$$n_p = n_i^2/N_a \tag{2.33}$$

in the neutral *n*-type region. Since $n_n = N_d$, (2.30) gives

$$p_n = n_i^2/N_d \qquad (2.34)$$

in the neutral *n*-type region. Equations (2.32)–(2.34) give the carrier concentrations in the neutral regions as functions of V_0 and the impurity concentrations. These equations will be important in our study of current flow through the junction.

We now turn our attention to the space-charge region of the junction. From the geometry of Figure 2.12, we see that electrical neutrality requires that

$$N_d x_n = N_a x_p \qquad (2.35)$$

where N_d and N_a are the densities of ionized donors and acceptors, respectively. At any point x, the space charge density $\varrho(x)$ is given by

$$\varrho(x) = -en(x) + ep(x) - eN_a + eN_d \qquad (2.36)$$

since ionized donors and acceptors have charges $+e$ and $-e$, respectively. Equation (2.36) gives us the space charge density $\varrho(x)$ at any point x of the space charge region. At any point x, Poisson's equation

$$\frac{d^2V}{dx^2} = -\frac{4\pi}{\varepsilon} \varrho(x) \qquad (2.37)$$

for the electrostatic potential $V(x)$ must be satisfied; in (2.37), ε is the dielectric constant. Substituting the expression (2.36) for the space charge density into Poisson's equation gives

$$\frac{d^2V}{dx^2} = -\frac{4\pi}{\varepsilon} \left[-en(x) + ep(x) - eN_a + eN_d \right] \qquad (2.38)$$

solution of which will give us $V(x)$ in the space charge region. We next make the reasonable approximation[10] that the *net* number $p(x) - n(x)$ of mobile free carriers in the space charge region is small compared to either N_a or N_d, the numbers, respectively, of fixed negative and positive charges. With this approximation, the space charge density $\varrho(x)$ becomes

$$\varrho(x) \cong -eN_a \qquad \text{for } -x_p \le x \le 0 \qquad (2.39)$$

and

$$\varrho(x) \cong +eN_d \qquad \text{for } 0 \le x \le x_n \qquad (2.40)$$

so Poisson's equation (2.38) reduces to

$$\frac{d^2V}{dx^2} = \frac{4\pi e}{\varepsilon} N_a \qquad \text{for } -x_p \leq x \leq 0 \qquad (2.41)$$

and

$$\frac{d^2V}{dx^2} = \frac{-4\pi e}{\varepsilon} N_d \qquad \text{for } 0 \leq x \leq x_n \qquad (2.42)$$

for the two parts of the space charge region. The charge distribution represented by (2.39) and (2.40) is shown in Figure 2.13. Since it is true that

$$N_a x_p = N_d x_n \qquad (2.43)$$

the two rectangular areas in Figure 2.13 are equal, thus demonstrating the required overall charge neutrality. We keep in mind that we are dealing with the ideal abrupt junction whose charge density distribution is given in Figure 2.13.

We now integrate Poisson's equation (2.41) and (2.42) for the electrostatic potential $V(x)$ and electric field $\mathscr{E}(x) = -dV/dx$ in the space charge region $-x_p \leq x \leq x_n$. Considering first the *n*-type side $0 \leq x \leq x_n$ of the space charge region, (2.42) can be written as

$$\frac{d^2V}{dx^2} = -\frac{d\mathscr{E}}{dx} = \frac{-4\pi e}{\varepsilon} N_d \qquad (2.44)$$

where $\mathscr{E} = \mathscr{E}(x)$ is the built-in electric field. Equation (2.44) can be integrated immediately to give

$$\mathscr{E}(x) = \frac{4\pi e}{\varepsilon} N_d x + C_1 \qquad (2.45)$$

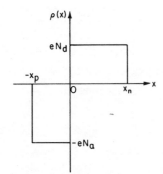

Figure 2.13. Space charge density $\varrho(x)$ given by equations (2.39) and (2.40) shown as a function of x.

The constant of integration C_1 is determined by the boundary condition, appropriate for the ideal junction model we are considering, that the built-in electric field is confined to the space charge region and vanishes outside of it. This condition is expressed as

$$\mathscr{E}(x_n) = 0 \qquad (2.46)$$

which, when applied to (2.45), leads to the value

$$C_1 = (-4\pi e/\varepsilon)N_d x_n \qquad (2.47)$$

for the constant of integration C_1. We thus obtain the expression

$$\mathscr{E}(x) = \frac{4\pi e}{\varepsilon} N_d(x - x_n) \qquad (2.48)$$

for the electric field in the *n*-type side $0 \le x \le x_n$ of the space-charge region, in which the space charge is positive. In a similar manner, considering the region $-x_p \le x \le 0$ of negative space charge on the *p*-type side of the junction, (2.41) becomes

$$\frac{d^2V}{dx^2} = -\frac{d\mathscr{E}}{dx} = \frac{4\pi e}{\varepsilon} N_a \qquad (2.49)$$

which, on integrating, gives

$$\mathscr{E}(x) = \frac{-4\pi e}{\varepsilon} N_a x + C_2 \qquad (2.50)$$

for the built-in electric field in the negative space charge region. The constant C_2 is determined by the boundary condition

$$\mathscr{E}(-x_p) = 0 \qquad (2.51)$$

a condition restricting the electric field to the space charge region and analogous to (2.46). The resulting value of $C_2 = (-4\pi e/\varepsilon)N_a x_p$ leads to

$$\mathscr{E}(x) = \frac{-4\pi e}{\varepsilon} N_a(x + x_p) \qquad (2.52)$$

for the built-in electric field in the region $-x_p \le x \le 0$ of negative space charge. We note from (2.48) and (2.52) that $\mathscr{E}(x)$ is negative for all x in the region $-x_p \le x \le x_n$, meaning that the built-in electric field is directed in the $-x$ direction, i.e., from the *n* side of the junction to the *p*

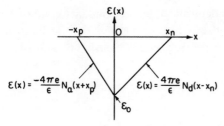

Figure 2.14. Electric field $\mathscr{E}(x)$ in the space charge region $-x_p \le x \le x_n$ shown as a function of x for the ideal abrupt *p–n* junction whose charge density distribution is that in Figure 2.13.

side. We see also that the electric field must be continuous at $x = 0$, so, using (2.48) and (2.52), we have

$$\mathscr{E}(0) = \frac{-4\pi e}{\varepsilon} N_d x_n = \frac{-4\pi e}{\varepsilon} N_a x_p \qquad (2.53)$$

in agreement with the condition (2.43) expressing the overall electrical neutrality of the space charge region. Figure 2.14 shows a plot of $\mathscr{E}(x)$ given by equations (2.48) and (2.52) as a function of distance x in the junction. From this figure, we see that the built-in electric field has its maximum magnitude $\mathscr{E}_0 \equiv |\,\mathscr{E}(0)\,|$, where

$$\mathscr{E}_0 = \frac{4\pi e}{\varepsilon} N_a x_p = \frac{4\pi e}{\varepsilon} N_d x_n \qquad (2.54)$$

at the point $x = 0$, i.e., at the position of the junction between the *n*-type and *p*-type regions. From the figure, we see that the magnitude of the electric field increases linearly from zero at $x = x_n$, reaches its maximum magnitude $\mathscr{E}_0$ at the junction, and then decreases linearly to zero at $x = -x_p$.

Since the electric field $\mathscr{E} = -dV/dx$, we now integrate again to obtain the electrostatic potential $V(x)$ in the space charge region. We will integrate (2.48) and (2.52) subject to the following boundary conditions which must be obeyed by $V(x)$. In the neutral *n*-type region, the electrostatic potential is equal to a constant, denoted as before by V_n, so we have

$$V(x_n) = V_n \qquad (2.55)$$

Similarly, in the neutral *p*-type region, the electrostatic potential is a constant, denoted by V_p, so

$$V(-x_p) = V_p \qquad (2.56)$$

Further, $V(x)$ must be continuous at the junction located at $x = 0$.

Integrating (2.48) written as $-dV/dx = (4\pi e/\varepsilon)N_d(x - x_n)$ gives

$$V(x) = (-4\pi e/2\varepsilon)N_d(x_n - x)^2 + C_3 \qquad (2.57)$$

and (2.55) yields the result that the constant $C_3 = V_n$, so we have

$$V(x) = (-2\pi e/\varepsilon)N_d(x_n - x)^2 + V_n \tag{2.58}$$

for the electrostatic potential in the region $0 \le x \le x_n$ of positive space charge. Similarly, integrating (2.52) written in the form

$$-dV/dx = (-4\pi e/\varepsilon)N_a(x + x_p)$$

and using the boundary condition (2.56), gives

$$V(x) = (2\pi e/\varepsilon)N_a(x + x_p)^2 + V_p \tag{2.59}$$

for the electrostatic potential $V(x)$ in the region $-x_p \le x \le 0$ of negative space charge on the *p*-type side of the junction. Finally, since the electrostatic potential must be continuous at $x = 0$, the two solutions (2.58) and (2.59) must have the same value at $x = 0$. This requirement leads to the condition

$$-\frac{2\pi e}{\varepsilon} N_d x_n{}^2 + V_n = \frac{2\pi e}{\varepsilon} N_a x_p{}^2 + V_p$$

which yields the equation

$$\frac{2\pi e}{\varepsilon} [N_d x_n{}^2 + N_a x_p{}^2] = V_n - V_p \equiv V_0 \tag{2.60}$$

where V_0 is the contact potential defined by equation (2.2). Equation (2.60) is a condition that must be obeyed by N_d, N_a, x_p, and x_n, and it, together with the requirement (2.6)

$$N_a x_p = N_d x_n \tag{2.6}$$

of electrical neutrality in the space charge region, constitutes a system of two equations in four unknowns. Equations (2.60) and (2.6) thus determine x_n and x_p as functions of N_a and N_d, so the total width $x_n + x_p$ of the space charge region is determined by the impurity densities N_a and N_d. These results are exhibited in Figure 2.15, which shows the electrostatic potential $V(x)$ given by equations (2.58) and (2.59) plotted as a function of distance x. We note that the features of the curve $V(x)$ in Figure 2.15 are qualitatively similar to the curve in Figure 2.6, but the variation of $V(x)$ in the space-charge region given by (2.58) and (2.59) is, of course, more detailed than that shown qualitatively in Figure 2.6. However, we note that both figures show the important difference V_0 in the electrostatic potential between the *n* and *p* sides of the junction.

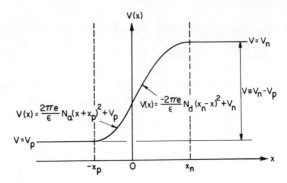

Figure 2.15. Electrostatic potential $V(x)$ given by equations (2.58) and (2.59) shown as a function of x. The figure is drawn for a donor–acceptor ratio $N_d/N_a = \frac{2}{3}$.

Just as we obtained the qualitative variation of electron energy as a function of distance shown in Figure 2.7, we may use Figure 2.15 to obtain the plot shown in Figure 2.16. This figure shows electron energy plotted as a function of distance. If we multiply the curve of $V(x)$ in Figure 2.15 by $-e$, we obtain $-eV(x)$, the potential energy of an electron in the junction. This quantity $-eV(x)$ corresponds, at any point x of the junction, to the valence band energy at that point. In this way, the variation with distance of the valence band electron energy shown in Figure 2.16 is obtained. Since the conduction band energy, at any point, is just the valence band energy plus the magnitude of the energy gap, we obtain also the variation with distance of the conduction band energies. While the variation of electron energy as a function of distance is only schematically shown in Figure 2.16, we could of course obtain analytical expressions for the valence band edge and conduction band edge energies as a function of distance

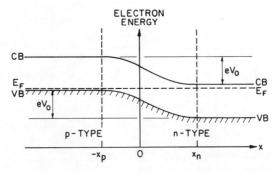

Figure 2.16. Electron energy as a function of distance x in the ideal abrupt *p–n* junction. (The variation in the space charge region is shown only schematically.)

from (2.58) and (2.59) by multiplying by $-e$, etc., as described above. However, the schematic curve shown in Figure 2.16 will be sufficient for our purposes. We note the similarity between Figure 2.16 and Figures 2.7 and 2.10, which we obtained in our earlier qualitative discussion. Figure 2.16 exhibits clearly the energy barrier of height eV_0 between conduction electrons on the n side and conduction electron states on the p side; holes in the valence band also encounter a barrier of height eV_0 between the p and n sides. We will see that the changes in the height of this energy barrier effected by an applied electric potential are the basis for the useful properties of the p–n junction. Figure 2.16 also indicates the position of the Fermi level as a function of distance through the junction in the equilibrium state in which no net currents flow.

We may now calculate the total width $W = x_n + x_p$ of the space charge region as a function of the impurity concentrations N_d and N_a. Considering equations (2.6) and (2.60), which must be true simultaneously, we can rewrite equation (2.60) as

$$V_0 = \frac{2\pi e}{\varepsilon} \left[N_d x_n \left(x_n + \frac{N_a}{N_d} x_p^2 x_n^{-1} \right) \right] \qquad (2.61)$$

Since, from (2.6), we have $x_p = (N_d/N_a)x_n$,

$$x_p^2 = (N_d/N_a)x_n x_p \qquad (2.62)$$

Substituting (2.62) into (2.61) yields

$$V_0 = \frac{2\pi e}{\varepsilon} N_d x_n (x_n + x_p) = \frac{2\pi e}{\varepsilon} N_d x_n W \qquad (2.63)$$

an equation expressing the space charge width W in terms of the contact potential V_0. Using (2.6), we have

$$W = (x_n + x_p) = x_n + \left(\frac{N_d}{N_a} \right) x_n = x_n \left(\frac{N_a + N_d}{N_a} \right)$$

so we obtain the result that

$$x_n = \left(\frac{N_a}{N_a + N_d} \right) W \qquad (2.64)$$

When this result is substituted into (2.63), we obtain

$$V_0 = \frac{2\pi e}{\varepsilon} \left(\frac{N_a N_d}{N_a + N_d} \right) W^2$$

which may be rewritten

$$W = \left[\frac{\varepsilon}{2\pi e} \left(\frac{N_a + N_d}{N_a N_d} \right) V_0 \right]^{1/2} \tag{2.65}$$

an equation giving the width of the space-charge region in terms of V_0, N_a, and N_d.

We may now express the contact potential V_0 in terms of N_a and N_d in order to have an expression for W in terms of the impurity concentrations alone. From equation (2.28), we have

$$V_0 = (k_B T/e) \ln(p_p/p_n) \tag{2.66}$$

where p_p and p_n are the hole concentrations in the neutral p and n regions, respectively. From (2.34),

$$p_n = n_i^2/N_d \tag{2.34}$$

and we know also that the hole concentration in the neutral p region is equal to the ionized acceptor density N_a, so

$$p_p = N_a \tag{2.67}$$

Combining (2.66), (2.67), and (2.34) yields

$$V_0 = \frac{k_B T}{e} \ln\left(\frac{N_a N_d}{n_i^2} \right) \tag{2.68}$$

which, when substituted in (2.65) gives the expression

$$W = \left[\frac{\varepsilon k_B T}{2\pi e^2} \left(\frac{N_a + N_d}{N_a N_d} \right) \ln\left(\frac{N_a N_d}{n_i^2} \right) \right]^{1/2} \tag{2.69}$$

for the total width W of the space charge region in terms of N_d and N_a. From this equation, W can be calculated for typical values of the parameters. Using $N_a = N_d = 10^{15}$ cm^{-3}, and the dielectric constant of silicon ($\varepsilon = 11.7$), one finds that W is about 1.3×10^{-4} cm, a typical value for a silicon $p–n$ junction.

It is often of interest to know the penetration distances x_n and x_p of the space charge region into the n and p sides of the junction. We can write, using (2.6),

$$W = x_p + x_n = x_p + \frac{N_a}{N_d} x_p = x_p \left(\frac{N_a + N_d}{N_d} \right) = x_p \left(1 + \frac{N_a}{N_d} \right)$$

leading to the relation

$$x_p = \frac{W}{1 + N_a/N_d} \tag{2.70}$$

Similarly, one can show that

$$x_n = \frac{W}{1 + N_d/N_a} \tag{2.71}$$

Equations (2.70) and (2.71) show that, the larger the ionized impurity density on a given side of the junction, the smaller the penetration of the space charge region into that side of the junction. In other words, lighter "doping" means that the space charge region extends further into that side of the junction. Heavily doped junctions have thin space charge layers, and vice versa. We note also, from (2.65), that the width of the space charge layer varies as the square root of the difference in electrostatic potential across the junction. In the equilibrium case we are considering, the electrostatic potential is the contact potential V_0, but, as we will see, applying a voltage across the junction changes this electrostatic potential difference and hence changes the space charge region width. Other useful equations for the width of the space charge region may be found in the literature.[11]

In summary, we see that our quantitative treatment of the abrupt *p–n* junction gave the same basic results as did the qualitative treatment, but with additional detail. It is useful at this point to reiterate that our model of a *p–n* junction is an idealized one in which the space charge region has abrupt boundaries, outside of which the semiconductor is assumed to be neutral. Further, we are tacitly assuming that we may use the results obtained earlier for carrier densities, Fermi levels, etc. in a nondegenerate semiconductor. (For details of these assumed approximations, the interested reader may consult the more advanced literature.[1,8]) Our model is a fairly good description of some alloyed and epitaxial junctions[12] but is often inadequate for discussing diffused junctions.[12] However, this idealized model provides enough physical insight for us to understand the action of a *p–n* junction under an applied potential, the subject which makes the junction of such great technological interest and use.

Before proceeding to do so, however, we digress briefly to discuss two topics which we will need in discussing the behavior of a *p–n* junction under an applied potential. These are the effect of an applied potential on the energy bands in a semiconductor and the diffusion and recombination of excess (i.e., nonequilibrium) carriers introduced into a semiconductor.

Effect of an Applied Potential on Electron Energy Bands

Consider a homogeneous semiconductor bar of length L and constant cross-section area. Let one end of the bar be at $x = 0$ and the other at $x = L$, and apply a constant electrostatic potential V across the sample. If we take the zero of potential, $V = 0$, at $x = 0$, then the applied potential $V(x)$ at point x (where $0 \leq x \leq L$) is

$$V(x) = Vx/L \tag{2.72}$$

At any point x, the effect of the applied potential $V(x)$ is to change the electron energy at that point by an amount

$$-eVx/L \tag{2.73}$$

If we look at electron energy as a function of distance x in the sample under the applied potential V, it will be as shown in Figure 2.17, which shows the conduction band (CB) and valence band (VB) of the semiconductor sample as a function of distance x. The application of the potential, corresponding in this case to a constant applied electric field of magnitude (V/L), lowers the electron energy levels, relative to the zero of potential at $x = 0$, as shown in Figure 2.17. At the end $x = L$ of the sample, the decrease in electric energy is equal to (eV), as shown in the figure. This effect is the basic effect of an applied electric field, and is sometimes called "tilting the bands."

We note specifically that, for our model of an abrupt $p–n$ junction, "band tilting" will occur only at points in the space charge region. This is because our model of a $p–n$ junction is one in which only the space charge region is depleted of free carriers and is of high electrical resistivity compared to the neutral n- and p-type regions. In our junction model, then, all of the applied potential will appear across the space charge region, so "band tilting" will occur only in that region.

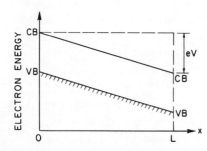

Figure 2.17. Electron energy bands as a function of distance x in a homogeneous semiconductor sample of length L with an applied electrostatic potential V such that, at point x, $V(x) = Vx/L$. The abbreviations CB and VB refer to the conduction and valence band edges, respectively.

Diffusion and Recombination of Excess Carriers[13,14]

We next want to consider excess carriers in a semiconductor. By "excess" is meant carriers introduced in some manner so that the density of such carriers has, at least temporarily, a value different from the equilibrium value. This is not an equilibrium situation, and the system will approach equilibrium as the excess carriers recombine and disappear and the carrier density approaches its equilibrium value.

We will discuss this situation by using the equation of continuity,[15] which, for electric charge, is

$$\frac{\partial \varrho}{\partial t} + \mathbf{\nabla} \cdot \mathbf{J} = G - R \tag{2.74}$$

where ϱ is the electric charge density, $\mathbf{J}$ is the current density, and G and R are the rates of generation and destruction, respectively, of electric charge per unit volume. We consider holes for concreteness. The current density $\mathbf{J}_p$ of holes will have, in general, a drift component given by equation (2.15) and a diffusion component given by (2.14), so

$$\mathbf{J}_p = -eD_p\mathbf{\nabla}p + ep\mu_p\mathscr{E} \tag{2.75}$$

where $\mathscr{E}$ is the electric field producing the drift current density. The electric charge density due to holes is

$$\varrho = ep \tag{2.76}$$

where p is the density of holes. Since we will be interested later in a situation in which drift currents are negligible,[16] we set $\mathscr{E} = 0$ in equation (2.75), which becomes, in one dimension,

$$J_p = -eD_p\frac{dp}{dx} \tag{2.77}$$

In one dimension, the equation of continuity (2.74) becomes, using (2.76),

$$e\frac{dp}{dt} - eD_p\frac{d^2p}{dx^2} = G_p - R_p \tag{2.78}$$

where the subscript p on G and R refers to holes. In equation (2.78), G_p is the rate of generation of charge (holes), per unit volume, due to external influences (e.g., photons), R_p is the *net* rate of destruction of charge (holes) due to recombination, and we include the effects of spontaneous thermal generation.[17]

We now want to consider an expression for R_p. Let p_0 be the equilibrium value of the hole concentration, so $p - p_0$ is the excess hole density in the nonequilibrium situation. For small concentrations of excess carriers, the rate of recombination of these excess carriers will be proportional[18] to the excess carrier density $p - p_0$. The rate of destruction of charge per unit volume will then be

$$R_p = e \frac{p - p_0}{\tau_p} \qquad (2.79)$$

where τ_p is a constant called the hole lifetime.[13,19] Substituting equation (2.79) into (2.78) gives

$$e \frac{dp}{dt} - eD_p \frac{d^2p}{dx^2} = G_p - e\left(\frac{p - p_0}{\tau_p}\right) \qquad (2.80)$$

as the equation governing the hole concentration $p(x, t)$. If the rate G_p of generation is equal to zero, (2.80) becomes

$$\frac{dp}{dt} = D_p \frac{d^2p}{dx^2} - \frac{1}{\tau_p} (p - p_0) \qquad (2.81)$$

We consider a situation in which holes are supplied to one end of a semi-infinite bar of semiconductor at a rate such that a steady state obtains. By this is meant that holes enter the bar, shown in Figure 2.18, at its left end $x = x_1$, and diffuse away to the right, recombining and disappearing as they go. In the steady state, the hole concentration p will be constant in time at every point $x \geq x_1$ of the bar. Then $dp/dt = 0$ in the steady state, and equation (2.81) becomes

$$\frac{d^2p}{dx^2} - \frac{1}{D_p\tau_p} (p - p_0) = 0 \qquad (2.82)$$

subject to the following boundary conditions. At the left end $x = x_1$ of the bar, the hole density has the constant value $p(x_1)$. As x approaches $+\infty$, the hole density must approach its equilibrium value p_0 as all of the

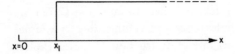

Figure 2.18. Semi-infinite bar of semiconductor with its left end at the point $x = x_1$.

excess holes recombine and disappear, so we must have $p(+\infty) = p_0$, where p_0 is a constant.

The quantity $(D_p \tau_p)^{1/2}$ has the dimensions of a length, and we define the diffusion length L_p for holes by

$$L_p{}^2 \equiv D_p \tau_p \tag{2.83}$$

The diffusion length is a measure of the distance a carrier diffuses in its lifetime. It can be shown[20] that the diffusion length is the average distance a carrier diffuses before recombining.

We rewrite (2.82) as

$$\frac{d^2}{dx^2}(p - p_0) = \frac{1}{L_p{}^2}(p - p_0) \tag{2.84}$$

where $dp_0/dx = 0$ because the equilibrium hole density p_0 is a constant. A solution of (2.84) is

$$p - p_0 = C_1 e^{x/L_p} + C_2 e^{-x/L_p} \tag{2.85}$$

where C_1 and C_2 are constants to be determined by the boundary conditions. Since p must approach p_0 as x approaches $+\infty$, C_1 must equal zero, so

$$p - p_0 = C_2 e^{-x/L_p} \tag{2.86}$$

The second boundary condition is that $p(x_1)$ be a constant, so (2.86) gives us

$$p(x_1) - p_0 = C_2 \exp(-x_1/L_p)$$

and

$$C_2 = [p(x_1) - p_0]e^{x_1/L_p} \tag{2.87}$$

which, substituted into equation (2.86) gives

$$p(x) - p_0 = [p(x_1) - p_0]\exp[-(x - x_1)/L_p] \tag{2.88}$$

as the solution for the spatially dependent hole concentration $p(x)$. If we define the *excess* hole density $\Delta p(x)$ at the point x by the equation

$$\Delta p(x) = p(x) - p_0 \tag{2.89}$$

then the solution (2.88) becomes

$$\Delta p(x) = [\Delta p(x_1)] \exp[-(x - x_1)/L_p] \tag{2.90}$$

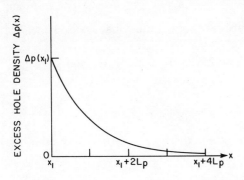

Figure 2.19. Exponential decrease of the excess hole density $\Delta p(x)$ with distance x, from equation (2.90). L_p is the diffusion length for holes and $\Delta p(x_1)$ is the value of $\Delta p(x)$ at $x = x_1$.

where $\Delta p(x_1)$ is the value of $\Delta p(x)$ at the left end $x = x_1$ of the bar. Equation (2.90) gives the space dependence of the excess hole density $\Delta p(x)$ and tells us that the excess hole density $\Delta p(x_1)$ created at the left end of the bar dies out exponentially with distance x into the bar, as shown in Figure 2.19. From that figure, we see that the *excess* hole density p approaches zero as x gets very large, while the *total* hole density approaches its equilibrium value p_0 in the same limit. As expected from equation (2.90), $\Delta p(x)$ falls to $1/e$ of its initial value $\Delta p(x_1)$ in a distance equal to the diffusion length L_p for holes.

In the same way, we can consider the equation of continuity (2.74) for electrons, and obtain the equation

$$n(x) - n_0 = [n(x_1) - n_0] \exp[-(x - x_1)/L_n] \tag{2.91}$$

describing the diffusion and recombination of excess electrons. In (2.91), $n(x)$ is the electron density at point x, n_0 is the equilibrium electron density, and the electrons are created at $x = x_1$ in a semi-infinite bar. The quantity L_n is the diffusion length for electrons, defined by

$$L_n^2 \equiv D_n \tau_n \tag{2.92}$$

where D_n is the diffusion coefficient for electrons and τ_n is the electron lifetime. Again, defining the excess electron density $\Delta n(x)$ by the relation

$$\Delta n(x) = n(x) - n_0 \tag{2.93}$$

equation (2.91) becomes

$$\Delta n(x) = [\Delta n(x_1)] \exp[-(x - x_1)/L_n] \tag{2.94}$$

the equation, describing the spatial dependence of the excess electron density $\Delta n(x)$, which is the analog of equation (2.90) for holes.

The process of creating excess carriers in a semiconductor is called *injection* of carriers. It can be done in a number of ways, among which two of the most important are electrical injection and optical injection. We will discuss injection further later. We introduce here some terminology which we will use throughout our study of semiconductor devices. Electrons in an *n*-type semiconductor or holes in a *p*-type semiconductor, are called *majority* carriers. Holes in an *n*-type semiconductor, or electrons in a *p*-type semiconductor, are called *minority* carriers. For example, if we were to create a local excess of concentration of holes in an *n*-type semiconductor, we would speak of the injection of minority carriers in the semiconductor. On the other hand, if we were to remove carriers from some region of the semiconductor, we would speak of carrier *extraction*.

Qualitative Discussion of a Junction under an Applied Potential

At equilibrium, the current densities of electrons and holes due to diffusion (i.e., the effect of concentration gradients) and due to drift (i.e., the effect of the built-in electric field) just cancel each other. This result was expressed in equations (2.7) and (2.8), which can be rewritten as

$$\mathbf{J}_{n1}(\text{diffusion}) + \mathbf{J}_{n2}(\text{drift}) = 0 \qquad (2.95)$$

$$\mathbf{J}_{p1}(\text{diffusion}) + \mathbf{J}_{p2}(\text{drift}) = 0 \qquad (2.96)$$

to emphasize that, for both electrons and holes, there are two different current densities, which move in opposite directions.

If an electrostatic potential V_a is applied across the junction, the situation is changed, equilibrium no longer holds, and net currents flow through the junction. In order to investigate this nonequilibrium situation, we make the assumption that all of the applied potential V_a appears across the space

Figure 2.20. Effect of an applied forward bias potential V_a and applied electric field $\mathscr{E}_a$ on a *p–n* junction; $\mathscr{E}$ is the built-in electric field; W is the width of the space charge region.

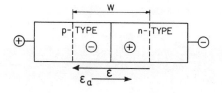

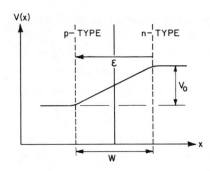

Figure 2.21. Electrostatic potential $V(x)$ (schematic) in the abrupt *p–n* junction, as a function of distance, for the equilibrium case; $\mathscr{E}$ is the built-in electric field, W is the width of the space charge region, and V_0 is the contact potential.

charge region of width W, which is depleted of carriers and hence is a region of higher electrical resistivity than the neutral *n*- and *p*-type regions. Recalling that the built-in electric field $\mathscr{E}$ in the junction at equilibrium is directed from the *n* side of the space charge region toward the *p* side, we see that applying a negative potential to the *n* region will *decrease* the built-in electric field. This is shown in Figure 2.20, in which $\mathscr{E}$ is the built-in electric field and $\mathscr{E}_a$ is the applied electric field, of magnitude V_a/W, due to the negative potential V_a applied to the *n*-type side of the junction. In this case, in which the *n* side is biased negatively, the resulting net electric field in the space charge region is equal to $\mathscr{E} - \mathscr{E}_a$ since $\mathscr{E}$ and $\mathscr{E}_a$ are antiparallel. If the *n* side is biased positively, then the applied electric field $\mathscr{E}_a$ is parallel to the built-in field $\mathscr{E}$, and the net resulting electric field in the space charge region has the increased value $\mathscr{E} + \mathscr{E}_a$.

There is a standard terminology by which the applied potential is described. If the *n*-type region is biased negatively, it is called *forward* bias. If the *n*-type region is biased positively, it is called *reverse* bias.

We need now to consider the effect of an applied potential V_a (or applied electric field $\mathscr{E}_a$) on the width W of the space charge region. Under an applied forward bias, the magnitude of the electric field in the space

Figure 2.~2. Electrostatic potential $V(x)$ (schematic) in the abrupt *p–n* junction, as a function of distance x, for the case of applied forward bias; V_a is the applied potential and $\mathscr{E}_a$ the applied electric field. The net electric field is $\mathscr{E} - \mathscr{E}_a$ and the potential difference is decreased to $V_0 - V_a$. The width W of the space charge layer has decreased.

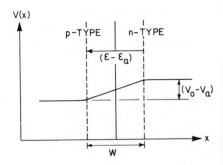

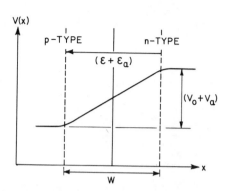

Figure 2.23. Electrostatic potential $V(x)$ (schematic) in the abrupt p–n junction, as a function of distance x, for the case of applied reverse bias; V_a is the applied potential and $\mathscr{E}_a$ the applied electric field. The net electric field is $\mathscr{E} + \mathscr{E}_a$ and the potential difference is increased to $V_0 + V_a$. The width W of the space charge layer has increased.

charge region decreases to $\mathscr{E} - \mathscr{E}_a$. This smaller electric field means that there must be fewer uncompensated ionized impurity atoms in the space charge region. This in turn means that the space charge region must be narrower in extent, so we see that an applied forward bias decreases the width W of the space charge region. Under reverse bias, the electric field increases in magnitude to $\mathscr{E} + \mathscr{E}_a$, so the width W of the space charge region is larger under reverse bias than it is at equilibrium.

To examine next the effect of the applied potential V_a on the difference in electrostatic potential between the n and p sides of the junction, we have seen that both the electric field and the width of the space charge region decrease under forward bias. Since the electrostatic potential $V(x)$ is equal to $W\mathscr{E}(x)$, the difference in electrostatic potential between the n and p regions is smaller under forward bias than it is at equilibrium. This result is shown in Figures 2.21 and 2.22. The sign of the applied potential is defined as follows. For forward bias, the applied potential is equal to V_a, where V_a is a positive quantity. For reverse bias, the applied potential is equal to $-V_a$. We see from these figures that, under forward bias, the potential difference between the n and p sides is decreased to $V_0 - V_a$, where V_0

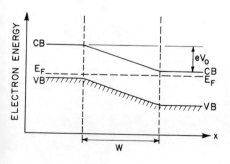

Figure 2.24. Electron energy band diagram (schematic) of the abrupt p–n junction at equilibrium. The energy barrier between n and p sides is eV_0 where V_0 is the contact potential, and W is the width of the space charge layer.

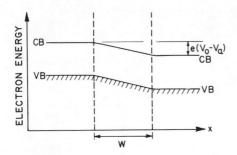

Figure 2.25. Electron energy band diagram (schematic) of the abrupt *p–n* junction under a forward bias potential V_a. The magnitude of the energy barrier has decreased to $e(V_0 - V_a)$, and the width W of the space charge layer has decreased.

is the contact potential at equilibrium. Under reverse bias, the opposite is true, and, as shown in Figure 2.23, the difference in potential between the n and p sides is increased to $V_0 + V_a$.

We may also plot the electron potential energy $-eV(x)$ as a function of distance x through the junction for forward and reverse bias. The energy barrier between the n and p sides of the junction is eV_0 at equilibrium, as shown in Figure 2.24, which corresponds to the potential diagram Figure 2.21. As seen in Figure 2.25, the application of a forward bias potential equal to V_a *decreases* the height of the energy barrier to $e(V_0 - V_a)$. The application of a reverse bias *increases* the height of the energy barrier to $e(V_0 + V_a)$, as shown in Figure 2.26.

The central, and most important, result above is that a forward bias applied to the junction decreases the magnitude of the energy barrier between the n- and p-type sides of the junction and a reverse bias increases the magnitude of that barrier. This fact is the basic physics underlying the behavior and operation of a semiconductor *p–n* junction under the influence of an applied electric potential. We shall see that this changing of the magnitude of the energy barrier explains the flow of current in a *p–n* junction under an applied potential, and is the basic physical reason why a semiconductor *p–n* junction acts as a diode rectifier of alternating current.

Figure 2.26. Electron energy band diagram (schematic) of the abrupt *p–n* junction under a reverse bias potential V_a. The magnitude of the energy barrier has increased to $e(V_0 + V_a)$, and the width W of the space charge layer has increased.

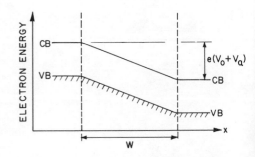

Qualitative Discussion of Current Flow in the Biased Junction

We have just seen that the application of an applied potential changes the magnitude of the energy barrier between the n and p sides of a p–n junction. We now consider the effect of these changes on the particle currents flowing in the junction.

Considering first the electron currents, we recall that there are two electron currents, which flow in opposite directions. The first, denoted J_{n1}(diffusion) is composed of those majority carrier electrons on the n side which have energy sufficient to surmount the energy barrier (of height eV_0 at equilibrium) between the n and p regions. The high-energy end of the energy distribution of majority electrons on the n side will have enough energy to do this. The diffusion current J_{n1}(diffusion) of majority electrons is indicated schematically in Figure 2.27a. The second current, which we earlier called J_{n2}(drift), is composed of minority electrons on the p side which diffuse into the space charge region and are swept down the energy barrier and into the n-type region. Because these minority electrons are generated by intrinsic thermal excitation across the energy gap, the current J_{n2} is usually called the generation current. This current J_{n2}(generation) is shown schematically in Figure 2.27(b); we will now use this terminology in place of the notation J_{n2}(drift) used earlier. The number of minority electrons comprising the current J_{n2}(generation) will be those created within a diffusion length L_n of the space charge region because electrons generated further into the p side will recombine and disappear before they can reach the space charge region to be swept down the energy barrier. These statements are also true of the diffusion current J_{p1}(diffusion) and the generation current J_{p2}(generation) of holes through the junction.

We consider first the electron currents. The diffusion current of electrons J_{n1} is made up of the small fraction of the majority electrons on the n side which have energy sufficient to surmount the energy barrier. The diffusion current J_{n1} will therefore depend on the magnitude of the energy barrier. The smaller the energy barrier, the larger the fraction of majority

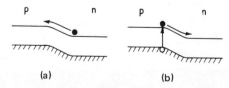

Figure 2.27. (a) Schematic representation of the diffusion current J_{n1} of electrons surmounting the energy barrier and going from the n side to the p side. (b) Schematic representation of the generation current J_{n2} of thermally generated electrons on the p side being swept down the energy barrier by the built-in electric field and entering the n-side.

electrons which can go over it and enter into the p side of the junction. Thus, if the height of the energy barrier is decreased, the magnitude of the diffusion current J_{n1} of majority electrons from n to p will increase.

The generation current of electrons J_{n2} from p to n is made up of all of the minority electrons which diffuse into the junction region from the p side. These are relatively few in number but essentially all of them are swept "down the energy hill" into the n side. To a good approximation, then, the generation current J_{n2} of minority electrons is independent of the height of the energy barrier because "it is just as easy for an electron to slide down a high energy hill as a low energy hill," so to speak. Essentially all of the small number of minority electrons that are generated within a diffusion length of the junction will thus enter into the generation current of electrons from p to n.

At equilibrium, the diffusion and generation currents are equal in magnitude and opposite in direction. The diffusion current from n to p is a small fraction of the large number of majority electrons, while the generation current from p to n is a large fraction of the small number of minority electrons.

We can now see how an applied potential affects the electron currents through the junction. Forward bias decreases the height of the energy barrier for electrons between n and p, so the diffusion current J_{n1} increases while the generation current J_{n2} is unchanged. The net electron current $J_{n1} + J_{n2}$ from n to p thus increases on forward bias. In the same way, we can see that an applied forward bias increases the net hole current $J_{p1} + J_{p2}$ moving from p to n. The total net current (composed of both electrons and of holes) moving through the junction increases with increasing forward bias applied to the junction.

To consider the effect of reverse bias, we have seen that reverse bias increases the height of the energy barrier. Thus the diffusion current J_{n1} of electrons from n to p is decreased by an applied reverse bias, while the generation current J_{n2} from p to n is more or less unchanged because the generation current is essentially independent of the barrier height. With increasing reverse bias, the diffusion current J_{n1} decreases until it is negligibly small with respect to the generation current J_{n2}. Eventually, with increasing reverse bias, we reach a state in which the total electron current through the junction is the generation current J_{n2} and is independent of the magnitude of the applied reverse bias. The same conclusions are true of the hole current through the junction. The total current (electrons and holes) through the junction decreases with increasing reverse bias until the current saturates with a magnitude equal to the value of the generation current $J_{n2} + J_{p2}$.

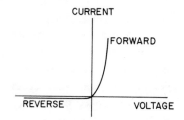

Figure 2.28. Current–voltage plot showing, qualitatively, the behavior expected for forward and reverse biases in a *p–n* junction.

This value of the current is the reverse saturation current of the junction.

From these qualitative conclusions, we expect that the total current J (both electrons and holes), where

$$J = J(\text{diffusion}) + J(\text{generation}) \qquad (2.97)$$

will behave approximately as follows. The total current J will be large and increasing with increasing forward bias, and will be small and constant for increasing magnitude of reverse bias; the situation is shown schematically in Figure 2.28.

Quantitative Treatment of Carrier Injection in the Junction

We now treat quantitatively some of the ideas discussed qualitatively about the *p–n* junction under an applied potential. We begin with a treatment of carrier injection and extraction. We have seen that decreasing the energy barrier between the n and p regions allows more majority carriers to diffuse across the junction; these then become minority carriers on the other side of the junction. When a forward bias is applied to the junction, electrons diffuse into the p region and holes diffuse into the n region, thereby increasing the minority carrier densities at the edges of the space charge region. This is the process of minority carrier injection.

On taking the reciprocal of equation (2.32) and making a slight change in notation, we have

$$\frac{p_{p0}}{p_{n0}} = \frac{n_{n0}}{n_{p0}} = \exp\left(\frac{eV_0}{k_B T}\right) \qquad (2.98)$$

where p_{n0} and p_{p0} are the hole concentrations in the neutral n and p regions, respectively, and n_{p0} and n_{n0} are the electron concentrations in the neutral p and n regions, all at *equilibrium*. Thus p_{n0} is the equilibrium minority hole concentration, p_{p0} is the equilibrium majority hole concentration, etc., and V_0 is the contact potential defined by equation (2.2). We saw that V_0, which appears in the exponential in equation (2.98), is the electrostatic potential difference across the junction space charge region at equilibrium. We would expect then, for the nonequilibrium situation in which an external potential V_a is applied across the junction, that the *total* potential difference across the space charge region would appear in the exponential in the equation, analogous to (2.98), appropriate for the *nonequilibrium* case. We would have, then, for an applied potential V_a, that

$$\frac{p_p}{p_n} = \frac{n_n}{n_p} = \exp\left[\frac{e(V_0 - V_a)}{k_B T}\right] \tag{2.99}$$

for the nonequilibrium case of an applied potential. In (2.99), the applied potential V_a is taken as positive for forward bias and is taken as negative for reverse bias. This is because the total potential difference across the junction has magnitude $V_0 - V_a$, with V_a positive, for forward bias, and magnitude $V_0 + V_a$ for reverse bias. Equation (2.99) is true for the carrier densities throughout the neutral n and p regions and is, in particular, true at the edges $x = x_n$ and $x = -x_p$ of the space charge region. We indicate this explicitly by rewriting (2.99) as

$$\frac{n_n(x_n)}{n_p(-x_p)} = \frac{p_p(-x_p)}{p_n(x_n)} = \exp\left[\frac{e(V_0 - V_a)}{k_B T}\right] \tag{2.100}$$

where $n_n(x_n)$ is the majority electron density at $x = x_n$, $n_p(-x_p)$ is the minority electron density at $x = -x_p$, $p_p(-x_p)$ is the majority hole density at $x = -x_p$, and $p_n(x_n)$ is the minority hole density at $x = x_n$, all for the nonequilibrium situation.

We next make the assumption that the *majority* carrier densities $n_n(x_n)$ and $p_p(-x_p)$ are, under the applied potential V_a, essentially unchanged from their equilibrium values. This assumption is reasonable because the majority carrier densities are so large relative to minority carrier densities in extrinsic semiconductors. This assumption is expressed by the equations

$$n_n(x_n) \cong n_{n0}, \qquad p_p(-x_p) \cong p_{p0} \tag{2.101}$$

which, when substituted in (2.100), give

$$\frac{n_{n0}}{n_p(-x_p)} = \frac{p_{p0}}{p_n(x_n)} = \exp\left[\frac{e(V_0 - V_a)}{k_B T}\right] \tag{2.102}$$

Next, we substitute equation (2.98) for the *equilibrium* majority carrier densities n_{n0} and p_{p0} into equation (2.102). The $\exp[eV_0/k_BT]$ term cancels, leading to the results for the nonequilibrium situation that

$$n_p(-x_p) = n_{p0} \exp(eV_a/k_BT) \qquad (2.103)$$

and

$$p_n(x_n) = p_{n0} \exp(eV_a/k_BT) \qquad (2.104)$$

Equations (2.103) and (2.104) give the minority electron density $n_p(-x_p)$ at the edge $x = -x_p$ of the space charge region and the minority hole density $p_n(x_n)$ at the edge $x = x_n$ of the space charge region, both under an applied potential V_a. In (2.103) and (2.104), n_{p0} is the *equilibrium* minority electron density in the neutral p region (including the point $x = -x_p$) and p_{n0} is the *equilibrium* minority hole density in the neutral n region (including $x = x_n$).

From equations (2.103) and (2.104), we see that, under forward bias, V_a is positive, and the minority carrier densities $n_p(-x_p)$ and $p_n(x_n)$ at the edges of the space charge region both increase to values greater than their equilibrium values n_{n0} and p_{p0}, respectively, because $\exp(eV_a/k_BT)$ is greater than unity for positive values of V_a. This increase in minority carrier density is the process of minority carrier injection. Similarly, under reverse bias, V_a is negative, and the minority carrier densities $n_p(-x_p)$ and $p_n(x_n)$ at the edges of the space charge region decrease to less than their equilibrium values n_{n0} and p_{p0}. This decrease in minority carrier density is the process of minority carrier extraction.

If we define the excess minority electron density Δn_p at the edge $x = -x_p$ of the p-type space charge region by

$$\Delta n_p \equiv n(-x_p) - n_{p0} \qquad (2.105)$$

we see that positive values of Δn_p represent minority carrier injection at $x = -x_p$, and negative values represent minority carrier extraction at that point. Substituting (2.103) into (2.105) we have

$$\Delta n_p = [n_{p0} \exp(eV_a/k_BT)] - n_{p0} \qquad (2.106)$$

$$\Delta n_p = n_{p0}[\exp(eV_a/k_BT) - 1] \qquad (2.107)$$

Equation (2.107) gives the density Δn_p of excess minority electrons injected or extracted (at the edge of the p-type space charge region) when the external potential V_a is applied to the junction. In a similar way, the excess minority hole density $\Delta p_n \equiv p(x_n) - p_{n0}$ at the edge $x = x_n$ of the n-type space charge region is given by

$$\Delta p_n = p_{no}[\exp(eV_a/k_BT) - 1] \tag{2.108}$$

which can be obtained from (2.104). Equation (2.108) gives the density Δp_n of excess minority holes injected or extracted (at the edge of the n-type space charge region) under the external applied potential V_a.

Having obtained the equations (2.107) and (2.108), it is useful at this point to summarize the physical picture of the processes taking place during minority carrier injection or extraction by an applied potential. At equilibrium, the diffusion current of, say, majority electrons from n to p over the energy barrier of magnitude eV_0 is just canceled by the opposing generation current of minority electrons moving from p to n. Under an applied forward bias V_a the magnitude of the energy barrier to the diffusion of majority electrons from n to p is decreased to $e(V_0 - V_a)$, thus increasing the diffusion current of majority electrons into the p region. However, at the same time, the opposing generation current of minority electrons from p to n remains unchanged by the applied potential. The result is an increase Δn_p in the density of minority electrons at the edge of the space charge region on the p-type side of the junction. This increase Δn_p in the minority carrier density on the p side is the process of minority carrier injection. Under an applied reverse bias, exactly the opposite takes place. The generation current of minority electrons from p to n predominates over the decreased diffusion current of electrons of n to p as the magnitude of the energy barrier is increased to $e(V_0 + V_a)$. In this case, Δn_p is negative, meaning that the minority electron density at the edge of the p-type space charge region has decreased. This process is that of minority carrier extraction. While we have used electrons for illustration, the same injection and extraction processes take place for minority holes.

Calculation of the Current through the Junction

We now want to calculate the current flowing through the junction. Denoting the electron current density by $\mathbf{J}_n$ and the hole current density by $\mathbf{J}_p$, we have, from equations (2.10)–(2.15), that

$$\mathbf{J}_n = \mathbf{J}_{n1} + \mathbf{J}_{n2} = eD_n\boldsymbol{\nabla}n + en\mu_n\mathscr{E} \tag{2.109}$$

$$\mathbf{J}_p = \mathbf{J}_{p1} + \mathbf{J}_{p2} = -eD_p\boldsymbol{\nabla}p + ep\mu_p\mathscr{E} \tag{2.110}$$

where the diffusion and drift current densities $\mathbf{J}_{n1}$ and $\mathbf{J}_{n2}$ of electrons and

J_{p1} and J_{p2} of holes, are given by equations (2.10), (2.13), (2.14), and (2.15). In equations (2.109) and (2.110), n and p are the electron and hole concentrations, μ_n and μ_p are the electron and hole mobilities, D_n and D_p are the electron and hole diffusion coefficients, and $\mathscr{E}$ is the electric field. We wish to find the total current density $\mathbf{J}$, where

$$\mathbf{J} = \mathbf{J}_n + \mathbf{J}_p \tag{2.111}$$

The current densities $\mathbf{J}_n$ and $\mathbf{J}_p$ must both satisfy the equation (2.74) of continuity for charge. Using ep and $(-e)n$ for the charge densities due to holes and electrons, respectively, the equations of continuity are

$$e\frac{dp}{dt} + \nabla \cdot \mathbf{J}_p = -e\left(\frac{p - p_0}{\tau_p}\right) \tag{2.112}$$

$$(-e)\frac{dn}{dt} + \nabla \cdot \mathbf{J}_n = +e\left(\frac{n - n_0}{\tau_n}\right) \tag{2.113}$$

In (2.112) and (2.113), the rates of generation G_p and G_n due to external agencies have been set equal to zero, and recombination terms R_p and R_n of the form (2.79) have been used. The symbol p is the hole density, p_0 is the equilibrium hole density, τ_p is the hole lifetime, n is the electron density, n_0 is the equilibrium electron density, and τ_n is the electron lifetime.

Equations (2.109), (2.110), (2.112), (2.113), and Poisson's equation

$$\nabla \cdot \mathscr{E} = 4\pi\varrho/\varepsilon \tag{2.114}$$

where ε is the dielectric constant and ϱ is the electric charge density, determine the hole and electron densities p and n. However, solution is difficult in most cases and it is necessary[21] to use a model that approximates the situation of interest in order to obtain physically useful information.

The model we shall consider is the idealized model[22] of an abrupt junction we have discussed all along. This junction was shown in Figure 2.12. The depletion layer extends from $x = x_n$ on the n side to $x = -x_p$ on the p side, with the abrupt junction itself located at $x = 0$. The homogeneous p-type neutral region is the region $-\infty \le x \le -x_p$ and the n-type neutral region is $x_n \le x \le +\infty$. We recall that the electrostatic potential was taken as constant in the neutral regions, so, in this model, the neutral regions are free of electric fields. We will thus assume that the electric field $\mathscr{E}$ is negligibly small outside the depletion layer. We assume also that the densities of minority carriers injected into the neutral regions are small compared to the majority carrier densities. This assumption of low-level injection means, along with the assumed small value of $\mathscr{E}$, that the drift currents in (2.109) and (2.110) may be neglected in the neutral regions near the junction. This is done by setting $\mathscr{E} = 0$ in those equations, which become, in one dimension,

$$J_n = eD_n \frac{dn}{dx} \qquad (2.115)$$

$$J_p = -eD_p \frac{dp}{dx} \qquad (2.116)$$

We now substitute (2.115) and (2.116) into the equations of continuity (2.112) and (2.113), and we consider explicitly minority carrier densities near the edges x_n and $-x_p$ of the depletion layer. We obtain

$$e \frac{dp_n}{dt} - eD_p \frac{d^2p_n}{dx^2} = -e\left(\frac{p_n - p_{n0}}{\tau_p} \right) \qquad (2.117)$$

$$(-e) \frac{dn_p}{dt} + eD_n \frac{d^2n_p}{dx^2} = +e\left(\frac{n_p - n_{p0}}{\tau_n} \right) \qquad (2.118)$$

In (2.117) and (2.118), p_n is the minority hole density in the neutral n region, p_{n0} is the equilibrium minority hole density in the neutral n region, n_p is the minority electron density in the neutral p region, n_{p0} is the equilibrium minority electron density in the neutral p region, and τ_p and τ_n are the minority hole and minority electron lifetimes, respectively.

In the steady state, the minority carrier densities are constant in time, so $dp_n/dt = 0$ and $dn_p/dt = 0$, and (2.117) and (2.118) become

$$\frac{d^2p_n}{dx^2} - \frac{p_n - p_{n0}}{L_p^{\ 2}} = 0 \qquad (2.119)$$

$$\frac{d^2n_p}{dx^2} - \frac{n_p - n_{p0}}{L_n^{\ 2}} = 0 \qquad (2.120)$$

Equations (2.119) and (2.120) describe the time-independent steady state minority carrier densities $p_n(x)$ and $n_p(x)$ which vary with position in the neutral regions. We see that, in this model which neglects drift currents in the neutral regions, the minority carrier currents at the edges of the neutral regions are taken as purely diffusive[16] in nature.

We next make the assumption[23] that there is no generation or recombination of carriers in the depletion layer. We are therefore considering an approximation[22] in which the electron and hole currents are constant across the depletion layer. Hence, if we find the electron or hole current at *any* point of the depletion layer, we know it at every point of the depletion layer. This feature of the model makes the calculation of the total current J through the junction relatively easy.[23] At any point x,

$$J(x) = J_n(x) + J_p(x) \qquad (2.121)$$

is the sum of the electron current $J_n(x)$ and the hole current $J_p(x)$ at that point. If we can calculate $J_n(x)$ and $J_p(x)$ at any point of junction, we know $J(x)$ at that point, and hence, at *every* point of the junction because the current through the junction is assumed uniform in space—J does *not* depend on x. In our model, at the edges x_n and $-x_p$ of the depletion layer, the only currents are minority carrier diffusion currents because we are neglecting all drift currents there. The result is that a calculation of the minority carrier diffusion currents at, say $x = x_n$, gives us the total current $J(x_n)$ at $x = x_n$, where

$$J(x_n) = J_n(x_n) + J_p(x_n) \qquad (2.122)$$

Then as discussed above, the total current J through the junction is equal to $J(x_n)$, so we have

$$J = J_n(x_n) + J_p(x_n) \qquad (2.123)$$

How can we calculate $J_n(x_n)$ and $J_p(x_n)$? We can find $J_p(x_n)$ by solving (2.119) for the minority hole density in the neutral n region, which includes the point $x = x_n$. We recall that equation (2.82), describing the steady state diffusion and recombination of holes from the point $x = x_1$, had the solution

$$p(x) - p_0 = [p(x_1) - p_0] \exp[-(x - x_1)/L_p] \qquad (2.88)$$

where $p(x)$ is the hole density at point x, and p_0 is the equilibrium hole density. We apply the solution (2.88) to our problem, the solution of (2.119), simply by substituting p_n for $p(x)$, p_{no} for p_0, and x_n for x_1. We obtain

$$p_n(x) - p_{no} = [p_n(x_n) - p_{no}] \exp[-(x - x_n)/L_p] \qquad (2.124)$$

as the equation giving the space-dependent minority hole density $p_n(x)$ in the neutral n region $x_n \leq x \leq +\infty$ of the junction. The minority hole diffusion current J_p is found from equation (2.116), so

$$J_p(x) = -eD_p \frac{dp_n}{dx}$$

and

$$J_p(x) = -eD_p[p_n(x_n) - p_{n0}](-1/L_p)\exp[-(x - x_n)/L_p] \qquad (2.125)$$

so

$$J_p(x_n) = e(D_p/L_p)[p_n(x_n) - p_{n0}] \qquad (2.126)$$

The quantity $p_n(x_n) - p_{n0}$ in equation (2.126) is just the excess minority hole density at the point x_n. However, from equation (2.104), we have

$$p_n(x_n) = p_{n0}\exp(eV_a/k_BT) \qquad (2.104)$$

where V_a is the potential applied to the junction. Substituting (2.104) into (2.126) gives

$$J_p(x_n) = e(D_p/L_p)p_{n0}[\exp(eV_a/k_BT) - 1] \qquad (2.127)$$

an expression which gives us the minority hole current at x_n in terms of the applied voltage V_a.

Next, equation (2.91) describes the diffusion and recombination of excess electrons created at the point $x = x_1$,

$$n(x) - n_0 = [n(x_1) - n_0]\exp[-(x - x_1)/L_n] \qquad (2.91)$$

where x_1 was assumed positive. If x and x_1 are negative, equation (2.91) becomes

$$n(x) - n_0 = [n(x_1) - n_0]\exp[(x - x_1)/L_n]$$

which, rewritten for the minority electron density $n_p(x)$ created at the edge $x_1 = -x_p$ of the neutral p region, becomes

$$n_p(x) - n_{p0} = [n_p(-x_p) - n_{p0}]\exp[(x + x_p)/L_n] \qquad (2.128)$$

Equation (2.128) describes the spatially dependent steady state minority electron density $n_p(x)$ in the neutral p region $-x_p \geq x \geq -\infty$ of the junction. The minority electron current $J_n(x)$ is given by (2.115),

$$J_n(x) = eD_n(dn_p/dx)$$

so, differentiating (2.128), we find that

$$J_n(x) = eD_n[n_p(-x_p) - n_{p0}](1/L_n)\exp[(x + x_p)/L_n] \qquad (2.129)$$

and, on evaluating $J_n(x)$ at $x = -x_p$, we have

$$J_n(-x_p) = e(D_n/L_n)[n_p(-x_p) - n_{p0}] \qquad (2.130)$$

The quantity $n_p(-x_p) - n_{p0}$ in equation (2.130) is the excess minority electron density at the point $-x_p$. However, from equation (2.103), we have

$$n_p(-x_p) = n_{p0} \exp(eV_a/k_BT) \qquad (2.103)$$

where, again, V_a is the applied potential. Substituting (2.103) into (2.130) gives

$$J_n(-x_p) = e(D_n/L_n)n_{p0}[\exp(eV_a/k_BT) - 1] \qquad (2.131)$$

for the minority electron current at the edge $-x_p$ of the neutral p region.

We now have, in equation (2.127), the term $J_p(x_n)$ in the expression (2.123) for the total current J through the junction. How do we find $J_n(x_n)$, the other term in the equation for J, when what (2.131) tells us is $J_n(-x_p)$, the electron current at a different point of the depletion layer? Our assumption that the electron and hole currents are constant across the depletion layer allows us to find $J_n(x_n)$. This is because this assumption says

$$J_n(x_n) = J_n(-x_p) \qquad (2.132)$$

since the electron current at one point $x = x_n$ of the depletion layer must be the same as the electron current at another point $x = -x_p$. We have then

$$J_n(x_n) = e(D_n/L_n)n_{p0}[\exp(eV_a/k_BT) - 1] \qquad (2.133)$$

Substituting equations (2.133) for the electron current at x_n and (2.127) for the hole current at x_n into equation (2.123) for the total current J gives

$$J = e\left(\frac{D_n}{L_n} n_{p0} + \frac{D_p}{L_p} p_{n0}\right)[\exp(eV_a/k_BT) - 1] \qquad (2.134)$$

Equation (2.134) is the celebrated Shockley diode equation which gives the total current density J through the ideal abrupt p–n junction as a function of the applied potential V_a and the various semiconductor parameters of the n and p sides of the junction.

We see from (2.134) that, for magnitudes of reverse bias (V_a negative) such that $|V_a| \gg k_BT/e$, the exponential term becomes very small, and the current density J approaches the value $-J(\text{generation})$ defined by

$$J(\text{generation}) \equiv e\left(\frac{D_n}{L_n} n_{p0} + \frac{D_p}{L_p} p_{n0}\right) \qquad (2.135)$$

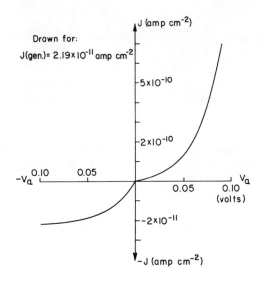

Figure 2.29. Plot of current density J as a function of applied voltage V_a from equation (2.134), drawn for J (generation) $= 2.19 \times 10^{-11}\,\text{A cm}^{-2}$. Forward bias corresponds to positive values of V_a and the direction of J (generation) is chosen as negative. Note the different scales for positive and negative values of current density J.

The magnitude of J(generation) defined by equation (2.135) is the reverse saturation current density of the *p–n* junction.

Figure 2.29 shows a plot of equation (2.134) using representative values[†] of the semiconductor parameters for silicon. As discussed above, the applied potential V_a is taken as positive for forward bias and negative for reverse bias. The direction of the current density J(generation), defined by equation (2.135), is chosen as negative. [A typical value of J(generation) is calculated in one of the problems at the end of this chapter.] From the figure, we see that the current density J increases rapidly with increasing magnitude of the forward bias and approaches saturation at a value equal to the reverse saturation current for increasing magnitude of the reverse bias. For forward bias voltages V_a such that $V_a \gg k_B T/e$, the exponential dominates in (2.134) and the current density increases exponentially with increasing magnitude of the forward bias voltage.

From equation (2.135) for the reverse saturation current density J(generation), we see that J(generation) is directly proportional to the

[†] The values used for silicon were $N_a = N_d = 10^{16}\,\text{cm}^{-3}$, with $n_i^2 = 4.6 \times 10^{19}\,\text{cm}^{-6}$ at 300 K, leading to $p_{n0} = n_{p0} = 4.6 \times 10^3\,\text{cm}^{-3}$. Values of $D_n = 33.8\,\text{cm}^2\,\text{sec}^{-1}$ and $D_p = 13.0\,\text{cm}^2\,\text{sec}^{-1}$ were calculated from the mobility values $\mu_n = 1300\,\text{cm}^2\,\text{v}^{-1}\,\text{sec}^{-1}$ and $\mu_p = 500\,\text{cm}^2\,\text{v}^{-1}\,\text{sec}^{-1}$. Diffusion lengths $L_n = 1.84 \times 10^{-3}\,\text{cm}$ and $L_p = 1.14 \times 10^{-3}\,\text{cm}$ were calculated using an assumed value of $10^{-7}\,\text{sec}$ for the electron and hole minority carrier lifetimes. The calculated reverse saturation current density of about $2 \times 10^{-11}\,\text{A/cm}^2$ is the magnitude of the values observed in silicon diodes at room temperature. (See A. S. Grove, Reference 4, page 178.)

carrier diffusion coefficients D_n and D_p, and also to the equilibrium minority carrier concentrations n_{p0} and p_{n0}. If N_d and N_a are the concentrations of ionized donors and acceptors, respectively, on the two sides of the junction, then we have

$$p_{n0} = n_i^2/N_d, \qquad n_{p0} = n_i^2/N_a \qquad (2.136)$$

From (2.136), we conclude that the larger are N_d and N_a, the smaller is the value of J(generation), all other things being equal. Since J(generation) is inversely proportional to both L_n and L_p, and L_n and L_p vary, respectively, as the square root of the minority carrier lifetimes τ_n and τ_p, the reverse saturation current J(generation) will vary as the $-\frac{1}{2}$ power of the minority carrier lifetimes, again assuming that all other parameters are held constant.

At this point, it is worthwhile to reiterate that the diode equation (2.134) was derived on the basis of an idealized model of a p–n junction. This model assumed (1) an ideal abrupt junction, in which (2) the electric field is confined to the depletion layer, (3) the injected minority carrier densities are small, and (4) there is no generation or recombination of carriers in the depletion layer. This idealized model does not describe real p–n junctions, and for a discussion of the validity of and deviations from this ideal model, the reader is referred to the literature, particularly the books by Moll[8] and by Streetman.[22] However, our discussion of the physics of the p–n junction will be confined to this idealized model.

Majority and Minority Carrier Components of the Junction Current

We calculated the total diode current J given in (2.134) by finding the minority carrier diffusion currents (2.133) and (2.127) at one point of the junction, and adding them. We neglected drift currents of minority carriers due to their small concentrations and neglected drift currents of majority carriers outside the junction because of our assumption that the electric field is confined to the space charge layer of the junction.

As remarked earlier, this assumption about the electric field is not really true because majority carriers in, say, the neutral n region must drift in from the external contact attached to the source of applied potential. These majority electrons fulfill three purposes[24] in the neutral n region of the junction. First, they recombine with the minority holes injected into the n region, thereby giving the familiar decrease in minority hole density with increasing distance into the neutral n region. Second, they provide space charge neutrality for the positive charges of the injected hole distribution.

Third, they supply the electrons for injection into the neutral p region as minority carriers there. Similar statements may be made concerning majority holes in the neutral p region.

We now wish to calculate[25] the majority and minority carrier components of the current in the two neutral regions and in the depletion layer. Consider first the n-type neutral region from $x = x_n$ to $x = +\infty$. The total current J is given by the diode equation (2.134), written in the form

$$J = J_{n0} + J_{p0} \tag{2.137}$$

where the convenient quantities J_{n0} and J_{p0} are defined by

$$J_{n0} = e(D_n/L_n)n_{p0}[\exp(eV_a/k_BT) - 1] \tag{2.138}$$

and

$$J_{p0} = e(D_p/L_p)p_{n0}[\exp(eV_a/k_BT) - 1] \tag{2.139}$$

We know also, from equation (2.125), that the minority hole current density $J_p(x)$ in the neutral n region is

$$J_p(x) = e(D_p/L_p)[p_n(x_n) - p_{n0}] \exp[-(x - x_n)/L_p] \tag{2.125}$$

where also, using equation (2.104),

$$p_n(x_n) = p_{n0} \exp(eV_a/k_BT) \tag{2.104}$$

we obtain

$$J_p(x) = e(D_p/L_p)p_{n0}[\exp(eV_a/k_BT) - 1] \exp[-(x - x_n)/L_p] \tag{2.140}$$

Equation (2.140) may be rewritten, using (2.139), as

$$J_p(x) = J_{p0} \exp[-(x - x_n)/L_p] \tag{2.141}$$

an equation that gives the space dependence of the steady state minority hole density in the neutral n region, where $x \geq x_n$.

With these results, we can find the *majority* electron current density $J_n(x)$ in the neutral n region because it must be true that

$$J_{n0} + J_{p0} = J = J_n(x) + J_p(x) \tag{2.142}$$

because the total current density J must be equal the sum of the minority and majority current densities. Substituting (2.141) for $J_p(x)$ into (2.142) gives

$$J_n(x) = J_{n0} + J_{p0} - J_{p0} \exp[-(x - x_n)/L_p]$$
$$J_n(x) = J_{n0} + J_{p0}\{1 - \exp[-(x - x_n)/L_p]\} \tag{2.143}$$

as the space dependence of the majority electron current density in the neutral n region. Together, equations (2.141) and (2.143) describe the space dependence of the minority and majority current densities in the neutral n-type region of the junction.

In the same way, we can obtain analogous expressions for the minority and majority carrier current densities in the neutral p-type region extending from $x = -x_p$ to $x = -\infty$. The results are that the minority electron current density is

$$J_n(x) = J_{n0} \exp[(x + x_p)/L_n] \tag{2.144}$$

and the majority hole current density is

$$J_p(x) = J_{p0} + J_{n0}\{1 - \exp[(x + x_p)/L_n]\} \tag{2.145}$$

Equations (2.145) and (2.144) are the analogs of equations (2.143) and (2.141), respectively, and give the spatial dependence of the currents in the neutral p region.

To find the electron current density *within* the depletion layer $-x_p \le x \le x_n$, we evaluate equation (2.143) at $x = x_n$, obtaining

$$J_n(x_n) = J_{n0} \tag{2.146}$$

and evaluating equation (2.144) at $x = -x_p$, obtaining

$$J_n(-x_p) = J_{n0} \tag{2.147}$$

Equations (2.146) and (2.147) tell us that the electron current density is constant across the depletion layer, an assumption built into the theory when we assumed no carrier generation or recombination in the depletion layer. Similarly, equations (2.141) and (2.145) give us

$$J_p(x_n) = J_{p0} \tag{2.148}$$

$$J_p(-x_p) = J_{p0} \tag{2.149}$$

exhibiting the constancy of the hole current across the depletion layer, also as required by our assumptions.

Equations (2.141), and (2.143)–(2.149) describe the spatial dependence of the electron and hole currents in the entire junction, composed of the depletion layer and two neutral regions. These results are exhibited in Figure 2.30, showing the total current $J = J_n(x) + J_p(x)$ through the junction, and its election component $J_n(x)$ and hole component $J_p(x)$ as func-

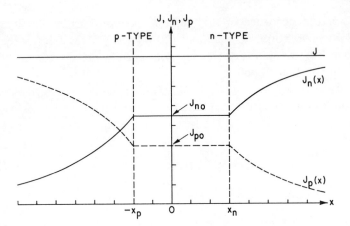

Figure 2.30. The total current $J = J_n(x) + J_p(x)$, where $J_n(x)$ is the electron current given by equations (2.143) and (2.144), and $J_p(x)$ is the hole current given by (2.141) and (2.145). The quantities J_{no} and J_{po} are given by (2.138) and (2.139), and are the values, assumed constant, of the electron and hole currents in the depletion layer of the junction. The figure has been drawn for $J_{no} = 1.5 J_{po}$, and $L_n = 1.3 L_p$, and where, for clarity and convenience, $L_n = |x_n|$.

tions of distance x, assuming that there is no recombination or generation of carriers in the space charge region. Far from the junction, at large values of $|x|$, the current will be entirely majority carriers, electrons on the n side and holes on the p side, and both will be drift currents driven by the small electric field maintained in the neutral regions. Nearer the edges of the space charge layer, on either side of the junction, there will be a minority carrier diffusion current and, for space charge neutrality, an accompanying diffusion current of majority carriers. At the edges of the space charge layer, majority carriers will be injected across the space charge region and into the opposite side of the junction, where they become minority carriers.

Summary of the Basic Physics of the *p–n* Junction

First, the fact that semiconductors like silicon can exist in both n and p types leads to the idea of a p–n junction. The simultaneous existence of electrons and holes in a semiconductor leads to currents, both diffusion and drift, of the two types of carriers.

Second, the difference in the Fermi level between the two sides of the junction leads to a difference in the electrostatic potential energy of an electron on the two sides of the junction. The fact that the magnitude of this energy barrier can be modulated by an applied potential results in a modulation of the diffusion currents in the junction.

Third, the small (and controllable, through doping) density of minority carriers on each side of the junction leads to a small reverse drift current which is essentially independent of the magnitude of the applied potential.

Fourth, the difference in the effect of an applied potential on the diffusion and drift currents in the junction leads to the great asymmetry of the current–voltage characteristic for forward and reverse bias. It is this asymmetric current–voltage characteristic that makes the *p-n* junction a useful device.

Reverse Breakdown in *p-n* Junctions

When the value of the reverse bias applied reaches a critical value, the magnitude of the reverse current increases sharply, as shown schematically in Figure 2.31. This figure indicates the increase in reverse current at the critical value V_B of the reverse voltage applied to the junction. This effect is called reverse breakdown[26,27] of the *p-n* junction. There are two different mechanisms that produce breakdown.

The first process is called Zener breakdown and is a quantum mechanical tunneling phenomenon. Figure 2.32 shows the band diagram of a *p-n* junction at equilibrium and under an applied reverse bias, showing that reverse bias lowers electron energies on the *n* side relative to electron energies on the *p* side. As shown in Figure 2.32b, there are, under sufficient reverse bias, filled electron states in the valence band on the *p* side at the same energy as empty conduction band states on the *n* side. This situation means that the possibility exists of an electron tunneling from the *p* side to the *n* side. If the width of the energy barrier (through which tunneling takes place) is denoted by *d* in Figure 2.32b, it may be shown[28] that *d*

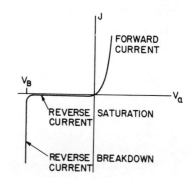

Figure 2.31. Reverse breakdown (schematic) in a *p-n* junction current–voltage characteristic. V_B is the critical breakdown voltage.

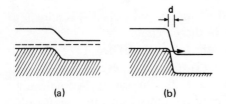

Figure 2.32. Band diagrams for a *p–n* junction (a) at equilibrium, and (b) under reverse bias. In (b), electron tunneling from *p* to *n* is indicated schematically by the arrow, and *d* is the width of the barrier to tunneling.

decreases with increasing magnitude of the applied reverse bias, assuming that the impurity doping is high enough that the width of the space charge region is narrow to begin with. As the magnitude of the reverse bias increases, *d* decreases and the magnitude of the electric field in the junction increases. The tunnel current of electrons from *p* to *n* increases with increasing electric field[29] and with decreasing barrier width, so, as the reverse bias approaches the critical value V_B, a significant tunnel current begins to flow. This flow of electrons from *p* to *n* constitutes a reverse current through the junction (just like the reverse current of thermally excited carriers making up the reverse saturation current). At the value V_B of reverse bias, the tunnel current dominates the ordinary *J*(generation) reverse saturation current, and the reverse breakdown shown in Figure 2.31 takes place.

The value of V_B for a particular junction will depend on the details of the junction type, doping, etc., but the critical value[30] of the electric field in silicon is about 10^6 V cm^{-1} for junction doping values in the 10^{17}–10^{19} cm^{-3} range. This critical field value corresponds to critical breakdown voltages V_B in the neighborhood of a few volts. It should be kept in mind, however, that Zener tunneling will take place only in narrow junctions with high doping.

A second process which can produce reverse breakdown in a *p–n* junction is called avalanche breakdown.[31] In this mechanism conduction electrons are accelerated by the electric field in the junction, thereby gaining kinetic energy. This kinetic energy may be large enough to produce free carriers by impact ionization. In this process, an electron with a kinetic energy at least as large as the energy gap in the semiconductor excites an electron from the valence band to the conduction band, thereby producing an electron–hole pair. These newly generated free carriers can also be accelerated by the electric field, producing more electron–hole pairs by impact ionization. The resulting multiple production of free carriers is called an avalanche and sharply increases the current through the junction, as shown in Figure 2.31.

The avalanche mechanism will produce reverse breakdown in the junction if the Zener tunneling mechanism is not dominant because the junction is not narrow enough for easy tunneling. Generally speaking, if Zener breakdown does not occur with a reverse bias of a few volts, avalanche breakdown will become dominant. Again, the value of the critical voltage for avalanche breakdown depends on the details of the junction.[31]

Other Topics on *p–n* Junctions

We have touched only on some of the main points of the physics of *p–n* junctions. Many other interesting topics are to be found in the literature. We have considered current flow in the case of low-level injection of minority carriers; the case of high injection levels, in which the density of minority carriers is comparable to that of the majority carriers, is discussed by Sze[32] and by Shockley.[33] We have discussed only the idealized abrupt junction; other, more realistic types of junction are treated by Moll,[34] Grove,[35] and Streetman.[36] The validity of the approximations made in the idealized model is discussed by Moll.[22] The high-resistivity depletion layer between the more conductive neutral regions has a capacitance associated with it. This topic is treated by, among other authors, Streetman,[37] as are questions of the response of the junction to transient and AC conditions. It is, however, useful to note here an important question. This is the time constant of the response[38–40] of a *p–n* junction to a sudden change of bias from forward to reverse. If the junction is forward biased, then minority carriers have been injected across the junction, giving the steady state excess carrier density distribution shown in Figure 2.19. If the bias voltage is suddenly switched from forward to reverse, the current through the diode will not immediately fall to its reverse-voltage value. The reason is that it takes a nonzero time interval for the excess minority carrier density to decrease from its large positive value under forward bias to its small negative value under reverse bias. This time interval is called the minority carrier storage time, and limits the speed with which the junction can be switched from a condition of forward bias to one of reverse bias.

Problems

2.1. *Depletion Layer Width.* To get an idea of the magnitudes involved, calculate the width of the depletion layer in an ideal abrupt silicon *p–n* junction at 300 K. Assume that both the *n* and *p* sides contain 10^{15} ionized impurities per cm³. Take $np \equiv n_i^2 = 2.2 \times 10^{18} \mathrm{cm}^{-6}$ and the dielectric constant equal to 11.7. Check to be sure that your answer has the correct units.

2.2. *Hole Currents in a Junction.* Make a simple band diagram of a *p–n* junction at equilibrium, showing schematically the diffusion and generation particle fluxes of holes.

2.3. *Currents in a Biased Junction.* Make sketches of the band diagram of a *p–n* junction under large ($| V_a | \gg k_B T/e$) forward and reverse bias. On each diagram, indicate with arrows the directions and relative magnitudes of all of the particle fluxes and all of the current densities involved. For the generation currents, show where the carriers are generated.

2.4. *Injected Hole Density.* Given a silicon *p–n* junction at 300 K in which the density of extrinsic carriers is 10^{15} cm^{-3} on either side of the junction, calculate the injected increase in minority holes on the *n* side for a forward bias of 0.6 V.

2.5. *Junction at Low Temperatures.* Given a silicon *p–n* junction, for which the total densities of donors and of acceptors are 10^{15} cm^{-3}, maintained at 30 K, discuss what current–voltage characteristic you would expect to observe.

2.6. *Carrier Densities in p Region.* Derive equations (2.144) and (2.145) for the minority electron and majority hole current densities in the neutral *p* region.

2.7. *Reverse Saturation Current Density.* Calculate the reverse saturation current density of a silicon *p–n* junction (using the ideal abrupt model). Assume that the doping on both sides of the junction is 10^{16} ionized impurities per cm^3, and that the lifetime of both carriers is 0.1 µsec. Note that carrier diffusion coefficients can be calculated from mobility values, which are $\mu_n = 1300$ cm^2 V^{-1} sec^{-1}, $\mu_p = 500$ cm^2 V^{-1} sec^{-1}. Compare your result with values observed for silicon diodes at room temperature. (See A. S. Grove, Reference 4, page 178.)

References and Comments

1. S. M. Sze, *Physics of Semiconductor Devices*, John Wiley, New York (1969), page 96.
2. C. Kittel, *Thermal Physics*, John Wiley, New York (1969), page 73.
3. B. G. Streetman, *Solid State Electronic Devices*, Prentice-Hall, New York (1972), Section 5.2.
4. A. S. Grove, *Physics and Technology of Semiconductor Devices*, John Wiley, New York (1967), pages 149–161.
5. C. Kittel, Reference 2, page 215.
6. C. Kittel, *Introduction to Solid State Physics*, Third Edition, John Wiley, New York (1966), page 323.
7. S. M. Sze, Reference 1, page 66, equations (93a) and (94a).
8. J. L. Moll, *Physics of Semiconductors*, McGraw-Hill, New York (1964), Chapter 7, especially pages 111, 117, 121–123, and 133–140.
9. B. G. Streetman, Reference 3, pages 199–201.
10. R. A. Smith, *Semiconductors*, Cambridge University Press, New York (1959), pages 265–267.
11. B. G. Streetman, Reference 3, page 158.

12. The important subject of the technology of p–n junction fabrication will, unfortunately, not be discussed here. See, for example, S. M. Sze, Reference 1, pages 75–78; A. S. Grove, Reference 4, Chapters 1, 2, 3; B. G. Streetman, Reference 3, pages 65–72.
13. S. Wang, *Solid State Electronics*, McGraw-Hill, New York (1966), pages 272–289.
14. S. M. Sze, Reference 1, pages 66–72.
15. See, for example, J. C. Slater and N. H. Frank, *Introduction to Theoretical Physics*, McGraw-Hill, New York (1933), page 187.
16. W. Shockley, *Electrons and Holes in Semiconductors*, Van Nostrand, New York (1950), page 313, equations (13) and (14).
17. W. Shockley, Reference 16, page 298.
18. C. Kittel, Reference 6, pages 322–324.
19. W. R. Beam, *Electronics of Solids*, McGraw-Hill, New York (1965), pages 190–200.
20. B. G. Streetman, Reference 3, page 128.
21. J. L. Moll, Reference 8, page 106.
22. J. L. Moll, Reference 8, pages 117, 133–140, discusses the validity of this ideal model, as does B. G. Streetman, Reference 3, pages 191–201.
23. N. W. Ashcroft and N. D. Mermin, *Solid State Physics*, Holt, Rinehart, and Winston, New York (1976), Chapter 29, page 609, gives a clear discussion.
24. B. G. Streetman, Reference 3, pages 171–175.
25. A. van der Ziel, *Solid State Physical Electronics*, Second Edition, Prentice-Hall, New York (1968), page 310.
26. S. M. Sze, Reference 1, pages 109–126.
27. B. G. Streetman, Reference 3, pages 175–180.
28. See, for example, B. G. Streetman, Reference 3, page 203, Problem 5-12.
29. S. M. Sze, Reference 1, page 111, equation (68).
30. A. S. Grove, Reference 4, page 193, Figure 6.27.
31. See, for example, B. G. Streetman, Reference 3, pages 177–180; S. M. Sze, Reference 1, pages 111–126.
32. S. M. Sze, Reference 1, pages 104–107.
33. W. Shockley, Reference 16, pages 328–333.
34. J. L. Moll, Reference 8, pages 121–123.
35. A. S. Grove, Reference 4, Chapter 6.
36. B. G. Streetman, Reference 3, pages 199–201.
37. B. G. Streetman, Reference 3, pages 187–191.
38. J. Millman and C. C. Halkias, *Integrated Electronics*, McGraw-Hill, New York (1972), pages 71–73.
39. S. Wang, Reference 13, pages 321–331.
40. R. S. Muller and T. I. Kamins, *Device Electronics for Integrated Circuits*, John Wiley, New York (1977), pages 179–187.

Suggested Reading

B. G. Streetman, *Solid State Electronic Devices*, Prentice Hall, New York (1972). Chapter 5 of this text discusses p–n junctions, and our treatment draws heavily on Streetman's Section 5.2 and 5.3. This treatment of the p–n junction is, in the author's opinion, the best available at the introductory level.

S. WANG, *Solid State Electronics*, McGraw-Hill, New York (1966). Chapter 5 of this advanced level text discusses the transport and recombination of excess carriers in semiconductors, while Chapter 6 discusses the physics of the p–n junction.

A. S. GROVE, *Physics and Technology of Semiconductor Devices*, John Wiley, New York (1967). Chapter 6 of this useful and practical book discusses, with many examples, the physics of the p–n junction.

J. L. MOLL, *Physics of Semiconductors*, McGraw-Hill, New York (1964). Chapter 6 of this text discusses nonequilibrium carrier densities, and Chapter 7 treats the p–n junction with special attention to the approximations made.

A. VAN DER ZIEL, *Solid State Physical Electronics*, Second Edition, Prentice-Hall, New York (1968). Sections 15.1 and 15.2 discuss the p–n junction in a terse and mathematical style.

S. M. SZE, *Physics of Semiconductor Devices*, John Wiley, New York (1969). Sections 2.7 and 3.1–3.4 of this large advanced treatise treat the mathematics of the p–n junction in detail.

N. W. ASHCROFT and N. D. MERMIN, *Solid State Physics*, Holt, Rinehart, and Winston, New York (1976). This textbook of solid state physics at the advanced level includes, in Chapter 29, a fine and careful discussion of the physics of the p–n junction.

W. SHOCKLEY, *Electrons and Holes in Semiconductors*, Van Nostrand, New York (1950). This is the book that began it all, written by one of the pioneers of semiconductor physics.

3

Semiconductor p–n Junction Devices

Introduction

The aim of this chapter is a discussion of the physics of a number of devices based on the semiconductor p–n junction. First is the p–n junction itself, used as a diode. Second is the important bipolar junction transistor. The emphasis is on a qualitative discussion of current amplification in this device, followed by a brief calculation of the currents involved. The tunnel diode, or degenerate diode, is treated next in a qualitative manner which obtains the negative dynamic resistance characteristic of this device. Finally, the junction field effect transistor is discussed, also qualitatively, and its function as a voltage-controlled amplifier is treated.

Semiconductor p–n Junction Diodes

Consideration of the highly asymmetric current–voltage characteristic shown in Figures 2.28 and 2.29 of Chapter 2 and expressed by equation (2.134) suggests the use of the semiconductor p–n junction as a diode. For use as a diode,[1,2] either as a rectifier or in other applications, the reverse saturation current of the diode should be small. From equation (2.135)

$$J(\text{generation}) \equiv e\left(\frac{D_n}{L_n}\, n_{p0} + \frac{D_p}{L_p}\, p_{n0}\right) \qquad (2.135)$$

we can see the factors[†] that govern the magnitude of the reverse saturation current J(generation). First, we see that small values of the equilibrium minority carrier densities n_{p0} and p_{n0} will favor a small magnitude of J(generation). We recall from equation (2.136) that

$$p_{n0} = n_i^2/N_d, \qquad n_{p0} = n_i^2/N_a \tag{3.1}$$

where the intrinsic constant n_i^2 is the product of the intrinsic electron and hole densities defined by equation (1.31). Therefore

$$n_i^2 \equiv 4(k_B T/2\pi\hbar^2)^3(m_e^* m_h^*) \exp(-E_g/k_B T) \tag{3.2}$$

where E_g is the semiconductor energy gap and m_e^* and m_h^* are, respectively, the electron and hole effective mass values. From equations (3.1) and (3.2) we see that a large semiconductor energy gap E_g will favor small values of the minority carrier densities and hence will favor small values of the reverse saturation current for given values of the impurity densities N_d and N_a and for a given value of the temperature. For this reason, silicon ($E_g = 1.11$ eV at 300 K) *p–n* junctions have a smaller reverse saturation current than germanium ($E_g = 0.67$ eV at 300 K) *p–n* junctions and hence are preferable as diodes.

We note also from (2.135) that small values of the ratio D/L for both electrons and holes favor small values of J(generation). Since $L^2 = D\tau$ from equation (2.83), and using the Einstein relation[(3)]

$$D = (k_B T/e)\mu \tag{3.3}$$

between the carrier mobility μ and its diffusion coefficient D, we find that

$$D/L = (k_B T/e)^{1/2}(\mu/\tau)^{1/2} \tag{3.4}$$

for both electrons and holes. Equation (3.4) tells us that small carrier mobilities and large minority carrier lifetimes will favor small values of D/L and hence small values of the reverse saturation current density. As discussed in Chapter 1, the mobility will be inversely proportional to the carrier effective masses, and effective masses are usually larger in semiconductors with larger energy gaps. Thus we see that a large value of E_g

---

[†] As pointed out in Chapter 2, equation (2.135) was derived using an idealized model of a *p–n* junction which neglected, among other factors, the generation of carriers in the space charge region. For this reason, equation (2.135) is not strictly true for silicon diodes, in which such generation is important. See References 22 and 35 of Chapter 2 for a discussion.

generally means lower carrier mobilities and hence lower values of D/L. Further, as also discussed in Chapter 1, an indirect band gap will favor longer carrier lifetimes; in this case, the lifetime of interest is the minority carrier lifetime. Finally, lower impurity densities in the junction will also favor longer minority carrier lifetimes, but this gain would have to be weighed against the fact that lower impurity densities will, as seen from equation (2.135), increase the equilibrium minority carrier densities.

In summary, we see that a larger energy gap will favor a smaller value of the reverse saturation current density at a given temperature. For this reason, silicon is preferred to germanium for making rectifying semiconductor diodes, and is also preferred for high-temperature operation.

Diodes are also used in switching circuits, in which applications they are alternately forward and reverse biased. The speed with which they can assume a new bias condition will determine their performance. One of the main factors determining this speed is the phenomenon of minority carrier storage,[4] discussed briefly in Chapter 2.

Finally, it should be mentioned that the critical voltage V_B for Zener breakdown[5] can be quite reproducible and is nondestructive. Such diodes[6] are called Zener diodes, and the reproducible breakdown voltage V_B serves as a voltage reference in many circuits.

The Bipolar Junction Transistor

We will study the bipolar junction transistor (BJT) because it is an important active amplifying device and its treatment is based on the p–n junction physics we developed in Chapter 2.

Consider first a p–n junction in which the p side is much more heavily doped than the n side, so, if N_a and N_d are the densities of ionized acceptors and donors, respectively, we are considering the situation for which

$$N_a \gg N_d$$

Since the intrinsic constant n_i^2 is a constant at a given temperature, the equilibrium minority hole density p_{n0} and minority electron density n_{p0} are, from equation (3.1), such that

$$p_{n0} \gg n_{p0} \tag{3.5}$$

meaning that there are more minority holes on the n side of the junction than there are minority electrons on the p side. In this case, then, the current density J, given by

$$J = e\left(\frac{D_p}{L_p} p_{n0} + \frac{D_n}{L_n} n_{p0}\right)(e^{eV_a/k_B T} - 1) \tag{3.6}$$

becomes, because of the condition (3.5),

$$J \cong e\left(\frac{D_p}{L_p} p_{n0}\right)(e^{eV_a/k_B T} - 1) \tag{3.7}$$

Equation (3.7) tells us that, in this kind of a junction, called a p^+n junction, the bulk of the injection current is carried by holes because, from (3.5), the density of minority electrons is negligibly low. (The symbol "+" in p^+n means that the p side of the junction is much more heavily doped than the n side.) Further, we can see, from equations (2.70) and (2.71),

$$x_p = \frac{W}{1 + N_a/N_d} \tag{2.70}$$

$$x_n = \frac{W}{1 + N_d/N_a} \tag{2.71}$$

where $W = x_n + x_p$ is the total width of the space charge region, that $x_n \cong W$ and $x_p \cong 0$ because $N_a \gg N_d$. As expected then, the space charge region of the p^+n junction is almost entirely in the n-type side of the junction.

From these considerations, we see that a forward-biased p^+n junction is a good way to inject minority holes into n-type material because equation (3.7) tells us that, in such a junction, many more holes are injected from p^+ to n than electrons are injected from n to p^+.

Suppose we now mentally construct a p–n–p junction transistor by taking a forward-biased p^+n junction and coupling to it a reverse-biased pn junction. The result is shown in Figure 3.1. The forward-biased p^+n junction, labeled E, is called the emitter junction, while the reverse-biased np junction, labeled C, is called the collector junction. The n-type region between the two junctions is called the base (labeled B) of the transistor. Figure 3.2 shows the band diagram of the pnp transistor with no external

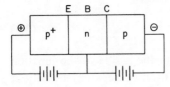

Figure 3.1. Schematic representation of a *pnp* junction transistor. The forward-biased p^+n emitter junction is labeled E, the reverse-biased np collector junction is labeled C, and the base is labeled B.

Figure 3.2. Band diagram of a *pnp* junction transistor with no applied bias. The shaded regions are filled valence band states, and electron energy increases upward. The Fermi level is denoted by E_F, and the junction labels are as in Figure 3.1.

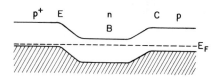

voltages applied to it, in the equilibrium state, where E, B, and C refer, respectively, to the emitter junction, base region, and collector junction. Figure 3.3 shows the same band diagram with forward bias applied to the emitter junction and reverse bias applied to the collector junction. Since the p^+n emitter junction E is forward biased, the energy barrier, between p^+ and n, for holes has decreased. Since the np collector junction C is reverse biased, the energy barrier to the diffusion of holes from p to n has increased. Figure 3.3 also shows, schematically, the injection of holes into the base B by the forward-biased p^+n emitter junction. The built-in electric field $\mathscr{E}$ at the collector junction is also shown.

We now consider the flow of particles in the *pnp* transistor; the directions of these particle flows are shown in Figure 3.4. In this figure, particle flow (1) consists of holes injected into the base by the emitter. Particle flow (2) shows the diffusion of these injected holes across the base region. At the collector junction, the built-in electric field there sweeps these holes into the collector, producing particle flow (3), as the holes fall down the energy hill for holes shown in Figure 3.3. Some of the holes diffusing across the base will recombine with majority electrons in the base; this is indicated as the hole flow (4). This means electrons, shown as particle flow (5), are supplied from the external circuit to replace those lost due to recombination with holes. Finally, back at the p^+n emitter junction, there will be some injection of electrons from n to p^+ even though the p^+ doping means that most of the current in the emitter junction will be due to holes. The electrons injected from n to p^+ are shown as particle flow (6) and are also replaced from the external circuit. (We are neglecting the small reverse current in the reverse-biased collector junction. This is composed of thermally generated electrons drifting from p to n and thermally generated holes drifting

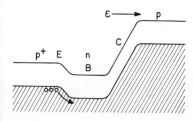

Figure 3.3. Band diagram of the *pnp* junction transistor with forward bias applied to the emitter junction E and reverse bias applied to the collector junction C. The electric field at the collector junction is $\mathscr{E}$, and the injection of holes ($\bigcirc$) into the base by the emitter is indicated schematically by an arrow.

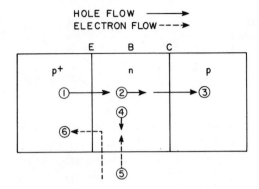

Figure 3.4. Particle flows in the biased *pnp* transistor. Hole flows are indicated by the solid arrows, and electron flows by the dashed arrows.

from *n* to *p*.) The same particle flows are shown on the transistor band diagram in Figure 3.5, in which the symbol ○ indicates a hole and the symbol ● indicates an electron, and where the particle flows are numbered as in Figure 3.4. The electric field $\mathscr{E}$ at the collector junction is also indicated.

We saw from Figure 3.4 that there is a flow of electrons into the base of the *pnp* transistor under bias. This flow comes from the external circuit and makes up the base current I_B in the transistor. [The direction of the current I_B is opposite to the direction of the electron flows (5) and (6) which constitute it.] The other currents are hole currents and are in the same direction as their respective particle flows. The hole flow (1) is the emitter hole current I_{E_p} and hole flow (3) is the collector hole current I_{C_p}. The emitter electron current I_{E_n} is due to electron flow (6) and is opposite in direction to that flow; I_{E_n} is thus in the same direction as the emitter hole current I_{E_p}. As mentioned earlier, we are neglecting the reverse current (both electrons and holes) at the collector junction.

In terms of these currents, we may redraw Figure 3.4 as shown in Figure 3.6. Also shown (at the "contacts" to the device, indicated by the symbol △) are the currents in the external circuit. The current I_E, where

$$I_E = I_{E_p} + I_{E_n} \tag{3.8}$$

Figure 3.5. Band diagram indicating the particle flows in the biased *pnp* transistor of Figure 3.3. Holes are represented by ○, and electrons by ●; $\mathscr{E}$ is the electric field at the collector junction.

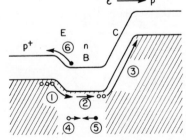

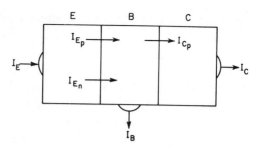

Figure 3.6. Current directions in the *pnp* transistor.

is the emitter current; the current I_C, where

$$I_C = I_{C_p} \tag{3.9}$$

is the collector current; and I_B is the base current. The currents I_E, I_C, and I_B are the currents in the external circuit of the transistor.

Amplification in the Bipolar Transistor

Given the currents flowing in the *pnp* transistor, it is useful to discuss physically why the device acts as a current amplifier. First, however, some comments about amplification in general[7] are appropriate here.

Generally, an amplifier is a device that increases the amplitude of an AC signal. Since the laws of thermodynamics tells us that we cannot get more energy out of our device than we originally put in, what is really going on in an amplifier is the conversion of energy, put into the device as DC current, into an output AC current which is proportional to the input AC current. Since the device has effectively "multiplied" the input current, we speak of the amplifier as having current gain, which is, of course, the constant of proportionality between the output and input AC signals. There are two types of amplifying devices. In the first, the output current is controlled by the input current. Such a device, of which the bipolar junction transistor is an example, is a current-controlled amplifier. In the second type of amplifying device, the output current is controlled by the input voltage. This type of device is called a voltage-controlled amplifier; an example (which we will discuss later) is the field-effect transistor.

We now examine the physical reasons why the bipolar junction transistor is a current-controlled amplifier, in which a small base current I_B controls a larger collector current I_C. Since the collector current I_C is a flow of holes across the base, and the base current I_B is a flow of electrons into the base, we can see why the electron flow I_B will have an effect on the hole flow I_C. The basic idea[8] is the existence of electrical neutrality in the base region of the transistor. If we increase the flow of electrons into the base, the flow of holes from the emitter (and hence to the collector) will increase in order to maintain space charge neutrality. If the flow of electrons into the base is reduced, the hole flow from the emitter will decrease. We see, then, why the electron flow into the base (the base current I_B) modulates the hole flow from emitter to collector (the collector current I_C) and we have a device in which the output current I_C is controlled by the input current I_B. The device is thus a current-controlled amplifier.

We must next examine why the bipolar junction transistor is such that the constant of proportionality between the output current and the input current can be greater than unity, i.e., why this amplifier can exhibit current gain greater than unity. For this to be true, one electron in the base or input current must "affect" or modulate *more* than one hole in the collector or output current. To see that this can be the case, suppose the lifetime τ_n of an electron in the base is longer than the time τ_t it takes an injected hole to diffuse from the emitter to the collector. In such a situation, the electron lives long enough to provide space charge electrical neutrality while several holes traverse the base. Each electron put into the base is able to electrostatically compensate for τ_n/τ_t holes passing through. Given the requirement of space charge neutrality in the base, we would expect the ratio of the hole current I_C in the base to the electron current I_B in the base to be equal to τ_n/τ_t (assuming for the moment that all of the holes injected into the base are collected). We see then that, if the device is such that τ_n is longer than τ_t, I_C/I_B will be greater than unity, and we will have current gain greater than unity.

Current Gain in the Bipolar Transistor

We now want to calculate the ratio I_C/I_B. First, we note that the collector current I_C will be the fraction of the injected hole current I_{E_p} that crosses the base without recombining. If we call this fraction the base transport factor B, then

$$I_C = BI_{E_p} \qquad (3.10)$$

The total emitter current I_E is made up of a hole current I_{E_p} and an electron current I_{E_n}, and was given by

$$I_E = I_{E_p} + I_{E_n} \qquad (3.8)$$

so we define the emitter injection efficiency γ by the relation

$$\gamma \equiv I_{E_p}/I_E = I_{E_p}/(I_{E_p} + I_{E_n}) \tag{3.11}$$

If the electron component of the emitter current were zero, then we would have $I_{E_n} = 0$ in (3.11) and the efficiency γ would be equal to unity, meaning that the current injected by the emitter was composed entirely of holes.

The base current I_B is calculated as follows. The base current is, as shown in Figure 3.4, composed of electron flow (6) injected from n to p^+ at the emitter junction and electron flow (5) entering the base to replace electrons lost due to recombination with holes. The first component (6) of the base current is thus equal to I_{E_n}, the electron current through the emitter junction. The second component (5) is numerically equal to the fraction of the injected hole current that does *not* traverse the base. This hole current is equal to $(1 - B)I_{E_p}$ because, from (3.10), BI_{E_p} is the fraction of the injected hole current that *does* cross the base. Adding the two components of the base current, we have the result that

$$I_B = I_{E_n} + (1 - B)I_{E_p} \tag{3.12}$$

We now define the base-to-collector current gain β as

$$\beta \equiv \frac{I_C}{I_B} = \frac{BI_{E_p}}{I_{E_n} + (1 - B)I_{E_p}} \tag{3.13}$$

where we have used (3.10) and (3.12). Substituting the definition (3.11) of γ into equation (3.13) gives

$$\beta = \frac{B\gamma}{1 - B\gamma} \tag{3.14}$$

for the base-to-collector current gain β. Since the base transport factor B is less than unity and the emitter injection efficiency γ is less than unity, $B\gamma$ is less than unity. If $B\gamma$ is close to unity, the current gain β will be large; for $B\gamma$ greater than 0.5, the current gain β will be greater than unity and current amplification will take place. A typical value of β for an ordinary transistor is $\beta = 100$.

We now consider the factors that influence B and γ. The base transport factor B is the fraction of injected holes that cross the base by diffusion and reach the collector. For B to be close to unity, one wants a narrow base width and a large value of the hole diffusion length L_p (where $L_p^2 = D_p\tau_p$, D_p is the minority hole diffusion coefficient, and τ_p is the minority hole lifetime in the base). The width of the base of the transistor is a question of fabrication technology[9,10] and of the doping of the base relative to the emitter and collector regions. If the base is lightly doped relative to the emitter and collector regions, then the space charge layers at the two junctions will extend relatively far into the base region, thereby making narrower the neutral n region which the minority holes must traverse. To have a long diffusion length L_p, one must have large values of the diffusion constant D_p and long values of the minority carrier lifetime τ_p. The value of D_p is fairly well determined by the semiconductor from which the transistor is made, but, from the Einstein relation (3.3), large values of the minority hole mobility will favor large values of D_p. While scattering of carriers by phonons is a process fixed by the fundamental characteristics of the material, light doping in the base will decrease carrier scattering by ionized impurities. Such light doping will thus serve to increase carrier mobility and hence will favor large values of D_p. The minority carrier lifetime τ_p is (as discussed in Chapter 1) favored by a band structure with an indirect energy gap; examples are germanium and silicon. Since electron–hole recombination in indirect gap semiconductors is predominantly via an intermediate state (e.g., a defect level), lower doping of the base also favors a longer minority carrier lifetime. In summary, a value of the base transport factor close to unity is favored by a long hole diffusion length and hence is favored by the conditions leading to a longer minority carrier lifetime.

The emitter injection efficiency γ can be shown[11] to approach unity as the ratio of the doping of the emitter to the doping of the base becomes large. The p^+n junction is therefore expected to have an injection efficiency close to unity.

We see then, from (3.14), that large values of the current gain β are favored by values of $B\gamma$ close to unity, and that values of $B\gamma$ close to unity are favored by, among other factors, a long minority carrier lifetime in the base and the use of a p^+n junction as an efficient hole injection at the emitter junction.

Circuit Configurations for Amplification with the Bipolar Transistor

Two circuit configurations for amplification[12] using the bipolar junction transistor are shown in Figures 3.7 and 3.8. In these configurations, the emitter junction is forward biased and the collector junction is reverse biased. This biasing scheme is called the forward active mode. The arrange-

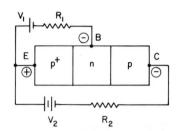

Figure 3.7. Common emitter circuit configuration for amplification with a *pnp* bipolar transistor.

ment in Figure 3.7 is called the common emitter configuration and that in Figure 3.8 is called the common base configuration.

In the common emitter circuit, the base current is determined by battery V_1 (e.g., 10 V) and resistor R_1 (e.g., $10^5 \, \Omega$), where we are neglecting the small voltage appearing across the forward-biased emitter junction. Almost all of the voltage between the collector and emitter appears across the relatively high resistance reverse-biased collector junction. The voltage V_2 might be 10 V and resistor R_2 5000 Ω. Note that, given the base current I_B, the collector current I_C is determined by the current gain β through equation (3.13),

$$I_C = \beta I_B \tag{3.13}$$

and not by the parameters V_2 and R_2 of the collector circuit, as long as the latter maintain the reverse bias at the collector junction. In the circuit in Figure 3.7, the voltage V_2 will be divided between the resistance R_2 and the resistance of the reverse-biased collector junction.

Using the typical values for the circuit elements quoted above, and using a current gain $\beta = 100$, one finds (with the approximations mentioned) that $I_B = 0.1$ mA and $I_C = 10$ mA. Suppose an AC current of the form $i_1 = i_0 \sin \omega t$, with $i_0 = 0.05$ mA, is applied as an input to the base, superposed on the DC bias current I_B. Then the total base current $i_B = I_B + i_1$ will have the form shown in Figure 3.9a. The total collector (output) current will have the form shown in Figure 3.9b, which is an AC component $i_2 = (\beta i_0) \sin \omega t$ superposed on the DC collector current I_C, where $I_C = \beta I_B$. This simple example exhibits the current gain by showing that

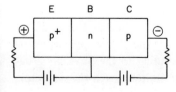

Figure 3.8. Common base circuit configuration for amplification with a *pnp* bipolar transistor.

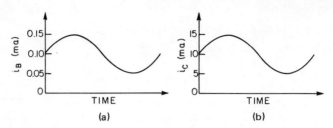

Figure 3.9. (a) Base (input) current i_B as a function of time. (b) Collector (output) current i_C as a function of time.

the output current i_C is equal to the input current i_B multiplied by the current gain β. Since the value of the input current i_B determines the value of the output current i_C, the transistor is, as expected from the discussion earlier, acting as a current-controlled amplifier.

In the common base configuration of Figure 3.8, the emitter-to-collector current amplification factor α is defined by

$$\alpha \equiv \frac{I_C}{I_E} = \frac{BI_{E_p}}{I_{E_p} + I_{E_n}} \tag{3.15}$$

using (3.10) and (3.8). Since the emitter injection efficiency γ is defined by (3.11), equation (3.15) becomes

$$\alpha = B\gamma \tag{3.16}$$

where B is the base transport factor. Since both B and γ are less than unity, the current amplification represented by α is amplification by a factor less than unity. However, the emitter-to-collector current is the transport of holes from the forward-biased, low-resistance emitter junction to the reverse-biased, high-resistance collector junction. Since this current flows in the high-resistance direction, the current flow represents power gain or amplification even though the current gain α is less than unity.[13] Finally, combining (3.16) with (3.14) gives

$$\beta = \frac{\alpha}{1 - \alpha} \tag{3.17}$$

as the relation between the base-to-collector current gain β and the emitter-to-collector current gain α. Equation (3.17) shows that, as α approaches its upper limit of unity, β becomes large, corresponding to the physical situation in which most of the holes injected by the emitter reach the collector.

Quantitative Discussion of the Bipolar Transistor

We now make a simple analysis of the bipolar transistor. This analysis will be sufficient for the situation in which the transistor is used as an amplifier, but is not the most general possible, in that a number of assumptions and approximations are made. The reader is referred to the literature for a discussion[14] of deviations from our idealized case, as well as a treatment of the important use of the bipolar transistor as a switch.[15,16] However, our discussion will provide, within the approximations made, interesting quantitative results.

We simplify the problem by making the following assumptions. First, holes diffuse from the emitter to collector in our *pnp* transistor. This is equivalent to assuming that no electric field, and hence no drift current of holes, exists in the base. (This assumption is the same as the one we made in Chapter 2 concerning the neutral regions of the *p–n* junction.) Second, the emitter current I_E is composed only of holes, so the emitter injection efficiency γ is taken as equal to unity. Third, we assume that the reverse saturation current in the collector junction can be neglected. Fourth, the transistor has a uniform cross section area, so the problem becomes one-dimensional. Fifth, we neglect generation and recombination of carriers in the space charge regions of the emitter and collector junctions. Sixth, all currents and voltages are taken to be steady state values.

Given these approximations, the calculation will be in three steps. First, we calculate the distribution in space of the excess holes injected into the base. Second, we calculate the emitter current I_E and the collector current I_C as diffusion currents due to the concentration gradients of injected holes at the emitter and collector boundaries of the base. The base current I_B is calculated from I_E and I_C. Third, we calculate the base-to-collector current gain β and the collector-to-emitter current gain α.

We set up the geometry of the problem as shown in Figure 3.10. The space charge regions of the emitter (E) and collector (C) junctions are delineated by vertical dashed lines. The neutral base region extends from $x = 0$ to $x = W_b$, so the base width is equal to W_b. We take $x = 0$ at

Figure 3.10. *pnp* junction transistor geometry. The space charge regions of the emitter junction E and the collector junction C are delineated by vertical dashed lines. The width of the neutral base region is W_b.

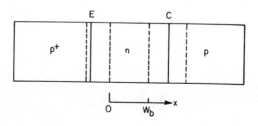

the edge of the emitter junction space charge region in the base and $x = W_b$ at the edge of the collector junction space charge region in the base. We use the symbol Δp_E for the injected excess hole density at the emitter junction at $x = 0$, and Δp_C for the excess hole density at the collector junction at $x = W_b$. If we use V_E for the forward bias voltage applied to the emitter junction and V_C for the reverse bias voltage applied to the collector junction, then equation (2.108) gives us

$$\Delta p_E = p_{n0}(e^{eV_E/k_BT} - 1) \tag{3.18}$$

$$\Delta p_C = p_{n0}(e^{eV_C/k_BT} - 1) \tag{3.19}$$

where p_{n0} is the equilibrium minority hole density in the base. Since the emitter junction is forward biased, V_E is positive, and, since the collector is reversed biased, V_C is negative.

The differential equation describing the diffusion of injected holes away from the emitter junction and into the base is, in the steady state, equation (2.84),

$$\frac{d^2}{dx^2}(p - p_0) = \frac{1}{L_p^2}(p - p_0) \tag{2.84}$$

where $p(x)$ is the nonequilibrium hole density at point x, p_0 is the (spatially constant) equilibrium hole density, and $p - p_0$ is the excess hole density at point x. As before, L_p is the diffusion length for minority holes in the base. The boundary conditions on equation (2.84) are given by the requirements that, at $x = 0$, $p - p_0$ must equal Δp_E, and, at $x = W_b$, $p - p_0$ must equal Δp_C. These conditions are expressed by the equations

$$p(0) = p_0 + \Delta p_E \tag{3.20}$$

$$p(W_b) = p_0 + \Delta p_C \tag{3.21}$$

A solution of (2.84) is

$$p(x) = p_0 + C_1 e^{x/L_p} + C_2 e^{-x/L_p} \tag{3.22}$$

where the boundary conditions determine the constants C_1 and C_2. From (3.20) and (3.21), we obtain

$$C_1 + C_2 = \Delta p_E \tag{3.23}$$

$$C_1 \exp(W_b/L_p) + C_2 \exp(-W_b/L_p) = \Delta p_C \tag{3.24}$$

Solving (3.23) and (3.24) for C_1 and C_2 gives

$$C_1 = [\Delta p_C - \Delta p_E e^{-W_b/L_p}](e^{W_b/L_p} - e^{-W_b/L_p})^{-1} \qquad (3.25)$$

$$C_2 = [-\Delta p_C + \Delta p_E e^{W_b/L_p}](e^{W_b/L_p} - e^{-W_b/L_p})^{-1} \qquad (3.26)$$

for the constants in equation (3.22) for $p(x) - p_0$, the excess hole density at point x in the base.

To find the diffusion current $J_p(x)$ of holes flowing from the emitter to the collector we use the result from (2.77) that

$$J_p(x) = -eD_p \frac{d}{dx}[p(x)] \qquad (2.77)$$

This leads, on differentiating $p(x)$ given by (3.22) and substituting the expressions (3.25) and (3.26) for C_1 and C_2, to the result

$$J_p(x) = \frac{-eD_p/L_p}{e^{W_b/L_p} - e^{-W_b/L_p}}$$
$$\times [\Delta p_C e^{x/L_p} - \Delta p_E e^{(x-W_b)/L_p} + \Delta p_C e^{-x/L_p} - \Delta p_E e^{-(x-W_b)/L_p}] \qquad (3.27)$$

for the diffusion current $J_p(x)$ of holes at any point x of the base.

To find the emitter current density J_E, we evaluate (3.27) at the edge $x = 0$ of the emitter junction space charge region (remembering that we assumed $\gamma = 1$ so J_E is composed only of holes), and find

$$J_E = J_p(0) = \frac{eD_p/L_p}{e^{W_b/L_p} - e^{-W_b/L_p}}[-2\Delta p_C + \Delta p_E(e^{W_b/L_p} + e^{-W_b/L_p})] \qquad (3.28)$$

Equation (3.28) can be rewritten using the identities

$$\coth x = (e^x + e^{-x})/(e^x - e^{-x}) \qquad (3.29)$$

and

$$2/(e^x - e^{-x}) = (\sinh x)^{-1} = \operatorname{csch} x \qquad (3.30)$$

so we obtain

$$J_E = e(D_p/L_p)[\Delta p_E \coth(W_b/L_p) - \Delta p_C \operatorname{csch}(W_b/L_p)] \qquad (3.31)$$

To find the collector current density J_C, we evaluate (3.27) at the edge $x = W_b$ of the collector junction space charge region in the base, so

$$J_C = J_p(W_b) = \frac{eD_p/L_p}{e^{W_b/L_p} - e^{-W_b/L_p}}[2\Delta p_E - \Delta p_C(e^{W_b/L_p} - e^{-W_b/L_p})] \qquad (3.32)$$

which, on rearranging to introduce hyperbolic functions, becomes

$$J_C = e(D_p/L_p)[-\Delta p_C \coth(W_b/L_p) + \Delta p_E \operatorname{csch}(W_b/L_p)] \quad (3.33)$$

To find the base current density J_B, we note that the current density J_E of holes into the base must equal the total current density $J_B + J_C$ of holes out of the base. Thus we have $J_B = J_E - J_C$, and

$$J_B = e(D_p/L_p)\{[\coth(W_b/L_p) - \operatorname{csch}(W_b/L_p)](\Delta p_E + \Delta p_C)\} \quad (3.34)$$

Using the identity $\coth x - \operatorname{csch} x = \tanh(x/2)$, equation (3.34) becomes

$$J_B = e(D_p/L_p)[(\Delta p_E + \Delta p_C) \tanh(W_b/2L_p)] \quad (3.35)$$

Equations (3.31), (3.33), and (3.35) give the emitter, collector, and base currents as functions [through Δp_E and Δp_C given by (3.18) and (3.19)] of the voltages V_E and V_C applied to the emitter and collector junctions. These expressions for J_E, J_C, and J_B are therefore correct for any bias voltages applied to the junctions of the transistor, and not just for the "usual" voltages in which the emitter junction is forward biased and the collector junction is reverse biased.

We now specialize equations (3.31), (3.33), and (3.35) to the particular case of "normal" or "active" biasing of the transistor, in which the emitter junctions is forward biased and the collector junction is reverse biased. For values of V_E and $|V_C|$ larger than $k_B T/e$, equations (3.18) and (3.19) become

$$\Delta p_E \cong p_{n0} \exp(eV_E/k_B T) \quad (3.36)$$

and

$$\Delta p_C \cong -p_{n0} \quad (3.37)$$

the negative sign in (3.37) meaning that carriers are extracted. Note that, in this situation, the total hole density $p_{n0} + \Delta p_C$ at the collector junction is approximately zero because of minority hole extraction. From (3.36) and (3.37), we see that, for the case of normal transistor bias, we have

$$\Delta p_E \gg \Delta p_C \quad (3.38)$$

and we can neglect the terms in Δp_C in the expressions (3.31), (3.33), and (3.35) for the currents in the device. We thus obtain

$$J_E = e(D_p/L_p) \Delta p_E \coth(W_b/L_p) \quad (3.39)$$

$$J_C = e(D_p/L_p) \Delta p_E \operatorname{csch}(W_b/L_p) \quad (3.40)$$

$$J_B = e(D_p/L_p) \Delta p_E \tanh(W_b/2L_p) \quad (3.41)$$

where Δp_E is given by (3.36), so (3.39)–(3.41) give the device currents as a function of the forward bias V_E on the emitter junction. We may exhibit that functional dependence explicitly by substituting (3.36) into (3.39) through (3.41), obtaining

$$J_E = e(D_p/L_p)[\coth(W_b/L_p)]p_{n0}\exp(eV_E/k_BT) \qquad (3.42)$$

$$J_C = e(D_p/L_p)[\operatorname{csch}(W_b/L_p)]p_{n0}\exp(eV_E/k_BT) \qquad (3.43)$$

$$J_B = e(D_p/L_p)[\tanh(W_b/2L_p)]p_{n0}\exp(eV_E/k_BT) \qquad (3.44)$$

We see from (3.42) through (3.44) that, in this approximation of neglecting Δp_C compared to Δp_E, the collector voltage V_C does not appear in the expressions for the currents. In particular, the collector current J_C is independent of the voltage across the collector junction in this approximation. Finally, we remind ourselves that the minority hole density p_{n0} in the base is equal to (n_i^2/N_d), where N_d is the doping in the base region.

We now use equations (3.42)–(3.44) to calculate the current gain factors α and β. From (3.15) and (3.13), we obtain

$$\alpha \equiv J_C/J_E = [\cosh(W_b/L_p)]^{-1} \qquad (3.45)$$

$$\beta \equiv J_C/J_B = [\operatorname{csch}(W_b/L_p)]/[\tanh(W_b/2L_p)] \qquad (3.46)$$

showing how the current gain varies with the ratio W_b/L_p of the base width W_b to the minority hole diffusion length L_p. Figure 3.11 is a plot of the collector to base current gain β as a function of L_p/W_b, showing how β increases as the ratio of the hole diffusion length L_p to the base width W_b

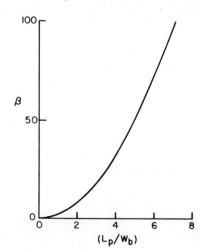

Figure 3.11. Current gain β, given by equation (3.46), plotted as a function of L_p/W_b, where L_p is the hole diffusion length and W_b is the width of the base.

increases. Since $L_p = (D_p\tau_p)^{1/2}$, this curve tells us that the current gain β increases with increasing minority hole lifetime τ_p.

Finally, for large values of L_p/W_b, W_b/L_p is less than unity, so we may expand the hyperbolic functions in (3.46) in series, where, for $x \ll 1$,

$$\tanh x \cong x - (1/3)x^3 + \cdots \tag{3.47}$$

$$(\sinh x)^{-1} \cong 1/x - (1/6)x + \cdots \tag{3.48}$$

Using the first term of the series expansions above in the expression (3.46) for β gives

$$\beta \cong 2L_p{}^2/W_b{}^2 = 2D_p\tau_p/W_b{}^2 \tag{3.49}$$

This result shows that, for $L_p \gg W_b$, the current gain β is proportional to the minority hole lifetime τ_p.

We note that the results (3.45) and (3.46) of our treatment predict that the current gains α and β are independent of the current densities. This is not actually true in real devices.[17]

Summary of the Physics of Amplification in the Bipolar Transistor

Several points may be mentioned. First, the action of the bipolar transistor as an amplifier depends directly on the simultaneous coexistence of electrons and holes in the base of the transistor. In a *pnp* transistor, this coexistence allows the base current of electrons to control the collector current of holes. Second, the bipolar transistor is a useful amplifier because a small base current can control a large collector current. This is equivalent to saying that large values of the current gain β are possible. This effect is, in turn, due to the fact that minority carriers in an indirect gap semiconductor like silicon have a lifetime long enough to yield usefully long values of the minority hole diffusion length, thereby allowing minority holes to cross the base of the transistor and be collected. Finally, the bipolar transistor is a minority carrier device, in that it depends on the existence of minority carriers in a semiconductor to produce amplification.

Tunnel Diodes

The tunnel diode is a particular type of semiconductor *p–n* junction which exhibits a negative differential conductivity. By this, it is meant that this device shows a decrease in current with increasing voltage over at

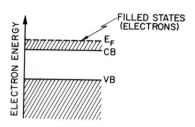

Figure 3.12. Band diagram showing free electrons in the conduction band of a degenerate *n*-type semiconductor. Filled electron states are shown as shaded; E_F is the Fermi level; CB and VB are the conduction and valence band edges, respectively.

least a part of its current–voltage characteristic. These negative resistance diodes, or tunnel diodes as they are more commonly called, have a number of useful circuit applications[18,19] because of their negative dynamic resistance.

Consider an *n*-type semiconductor so heavily doped with impurities that it is degenerate. Then, as discussed in Chapter 1 and exhibited in Figures 1.15 and 1.16, the Fermi level is above the conduction band minimum and there is a density of free electrons in the conduction band. For a degenerate *p*-type semiconductor, there is a density of free holes in the valence band. Band diagrams exhibiting these effects are showing in Figures 3.12 and 3.13. Consider next a *p–n* junction, at equilibrium, in which both the *n*-type and *p*-type sides are degenerate. The band diagram of such a "degenerate junction," at equilibrium, is shown in Figure 3.14. The junction is assumed to be ideally abrupt and, since the impurity density is high in both *n*-type and *p*-type regions, the width W of the space charge region will be small for this type of junction. A representative value[18,20] of W of the order of 100 Å for a degenerate silicon *p–n* junction with donor and acceptor densities of 10^{19} cm^{-3}.

If we consider the possibility of the tunneling of electrons from one side of the junction to the other, the width W of the depletion layer will be the width of the energy barrier to tunneling. Since W will be small for the degenerate junction, electron tunneling is possible, and we examine next the factors affecting tunneling.

What factors will affect the probability of tunneling by an electron? We expect first that the current of tunneling electrons will increase as the

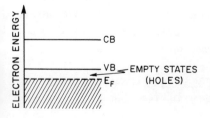

Figure 3.13. Band diagram showing free holes in the valence band of a degenerate *p*-type semiconductor. Filled electron states are shown as shaded; E_F is the Fermi level; CB and VB are the conduction and valence band edges, respectively.

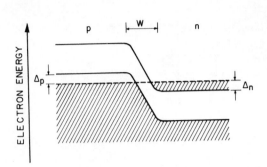

Figure 3.14. Degenerate *p–n* junction at equilibrium. The junction is assumed to be abrupt and the width W of the space charge region is exaggerated. The Fermi level is the dashed line and filled electron states are shown as shaded. The energies to which the conduction and valence bands are filled with electrons and holes are denoted by $\varDelta_n$ and $\varDelta_p$, respectively.

width W of the barrier decreases since we expect[21] that the transmission coefficient for the tunneling process will behave in this manner. While the calculation of the tunneling probability will depend on the details of the energy barrier,[22] we can see qualitatively (say from[23] Fermi's golden rule number 2) that the probability of tunneling per unit time will be directly proportional to the density of final states available to the electron. For a given tunneling probability per unit time, we expect that the number of electrons tunneling will be proportional to the number of initial states occupied by electrons. In summary, we expect the current of tunneling electrons to increase with increasing density of empty final states, with increasing number of occupied initial states from which tunneling may take place, and with decreasing barrier width.

Given the degenerate junction band diagram shown in Figure 3.14 how will a voltage V_a applied to the junction affect the factors above which influence the magnitude of the tunnel current? One would expect that forward bias would decrease the width of the space charge region somewhat, but not greatly because W is already relatively small owing to the high doping. Forward bias will, however, increase the electron energies on the *n*-type side of the junction, thus "raising" the electron energy states on the *n*-side relative to the *p*-side. This effect is exhibited in Figure 3.15, which shows the band diagram with an applied forward bias. The symbols $\varDelta_n$ and $\varDelta_p$ are used, as in Figure 3.14, for the energies to which the conduction and valence bands are, respectively, filled with electrons and holes. Figure 3.15 is drawn for an applied forward bias $V_a = \varDelta_n/e$. We can see from Figure 3.15 that, since the tunneling process takes place at constant electron energy, forward bias will increase both the number of possible initial states and the number of available final states.

Beginning with an applied potential $V_a = 0$, as shown in Figure 3.14, we see that no filled state on the *n* side is at the same energy as an empty state on the p side. Further, no filled state on the p side is at the same energy as an empty state on the n side. At $V_a = 0$, then, no tunnel current of electrons flows. Figure 3.15 shows the application of a nonzero forward bias V_a. Since the electron states on the n side are "raised," there are now filled electron states at the same energy as empty states on the p side. Electrons can now tunnel from n to p, and a tunnel current of electrons flows through the junction.

At some value of the applied potential V_a, a maximum in the joint "overlap" of filled states on the n side with empty states on the p side will take place. Calling this value V_a', we can see intuitively that the value of V_a' will be related to the fillings Δ_n and Δ_p of the bands in the junction. Even though V_a' does not turn out to be a simple function[24] of Δ_n and Δ_p, we do expect that the tunnel current of electrons from n to p will reach a maximum at some value V_a' of the applied forward bias. At values of the forward bias larger than V_a', the joint "overlap" of filled states on the n side and empty states on the p side begins to decrease, until this "overlap" becomes zero at a value V_a'' of forward bias, where

$$eV_a'' = \Delta_n + \Delta_p \qquad (3.50)$$

We expect, then, that the tunnel current of electrons from n to p will go to zero at a value of forward bias V_a equal to V_a''. A schematic plot of electron tunnel current density J_t as a function of forward bias V_a is shown in Figure 3.16. We note that, for values of the forward bias between V_a' and V_a'', the dynamic conductance dJ_t/dV_a is negative. This means that, for $V_a' \leq V_a \leq V_a''$, the device current decreases for increasing values of the applied forward bias V_a.

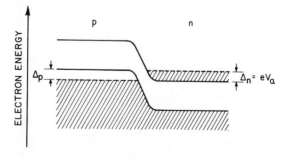

Figure 3.15. Band diagram of a degenerate p–n junction under an applied forward bias V_a. The filled electron states are shown as shaded and the energies to which the conduction and valence bands are filled with electrons and holes are Δ_n and Δ_p, respectively. The figure is drawn for an applied forward bias $V_a = \Delta_n/e$.

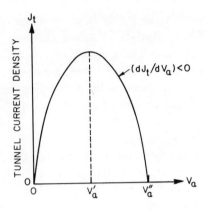

Figure 3.16. Variation (schematic) of the tunnel current density J_t with applied forward bias V_a. The voltage V_a'' at which J_t goes to zero is given by equation (3.50), and the dynamic conductance dJ_t/dV_a is negative for the voltage range $V_a' < V_a \leq V_a''$.

For all values of the applied forward bias, there will also be a forward injection current of electrons from *n* to *p* and holes from *p* to *n*, over their respective energy barriers. The total current density J will be the sum of the tunnel and injection currents and the behavior of J with applied forward bias is shown schematically in Figure 3.17. The total current–voltage characteristic has a minimum at a value V_V of the applied forward bias. A typical value[25] of the "valley voltage" V_V is 0.5 V for a GaAs tunnel diode. The ratio of the current density J_P (at the peak of the curve) to the current density J_V (at the valley voltage V_V) is typically[25] 15 for a GaAs device.

The discussion of tunneling here has tacitly assumed that the semiconductor has a direct band gap (e.g., GaAs) and that the electron tunneling process takes place between band extrema at the same point in the Brillouin zone. For indirect tunneling, in which the initial and final electron states are at different points in the Brillouin zone (e.g., silicon), the plot[26] of tunnel current as a function of forward bias is qualitatively similar to that for direct tunneling.

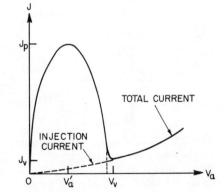

Figure 3.17. Schematic variation of the total current density J, as a function of applied forward bias V_a, for a tunnel diode.

For values of the forward bias larger than V_a'', one would expect only injection current to flow in the tunnel diode. However, the observed currents at such values of forward bias are larger than the expected injection current. This excess is referred to as the "excess current" in the tunnel diode and is, at least partially, due[27] to carrier tunneling through intermediate states in the semiconductor band gap.

Finally, for reverse bias, one can see from Figure 3.14 that there is "overlap" of states permitting tunneling of electrons from p to n for all values of the reverse bias. Since the "overlap" increases monotonically with increasing magnitude of the reverse bias, we expect that the reverse current–voltage characteristic will show a monotonic increase in the current with increasing reverse bias. This characteristic resembles that of a resistor.

It is also appropriate to remark on the fact that the process of tunneling by an electron is fast[28] compared to the drift or diffusion motion of an electron. This makes the tunnel diode a device that can shift rapidly from one point to another on its current–voltage characteristic, so it is very useful in high-speed circuits,[29] including switching circuits and high-frequency oscillators. This utility in high-speed applications, in addition to its negative dynamic resistance,[30] makes the tunnel diode a useful device.

Finally, it is useful to summarize how the physics of the heavily doped p–n junction gives the tunnel diode its useful properties of high speed and negative dynamic resistance. First, it is the rapidity of electron tunneling, compared to electron drift or diffusion, that gives the device its high speed. Second, the high doping in the junction does two things. It creates empty states in the valence band on the p side and filled states in the conduction band on the n side, thereby making possible electron tunneling from n to p under forward bias. (If this were not true, only electron tunneling from p to n under reverse bias would be possible.) The heavy doping also decreases the width of the energy barrier to tunneling by making the width of the space charge region small. Third, the negative resistance portion of the tunnel diode current–voltage characteristic comes about because the probability of tunneling decreases for values of applied forward bias greater than a critical value, thereby decreasing the device tunnel current for increasing magnitude of the forward bias.

The Junction Field Effect Transistor (JFET)

We consider next another solid state device that acts as an amplifier, the junction field-effect transistor[31] (or JFET). This device is one in which

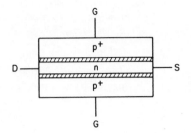

Figure 3.18. Schematic structure of a junction field-effect transistor. The shaded areas are the depletion layers of the reverse-biased p^+n junctions; S is the source, D the drain, and G the gate electrodes of the device.

the device current is controlled by an applied voltage, so it is a voltage-controlled amplifier.

The basic idea of the junction field-effect transistor is to vary the spatial extent of the depletion layer of a reverse-biased *p–n* junction by varying the magnitude of the applied reverse bias. The spatial extent of the depletion layer, in turn, controls the conductance of a channel in a semiconductor, thereby controlling the current flowing through this channel. Figure 3.18 shows a schematic picture of a JFET structure. In Figure 3.18, the depletion layers of the reverse-biased p^+n junctions are shown as shaded, and the source, drain, and gate electrodes of the device are marked S, D, and G, respectively. The *n*-type region is called the channel; its conductivity is determined by its doping. If the p^+n junctions are reverse biased, the width of the depletion layer (almost all of which is on the *n* side because of the high doping on the *p* side) will be increased. The shaded regions of Figure 3.18 are depleted of electrons and are thus of high resistance, thereby decreasing the effective cross-section area of the *n*-type channel and increasing the resistance of the channel. It is clear that the current through the channel will decrease with increasing magnitude of the reverse bias applied to the p^+n junctions. In this manner the voltage applied to the gate electrode controls the current through the device, so the junction field-effect transistor is a voltage-controlled amplifier.

Physical Basis of the Current–Voltage Characteristic of the JFET

Consider a JFET connected as shown in Figure 3.19. The potential of the drain relative to the source is $(+V_{DS})$ and V_{GS} is the potential of the gates relative to the source. The *n*-type channel is of length L, and the shaded regions represent the equilibrium depletion layers in the *n* channel due to the reverse-biased p^+n junctions. We consider first the situation when $V_{GS} = 0$. The reverse bias applied to the junctions is then of magnitude

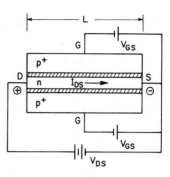

Figure 3.19. Junction field-effect transistor with circuit connections; the gates G are connected to the source S. The shaded areas are the equilibrium depletion layers of the p^+n junctions; the n-type channel is of length L.

$|V_{DS}|$. As shown in Figure 3.19, the channel current I_{DS} flowing from drain to source is composed of electrons flowing from source to drain.

What happens to the current I_{DS} as the magnitude of V_{DS} increases? We assume that the heavily doped p^+ gate regions have such a high conductivity that they are essentially metallic, so that the electrostatic potential is the same throughout the gate regions. Since the gates in Figure 3.19 are connected to the source, the potential throughout the gates is the source potential, which we choose to be zero. Since the n channel is lightly doped, the channel will act as a distributed resistor. For low currents, voltage V_{DS} varies approximately linearly across the length L of the channel. If we call $V(x)$ the potential at point x of the channel (where the source is at $x = L$ and the drain at $x = 0$), $V(L) = 0$ at the source and $V(0) = +V_{DS}$ at the drain. We see that $V(x)$ will have a larger magnitude near the drain than near the source. In this configuration (with no bias applied to the gates G) the voltage $V(x)$ is the magnitude of the reverse bias applied at point x to the p^+n junctions. The conclusion is that a larger magnitude reverse bias is applied to the p^+n junctions near the drain D than is applied near the source S.

The result of this conclusion is that the junction depletion layers extend further into the channel near the drain than they do near the source. For a given value of the voltage V_{DS}, the depletion layers in the JFET look (schematically) as shown in Figure 3.20, where the depletion layers are

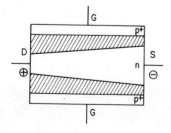

Figure 3.20. Depletion layers (schematic) of JFET for a given value of V_{DS} and no voltage applied to the gates G ($V_{GS} = 0$). The depletion layers are shown as shaded.

the shaded areas. As the magnitude of V_{DS} is increased the depletion layers extend further into the channel, increasing the resistance R of the channel. Since $R = dV_{DS}/dI_{DS}$ and R increases with increasing V_{DS}, a plot of V_{DS} as a function of I_{DS} will have a positive slope which increases with increasing V_{DS}. A plot of I_{DS} as a function of V_{DS} will then have a slope $1/R = dI_{DS}/dV_{DS}$ which decreases with increasing V_{DS}, as shown in Figure 3.21. This figure shows the current–voltage characteristic of a JFET with no voltage applied to the gates.

Figure 3.21 shows the device current I_{DS} saturating with increasing magnitude of the voltage V_{DS}. Is that result physically reasonable? The answer is yes, as can be seen from the shape of the depletion layers in the drawing in Figure 3.20. As V_{DS} is increased, the depletion layers extend further into the channel until, at a particular value of V_{DS}, called V_P, the two depletion layers meet near the drain. This effect is called "pinching-off" the conducting part of the channel, and V_P is called the "pinch-off" voltage. (The physics involved in pinch-off may be more complicated than simple merging of the depletion layers.[32]) When pinch-off takes place, the current I_{DS} does not increase significantly with increasing V_{DS} above the pinch-off voltage V_P, resulting in the approximate saturation of the device current shown in Figure 3.21.

Consider next the application of a negative potential $V_{GS} = -V_1$ between the gates G and the source S. Since this potential makes the gates (p^+ regions) more negative with respect to the source, application of a negative voltage to the gates increases the magnitude of the reverse bias applied to the p^+n junctions (for a given value of V_{DS}). The result is an increase in the penetration of the junction depletion layer into the channel for a given value of the voltage V_{DS}. Pinch-off thus takes place at a smaller values of V_{DS} when a negative bias is applied to the gates of the device. This effect is shown in Figure 3.22, where we see that pinch-off occurs at smaller values of V_{DS} for increasing magnitude ($V_2 > V_1$) of the negative potential V_{GS} applied to the gates. The locus of the points showing the

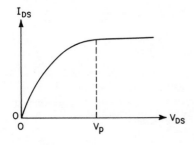

Figure 3.21. Current I_{DS} as a function of voltage V_{DS} for JFET with $V_{GS} = 0$. The slope of the curve decreases with increasing V_{DS} until the curve saturates at a value of $V_{DS} = V_P$, the pinch-off voltage.

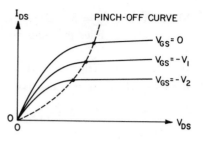

Figure 3.22. Current I_{DS} as a function of voltage V_{DS} for JFET for $V_{GS} = 0, -V_1, -V_2$, where V_2 is greater than V_1. The black dots indicate the pinch-off voltage V_p at each value of V_{GS}; the pinch-off curve is shown as a dashed curve.

pinch-off voltage V_P on the I_{DS}–V_{DS} curves is called the pinch-off curve of the device. We can see that the channel resistance $R = dV_{DS}/dI_{DS}$ is larger (for a given value of V_{DS}) for larger magnitudes $|V_{GS}|$ of the negative potential applied to the gates. Since the slope dI_{DS}/dV_{DS} of the curves in Figure 3.22 is equal to $1/R$, this fact means that the curves in Figure 3.22 have smaller slopes (at a given value of V_{DS}) for larger magnitudes $|V_{GS}|$ of the gate voltage.

Examining the current–voltage plots of Figures 3.22 for values of V_{DS} above the pinch-off voltage, we see that the JFET is a device in which the current I_{DS} is (for a given value of V_{DS}) controlled by the magnitude of the gate voltage V_{GS}. The JFET is thus a voltage-controlled amplifier in which the gate voltage V_{GS} controls the device current I_{DS} through the effect of the reverse bias on the channel resistance.

The JFET device current I_{DS} is composed only of majority carriers which, in this example of an n channel, are electrons. The JFET is therefore a majority carrier device, as opposed to the bipolar junction transistor, which depends for its operation on the coexistence of majority and minority carriers in the base.

Since the gate voltage V_{GS} is applied across a reverse-biased p–n junction, the input impedence of a JFET is high, typically many megohms. The control of I_{DS} by the gate voltage V_{GS} is represented by the mutual transconductance g_m of the device, where

$$g_m = (dI_{DS}/dV_{GS})_{V_{DS}} \tag{3.51}$$

The amplification factor μ of the JFET is defined by

$$\mu = g_m r_d \tag{3.52}$$

where the drain resistance $r_d = (dV_{DS}/dI_{DS})_{V_{GS}}$ is what we have been calling the channel resistance R. Typical values[33] of these parameters are $g_m =$

10^{-3} A/V and $r_d = 5 \times 10^5 \, \Omega$, giving a value of $\mu = 500$. The use of the JFET as an amplifier is discussed in various electrical engineering texts;[34] however, the JFET is not as important commercially as another type of field-effect transistor (the metal-oxide–semiconductor field effect transistor, or MOSFET), which we will discuss in a later chapter.

Problems

3.1. *Band Diagram of npn Transistor.* (a) Draw band diagrams for an *npn* transistor at equilibrium and under the normal active bias conditions for amplification. (b) Draw a figure showing the currents flowing and use it to discuss the injection and collection of electrons. (c) What are the components of the base current in an *npn* transistor?

3.2. *Transistor at High Temperature.* A silicon *pnp* transistor is operated at 150°C. Discuss qualitatively the changes you would expect in its characteristics relative to those at room temperature.

3.3. *Reverse Current of a Tunnel Diode.* Using a band diagram of a degenerate junction, discuss the current flow in a tunnel diode under reverse bias. Sketch qualitatively the current–voltage characteristic for reverse bias.

3.4. *High-Temperature Rectifier.* Discuss the design of a possible *p–n* junction diode for use as a high-temperature rectifier. Suggest a specific material for a diode with a reverse saturation current density no greater than 10^{-12} A/cm² at 150°C (for reverse voltages less than the breakdown value). Discuss doping and other variables for such a design.

3.5. *Depletion Layer Width in a Tunnel Diode.* Calculate the width of the depletion layer in a degenerate *p–n* junction in silicon (of dielectric constant 11.7). Take the donor and acceptor densities as equal to 2×10^{19} cm⁻³.

References and Comments

1. W. G. Oldham and S. E. Schwarz, *An Introduction to Electronics*, Holt, Rinehart, and Winston, New York (1972), Chapter 5.
2. J. Millman and C. C. Halkias, *Integrated Electronics*, McGraw-Hill, New York (1972), Chapter 4.
3. See, for example, C. Kittel, *Introduction to Solid State Physics*, Third Edition, John Wiley, New York (1966), page 323.
4. See References 38–40 of Chapter 2.
5. See References 26 and 27 of Chapter 2.

6. J. Millman and C. C. Halkias, Reference 2, pages 73–77.
7. B. G. Streetman, *Solid State Electronic Devices*, Prentice-Hall, New York (1972), pages 278–285.
8. B. G. Streetman, Reference 7, pages 306–308.
9. B. G. Streetman, Reference 7, pages 308–313.
10. A. S. Grove, *Physics and Technology of Semiconductor Devices*, John Wiley, New York (1967), pages 1–88 and 208–209.
11. S. M. Sze, *Physics of Semiconductor Devices*, John Wiley, New York (1969), pages 270–272.
12. B. G. Streetman, Reference 7, pages 306–308; W. G. Oldham and S. E. Schwarz, Reference 1, page 438; J. Millman and C. C. Halkias, Reference 2, pages 126–134.
13. See, for example, C. Kittel, *Introduction to Solid State Physics*, Second Edition, John Wiley, New York (1956), pages 397–398; W. G. Oldham and S. E. Schwarz, Reference 1, page 514.
14. S. Wang, *Solid State Electronics*, McGraw-Hill, New York (1966), pages 340–349; see also A. S. Grove, Reference 10, pages 222–228.
15. W. G. Oldham and S. E. Schwarz, Reference 1, pages 319–323; S. M. Sze, Reference 11, pages 302–309; B. G. Streetman, Reference 7, pages 325–340.
16. R. S. Muller and T. I. Kamins, *Device Electronics for Integrated Circuits*, John Wiley, New York (1977), pages 218–222.
17. S. M. Sze, Reference 11, page 272; S. Wang, Reference 14, page 346; J. Millman and C. C. Halkias, Reference 2, page 133.
18. J. Millman and C. C. Halkias, Reference 2, pages 77–79.
19. K. K. N. Chang, *Parametric and Tunnel Diodes*, Prentice-Hall, New York (1964), pages 147–212; H. C. Okean in *Semiconductors and Semimetals*, R. K. Willardson and A. C. Beer (editors), Academic Press, New York (1971), Volume 7, Part B, pages 473–624.
20. A. S. Grove, Reference 10, page 163, Figure 6.9.
21. L. I. Schiff, *Quantum Mechanics*, Third Edition, McGraw-Hill, New York (1968), page 103, equation (17.7).
22. S. M. Sze, Reference 11, pages 156–169; S. Wang, Reference 14, pages 368–372.
23. L. I. Schiff, Reference 21, page 285, equation (35.14).
24. See S. M. Sze, Reference 11, page 167, for a particular case; see also S. Wang, Reference 14, pages 378–379.
25. J. Millman and C. C. Halkias, Reference 2, page 79.
26. S. M. Sze, Reference 11, page 168, Figure 11(a).
27. S. M. Sze, Reference 11, pages 169–172, discusses excess current.
28. K. K. Thornber, T. C. McGill, and C. A. Mead, *Journal of Applied Physics*, **38**, 2384–2385 (1967) for a discussion of the tunneling time of an electron.
29. J. Millman and C. C. Halkias, Reference 2, page 78.
30. J. J. Brophy, *Basic Electronics for Scientists*, Third Edition, McGraw-Hill, New York (1977), pages 266–270.
31. B. G. Streetman, Reference 7, pages 285–293; S. M. Sze, Reference 11, pages 340–346; A. S. Grove, Reference 10, pages 243–252.
32. See B. G. Streetman, Reference 7, page 288, and further references on his page 313.
33. J. Millman and C. C. Halkias, Reference 2, pages 318–321.
34. J. Millman and C. C. Halkias, Reference 2, Chapter 10; J. J. Brophy, Reference 30, Chapter 6.

Suggested Reading

B. G. STREETMAN, *Solid State Electronic Devices*, Prentice-Hall, New York (1972). Our discussion of junction devices parallels that of Streetman to some extent. This text is excellent as an introduction to semiconductor devices.

S. M. SZE, *Physics of Semiconductor Devices*, John Wiley, New York (1969). Once you have learned the basic physical ideas behind a semiconductor device, this formidable treatise will furnish the details and the mathematics, in addition to references to the original literature.

A. S. GROVE, *Physics and Technology of Semiconductor Devices*, John Wiley, New York (1967). Recommended for both devices physics and fabrication technology, especially for discrete silicon devices.

J. MILLMAN and C. C. HALKIAS, *Integrated Electronics*, McGraw-Hill, New York (1972). This modern electrical engineering textbook covers circuit applications of all sorts of devices. (A revised version, entitled *Microelectronics*, by J. Millman, was published in 1979 by McGraw-Hill.)

W. G. OLDHAM and S. E. SCHWARZ, *An Introduction to Electronics*, Holt, Rinehart, and Winston, New York (1972). This electronics text is also recommended for device applications.

J. J. BROPHY, *Basic Electronics for Scientists*, Third Edition, McGraw-Hill, New York (1977). The two textbooks immediately above are primarily for electrical engineers who will be concerned with the design of devices and circuits. This electronics text is aimed at the science student who wishes to understand the operation of electronic instruments, and therefore does not require as intensive a familiarity as does the electrical engineer. This book is recommended to physics students wishing some knowledge of the electronics they use in research.

4

Physics of Metal–Semiconductor and Metal–Insulator–Semiconductor Junctions

Introduction

This chapter discusses some topics in the physics of metal–semiconductor junctions and metal–insulator–semiconductor (MIS) junctions. The former topic is treated first in a discussion of the metal–semiconductor junction band diagram at equilibrium and under an applied potential. The results are used to show that the current–voltage characteristic is asymmetric and to treat metal contacts to semiconductors. The latter topic is then discussed using the band diagram of an idealized MIS structure. The key result is the formation of an inversion layer at the insulator–semiconductor interface when an appropriate potential is applied. This result will be used in the next chapter to study two important devices, the induced-channel field-effect transistor and the charge-coupled device.

The Metal–Semiconductor Junction at Equilibrium

We consider a junction between a metal and a semiconductor,[1-4] often called a Schottky junction or a Schottky diode. These are useful solid state devices and this topic will also give us some information on metal contacts to semiconductors.

In the following discussion we neglect the existence of surface states at the semiconductor surface. (In Chapter 6, a brief discussion of surface physics is presented, and it will consider the effect of surface states on the

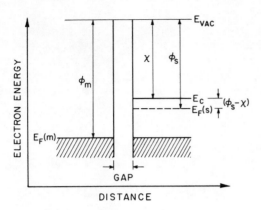

Figure 4.1. Band diagrams of a metal and an n-type semiconductor before contact is made, and with a gap or space between them. The work function of the metal is ϕ_m, the work function of the semiconductor is ϕ_s, χ is the electron affinity of the semiconductor, and ϕ_m is larger than ϕ_s. The semiconductor Fermi level is $E_F(s)$, E_C is the energy of the conduction band edge, E_V is the energy of the valence band edge, and $E_F(m)$ is the Fermi level in the metal. Filled electron states are shaded, and the position of $E_F(s)$ relative to E_C is exaggerated for clarity.

properties of metal–semiconductor contacts.) Figure 4.1 shows the band diagrams of a metal and of an n-type semiconductor before contact is made and with a small gap or space between them. The notation used in Figure 4.1 is that $E_F(m)$ denotes the Fermi level of the metal, $E_F(s)$ is the Fermi level of the semiconductor, and E_{VAC} is the common vacuum energy level for both before contact. The work function ϕ_m of the metal is defined by

$$\phi_m \equiv E_{\mathrm{VAC}} - E_F(m) \tag{4.1}$$

and is the energy necessary to remove an electron from the top of the Fermi distribution in the metal to the vacuum level. Similarly, the work function ϕ_s of the semiconductor is defined by

$$\phi_s \equiv E_{\mathrm{VAC}} - E_F(s) \tag{4.2}$$

and is the energy difference between the Fermi level in the semiconductor and the vacuum level. Since the position of the semiconductor Fermi level $E_F(s)$ depends on the doping-impurity content, the work function ϕ_s is also a function of the doping. The electron affinity χ of the semiconductor is defined as

$$\chi \equiv E_{\mathrm{VAC}} - E_C \tag{4.3}$$

where E_C is the energy of the conduction band edge. The electron affinity is the energy required to remove an electron from the bottom of the conduction band to the vacuum level. If we choose the zero of energy at the conduction band edge, so $E_C = 0$, then equations (4.2) and (4.3) can be combined to give

$$\phi_s = \chi - E_F(s) \qquad (4.4)$$

where $E_F(s)$ is negative. Both the semiconductor[5] and metal[6] work functions and the semiconductor electron affinity[5] are of the order of a few electron volts.

We now consider the formation of a metal–n-type-semiconductor junction for the case, shown in Figure 4.1, for which the metal work function ϕ_m is larger than the semiconductor work function ϕ_s. We (mentally) bring together the metal and the semiconductor until the gap between them is very small and a flow of electrons from the semiconductor into the metal now takes place. This electron transfer creates a region in the semiconductor, of width W, that is depleted of electrons and hence bears a positive space charge due to uncompensated fixed donors. The electrons flowing into the metal form a surface density of negative charge. These effects are shown in Figure 4.2. The charge densities in the semiconductor and on the metal surface produce an electric field $\mathscr{E}$ as shown, located almost entirely in the semiconductor depletion layer. The electric field $\mathscr{E}$ corresponds to a gradient of electrostatic potential and of electron potential energy across the depletion layer, also as shown (schematically) in Figure 4.2. The energy of an electron at the conduction band edge is higher at the semiconductor surface than it is in the bulk of the semiconductor, outside the depletion

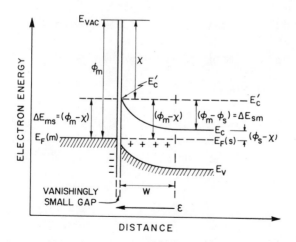

Figure 4.2. Band diagram of a metal–n-type-semiconductor junction with an infinitesimally small gap between them. The width of the positive space charge region in the semiconductor is W, $\mathscr{E}$ is the electric field, ΔE_{ms} is the energy barrier between metal and semiconductor, and ΔE_{sm} is the energy barrier between semiconductor and metal.

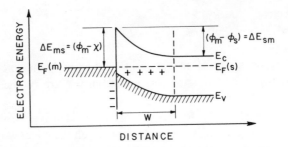

Figure 4.3. Simplified version of the metal-*n*-type-semiconductor junction band diagram shown in Figure 4.2, at equilibrium, with the gap reduced to zero, for the case in which ϕ_m is larger than ϕ_s.

layer. The transfer of electrons from semiconductor to metal continues until the Fermi energy in the bulk of the semiconductor is the same as the Fermi energy in the metal.

In the limit, as the gap between metal and semiconductor vanishes, the value of the energy difference, *at the surface*, between the vacuum level and the conduction band edge is equal[†] to χ. This is shown in Figure 4.2, and the energy barrier ΔE_{ms} between metal and semiconductor is thus $\phi_m - \chi$. As seen further from Figure 4.2, $\phi_m - \chi$ is also the energy difference between the Fermi energy $E_F(s)$ in the bulk semiconductor and the energy E_C' of the conduction band edge at the semiconductor surface.

In the bulk of the semiconductor, we see from Figure 4.2 that the energy difference between the conduction band edge E_C and $E_F(s)$ is equal to $\phi_s - \chi$. Therefore, the energy barrier ΔE_{sm} for an electron going from an energy E_C in the bulk semiconductor into the metal is given by

$$\Delta E_{sm} = (\phi_m - \chi) - (\phi_s - \chi) = \phi_m - \phi_s \qquad (4.5)$$

We conclude from these considerations that the energy barrier ΔE_{ms} for an electron going from metal to semiconductor is $\phi_m - \chi$, while the barrier ΔE_{sm} for an electron going from the semiconductor bulk into the metal is $\phi_m - \phi_s$. Figure 4.3 shows, in the conventional manner, the band diagram based on these results, with the gap or space between metal and semiconductor reduced to zero. The width W of the depletion layer may be found by solving[7] Poisson's equation for the charge density $+eN_d$ in the space

---

[†] This is equivalent to the statement[1] that, as the gap vanishes, the value of the semiconductor work function, at the surface, is equal to the metal work function less the semiconductor electron affinity.

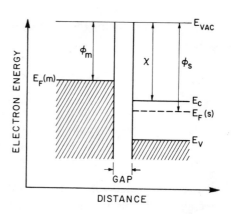

Figure 4.4. Band diagrams of a metal and an n-type semiconductor before contact is made, and with a gap or space between them. The metal work function ϕ_m is smaller than the semiconductor work function ϕ_s in this case. The notation is the same as that of Figure 4.1.

charge region, where N_d is the density of ionized donors in the semiconductor.

We consider next the metal–n-type-semiconductor junction for the situation in which the semiconductor work function ϕ_s is larger than the metal work function ϕ_m. (The previous discussion treated the opposite case, in which ϕ_m is the larger.) Figure 4.4 shows the band diagram before contact, with a gap between the metal and the semiconductor; the symbols are the same as those used in Figures 4.1–4.3. The metal Fermi energy $E_F(m)$ is now higher than the Fermi level $E_F(s)$ in the semiconductor, so, as the gap between metal and semiconductor is decreased, electrons are transferred to the semiconductor. This is shown in Figure 4.5, in which the Fermi level in the semiconductor is raised, and is equal to the Fermi energy in the metal.

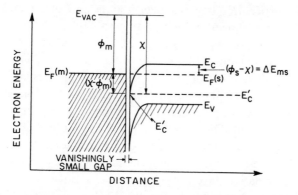

Figure 4.5. Band diagram of a metal–n-type-semiconductor junction, with an infinitesimally small gap between them, for the situation in Figure 4.4. The energy barrier between metal and semiconductor is $\Delta E_{ms} = \phi_s - \chi$.

At the semiconductor surface, the energy E_C' of the conduction band edge is still an energy χ below the vacuum level, as shown in Figure 4.5. Thus we have

$$\chi = E_{\text{VAC}} - E_C' \tag{4.6}$$

at the surface. Since equation (4.1) is also still true at the metal surface, we have, combining equations (4.6) and (4.1),

$$\chi - \phi_m = E_F(m) - E_C' \tag{4.7}$$

at the joint metal–semiconductor interface, so the conduction band edge energy E_C' at the interface is an energy $\chi - \phi_m$ below the Fermi level in the metal.

We now calculate the energy barrier

$$\Delta E_{ms} \equiv E_C - E_F(m) \tag{4.8}$$

for an electron going from the metal to the conduction band edge in the bulk semiconductor, where its energy will be E_C. Since, at equilibrium, $E_F(s)$ equals $E_F(m)$, equation (4.8) becomes

$$\Delta E_{ms} = E_C - E_F(s) = -E_F(s) \tag{4.9}$$

since we defined E_C (in the bulk) as the energy zero, so $E_C = 0$, and $E_F(s)$ is negative. Substituting equation (4.4) into (4.9) gives

$$\Delta E_{ms} = (\phi_s - \chi) \tag{4.10}$$

as the energy barrier ΔE_{ms} between metal and the bulk of the semiconductor. The result (4.10) is shown in Figures 4.5 and 4.6; the latter figure shows the results of Figure 4.5 as the gap between metal and semiconductor goes to zero. The band diagram of Figure 4.6 is that of a metal–n-type-semiconductor junction, at equilibrium, for the case for which ϕ_s is larger than ϕ_m.

Since electrons are transferred from metal to semiconductor in the situation in which ϕ_m is less than ϕ_s, a negative surface charge develops on the semiconductor. The negative charge is a surface charge because, as shown in Figure 4.6, the semiconductor Fermi level $E_F(s)$ crosses the conduction band edge near the semiconductor–metal junction. This produces an accumulation layer of filled electron states at those electron energies below the Fermi level, and the electron accumulation layer is, as shown in Figure 4.6, confined in space to the region of the semiconductor–metal junction surface. (This should be compared with the extended positive

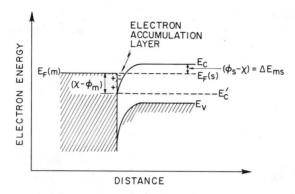

Figure 4.6. Simplified version of the metal–n-type-semiconductor junction band diagram shown in Figure 4.5, at equilibrium, with the gap reduced to zero, for the case in which ϕ_m is smaller than ϕ_s. An electron accumulation layer has formed in the semiconductor near the junction. The spatial extent of the accumulation layer is schematic and has been exaggerated for clarity.

space charge region in the semiconductor shown in Figure 4.3 for the case in which ϕ_m is greater than ϕ_s.) This negative surface charge is compensated by a positive surface charge on the metal.

We summarize these results concerning the metal–n-type-semiconductor junction at equilibrium. If the metal work function ϕ_m is larger than the semiconductor work function ϕ_s, then the band diagram of the junction at equilibrium is shown in Figure 4.3. If ϕ_m is smaller than ϕ_s, then the band diagram is shown in Figure 4.6, in which an electron accumulation layer is formed in the semiconductor near the junction. The corresponding results for the metal–p-type-semiconductor junction are left as an exercise.

Effect of an Applied Potential on the Metal–Semiconductor Junction

We next look into the effect of an applied potential on the energy barrier ΔE_{ms} to electron flow from the metal to the semiconductor and the energy barrier ΔE_{sm} to electron flow from the semiconductor to the metal. We consider the metal–n-type-semiconductor junction, so forward bias is a negative potential of magnitude V_a applied to the semiconductor and reverse bias is a positive potential V_a applied to the semiconductor.

In the case in which ϕ_m is larger than ϕ_s, there is a positively charged depletion layer in the semiconductor, as shown in Figure 4.3. We assume that all of the applied potential V_a appears across the depletion layer, so no "band tilting" takes place in the bulk semiconductor. Figure 4.7 shows

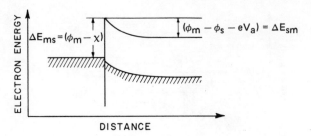

Figure 4.7. Band diagram of the metal–n-type-semiconductor junction in Figure 4.3 (for which ϕ_m is larger than ϕ_s) under an applied forward bias (semiconductor negative) of magnitude V_a.

the effect on the band diagram of Figure 4.3 of applying a forward bias V_a. The energy barrier ΔE_{sm} to the flow of electrons from semiconductor to metal is decreased from its equilibrium value $\phi_m - \phi_s$ to the smaller value $\phi_m - \phi_s - eV_a$. The energy barrier ΔE_{ms} to electron flow from metal to semiconductor is unchanged from its equilibrium value $\phi_m - \chi$ by forward bias. Figure 4.8 shows the effect of a reverse bias of magnitude V_a on the band diagram in Figure 4.3. The energy barrier ΔE_{ms} is again unchanged, while the energy barrier ΔE_{sm} is increased to $\phi_m - \phi_s + eV_a$.

We consider next the net currents through the junction. These currents are primarily majority carriers[8] in a metal–semiconductor junction, so we neglect any hole currents[9] in our treatment of the metal–n-type-semiconductor junction. The net electron current through the junction is the difference between two components, J_1 and J_2. The current J_1 is the electron current from metal to semiconductor, and J_2 is the electron current from semiconductor to metal. At equilibrium, when the applied potential $V_a = 0$, the net current $J = J_1 + J_2$ must vanish. Choosing the direction from semi-

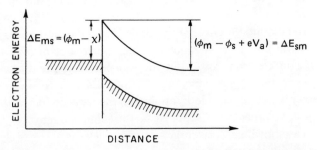

Figure 4.8. Band diagram of the metal–n-type-semiconductor junction in Figure 4.3 (for which ϕ_m is larger than ϕ_s) under an applied reverse bias (semiconductor positive) of magnitude V_a.

conductor to metal as positive, we write

$$J_2 = J_0, \qquad J_1 = -J_0 \qquad\qquad (4.11)$$

and at equilibrium, $J = J_1 + J_2 = 0$, as required.

To find the effect of V_a on the electron current J, we note from the lack of effect of V_a on the energy barrier ΔE_{ms} that the electron current J_1 will be essentially unchanged by the applied potential V_a. Since V_a decreases the barrier ΔE_{sm}, we expect that the component J_2 will increase exponentially with increasing magnitude of the forward bias V_a. These results are expressed as

$$J_1 = -J_0$$
$$J_2 = J_0 \exp(eV_a/k_BT)$$

giving the effect of the applied potential V_a on the currents J_1 and J_2. The total current $J = J_1 + J_2$ is then

$$J = J_0[\exp(eV_a/k_BT) - 1] \qquad\qquad (4.12)$$

giving the total current J as a function of applied bias V_a. We note that for large magnitudes of reverse bias, $J = -J_0$, so J_0 is a reverse saturation current. The overall result is that the total electron current is large for forward bias applied to the metal–semiconductor junction, and is small for reverse bias. The $J - V_a$ characteristic expressed by equation (4.12) is highly asymmetric,[10] just as it is for the semiconductor p–n junction. The calculation of the reverse current J_0 for the metal–semiconductor junction is more complex than it is for the semiconductor p–n junction, and depends on the width of the space charge region relative to the carrier mean free path. For a discussion of the calculation of J_0 for metal–semiconductor junctions, the reader is referred to the literature.[11]

We consider next the effect of an applied potential V_a on a metal–n-type-semiconductor junction in which ϕ_m is smaller than ϕ_s, so the band diagram at equilibrium is that of Figure 4.6. In this case[12] there is no extended space charge layer in the semiconductor (as there was for the situation in which ϕ_m is larger than ϕ_s). We must therefore consider the applied potential V_a as appearing across the entire bulk of the semiconductor when discussing the effect of V_a on the band diagram in Figure 4.6. If V_a is a negative potential applied to the n-type semiconductor (relative to the metal), the result is an increase in the electron potential energy in the semiconductor, i.e., the "band tilting" discussed in Chapter 2. The result of such an applied bias on the band diagram of Figure 4.6 is shown in

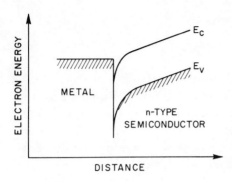

Figure 4.9. Band diagram of a metal–n-type-semiconductor junction (for which ϕ_m is smaller than ϕ_s) under an applied bias in which the semiconductor is negative with respect to the metal.

Figure 4.9. From the figure, we see that there is no energy barrier to electron flow from the semiconductor to the metal when the n-type semiconductor is negative. The band diagram under an applied bias with the n-type semiconductor positive is shown in Figure 4.10; here the electron energy in the semiconductor is decreased. The flow of electrons is from metal to semiconductor when the semiconductor is positive, over the small energy barrier shown in Figure 4.10. The barrier is expected to be small if the quantity $\phi_s - \chi$ is small, as it usually is in a doped n-type semiconductor. We conclude that there will be little difference in the electron flow for the two conditions of bias for the metal–n-type-semiconductor junction in which ϕ_m is smaller than ϕ_s. We expect therefore that such a metal–semiconductor junction will form an ohmic (i.e., nonrectifying) contact to the n-type semiconductor.

We may combine our results on the metal–n-type-semiconductor junction as follows. If the metal work function ϕ_m is larger than the semiconductor work function ϕ_s, then the current–voltage characteristic is asymmetric. We therefore expect a metal–n-type-semiconductor contact to be rectifying if ϕ_m is larger than ϕ_s. Conversely, we expect such a contact to be ohmic if ϕ_m is smaller than ϕ_s. These results are an approximate "rule" for the kind of contact, ohmic or rectifying, formed by a metal of work function

Figure 4.10. Band diagram of a metal–n-type-semiconductor junction (for which ϕ_m is smaller than ϕ_s) under an applied bias such that the semiconductor is positive.

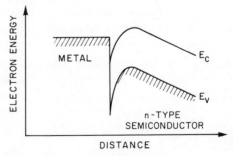

ϕ_m on an n-type semiconductor of work function ϕ_s. One may show[13] that this "rule" is reversed for metal contacts to a p-type semiconductor. In that case, the contact is ohmic if ϕ_m is larger than ϕ_s, and is rectifying if ϕ_m is smaller than ϕ_s. It should be noted that this "rule" is only approximate and does not always predict the type of contact obtained experimentally. One reason[14] for this is the neglect of the effect of surface states at the semiconductor surface. This point will be treated in Chapter 6 when surface states are discussed.

Physics of the Metal–Insulator–Semiconductor Structure

We now discuss the physics[15-18] of a structure composed of a metal, an insulator, and a semiconductor, as shown schematically in Figure 4.11. This structure is usually referred to as an MIS structure or junction. While the physics of this structure is interesting for its own sake, the MIS junction is the basis of important solid state devices, including charge coupled devices (CCD) and the insulated-gate field effect transistor (IGFET), both of which will be discussed later.

We consider first the band diagram, shown in Figure 4.12, of a simple metal–insulator–semiconductor of the type shown in Figure 4.11. We consider the case in which the semiconductor is n type. Figure 4.12 shows the equilibrium band diagram of an "*ideal*" MIS structure, in which the work function ϕ_m of the metal is equal to the work function ϕ_s of the semiconductor. While this "ideal" case is oversimplified,[19,20] it will give us the central physical results necessary for the applications we are interested in. A further idealization is made by neglecting surface states at the interfaces of the structure.

We may now make the following points regarding the equilibrium band diagram in Figure 4.12. The metal and semiconductor Fermi energies are equal at equilibrium in this ideal case, so there is no charge flow between metal and semiconductor. There are assumed to be no free charges in the insulator. Thus, even with an applied DC potential difference existing between the metal and the semiconductor, the only mobile charges present are those in the semiconductor and those at the metal–insulator interface. Since there is no charge transfer between the metal and the semiconductor,

Figure 4.11. Metal–insulator–semiconductor (MIS) structure (schematic). The relative thicknesses of the metal and insulator layers are exaggerated.

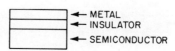

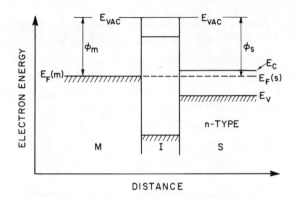

Figure 4.12. Band diagram at equilibrium of an "ideal" MIS structure, in which the work function ϕ_m of the metal (M) is equal to the work function ϕ_s of the semiconductor (S). The notation is the same as that used in Figure 4.1; the insulator is denoted by I, and its bands and Fermi level are not explicitly labeled.

the Fermi level in the semiconductor remains "flat."[21] However, the mobile charges in the semiconductor may rearrange themselves under the influence of an applied electric field and the semiconductor energy bands may bend up or down (relative to the "flat" Fermi level) to reflect such a rearrangement.

We now consider applying a negative potential $-V_a$ to the metal, relative to the semiconductor. Choosing the potential of the metal as zero, the semiconductor is now at a potential $+V_a$ relative to the metal. This means that there is now a difference of potential energy equal to $(-e)(V_a)$ $= -eV_a$ between electrons in the semiconductor and electrons in the metal. The Fermi level in the semiconductor is now an energy eV_a below the Fermi level in the metal. This is shown in Figure 4.13 depicting the situation before

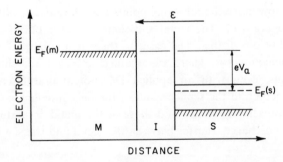

Figure 4.13. Band diagram of the ideal MIS structure of Figure 4.12 with a negative potential $-V_a$ applied to the metal. The diagram is shown before the rearrangement of the mobile charges in the semiconductor under the influence of the electric field $\mathscr{E}$.

any rearrangement of charge in the semiconductor. Figure 4.13 shows the Fermi level in the semiconductor an energy eV_a below the Fermi level in the metal. The electron energy bands in the insulator are ignored because there are assumed to be no free electrons in the insulator. Figure 4.13 also shows the electric field $\mathscr{E}$, due to the applied potential, which is directed from the semiconductor toward the metal.

Even though no charge flows between the metal and the semiconductor, the mobile charges (i.e., mostly majority electrons) in the semiconductor rearrange themselves under the influence of the electric field $\mathscr{E}$ which exists in the semiconductor and insulator. Electrons are repelled from the insulator–semiconductor interface, producing a region depleted of electrons; part of the potential difference V_a appears across this depletion layer in the semiconductor and part across the insulator. The fact that electrons are moving away from the insulator–semiconductor interface means that the applied potential makes the electron energy higher at that interface than it is in the bulk of the semiconductor. This means that the energy bands in the semiconductor bend upward at the interface with the insulator, as shown in Figure 4.14. Also shown in Figure 4.14 is the depletion layer of width x_d in the semiconductor, and the negative surface charge at the metal–insulator interface.

The space charge density $\varrho(x)$ in the depletion layer is given by

$$\varrho(x) = +eN_d \tag{4.13}$$

where N_d is the density of donors in the n-type semiconductor. Figure 4.15 shows a plot of $\varrho(x)$ as a function of x; $x = 0$ is taken at the insulator–semiconductor interface. The negative surface charge density at the metal–

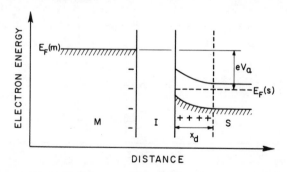

Figure 4.14. Band diagram of the ideal MIS structure of Figure 4.13 after rearrangement of electrons in the semiconductor. The negative potential $-V_a$ applied to the metal produces a positive space charge region of width x_d in the semiconductor and a corresponding negative surface charge density at the metal–insulator interface.

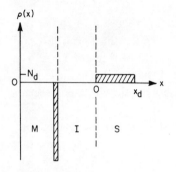

Figure 4.15. Space charge density $\varrho(x)$ as a function of distance x for the ideal MIS structure (n-type semiconductor) with a negative potential applied to the metal. The quantity x_d is the extent of the positive space charge density in the semiconductor and N_d is the ionized donor density. The space charge at the metal–insulator interface represents a surface density of negative charge.

insulator interface is also indicated in Figure 4.15. We could find the magnitude x_d of the depletion layer in the semiconductor by solving Poisson's equation [for the electrostatic potential $V(x)$ in the depletion layer and the insulator] using the charge density (4.13). However, we will merely note that the solution would give V as a function of x, and vice versa, so we expect that $x = x(V)$, meaning that the extent x_d of the depletion layer will depend on the magnitude $|-V_a|$ of the negative potential applied to the metal. If the magnitude $|-V_a|$ is increased, the extent of the depletion layer will increase (just as for a reverse-biased p–n junction), to the larger value x_d'. Increasing the magnitude $|-V_a|$ further will increase the energy difference eV_a between the metal and the semiconductor. We thus expect the bands in the semiconductor depletion layer to bend further upward when the magnitude $|-V_a|$ is increased. These effects are shown in Figure 4.16, in which the depletion layer in the semiconductor is now of width

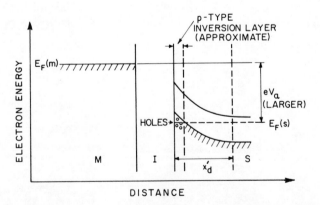

Figure 4.16. Band diagram of the ideal MIS structure of Figure 4.14 with a negative potential $-V_a$ of larger magnitude applied to the metal. The width of the positive space charge layer in the semiconductor has increased to x_d' and a p-type inversion layer has formed at the insulator–semiconductor interface.

x_d'. The upward bending of the semiconductor energy bands may, with increasing magnitude $|-V_a|$, produce a situation in which the valence band edge at the semiconductor surface comes close to, or even crosses, the Fermi level $E_F(s)$.

This crossing of the Fermi level by the valence band edge at the surface means that the semiconductor becomes effectively p type near the surface, forming an inversion layer of p-type material in the n-type semiconductor. If there are sufficient minority holes available, they will accumulate near the surface, thereby forming the inversion layer (if there are not sufficient minority holes available, the inversion layer will not form) as minority holes diffuse to, and accumulate in, the potential energy well formed by the band bending at the surface.[22] The formation of the inversion layer modifies the plot of space charge density $\varrho(x)$ in Figure 4.15. There would now be an additional component of $\varrho(x)$, extending the width of the inversion layer from $x = 0$, representing the positive space charge density due to accumulated holes in the surface inversion layer.[23]

We may summarize the physical results embodied in the Figure 4.16 as follows. If a negative potential $-V_a$ is applied to a metal–insulator–n-type-semiconductor structure, a depletion layer of positive space charge will form in the semiconductor. For sufficiently large values of $|-V_a|$, a p-type inversion layer will form at the surface of the n-type semiconductor if sufficient minority holes are available.

In the following chapter, we discuss a number of electronic devices based on metal-semiconductor and metal–insulator–semiconductor junctions.

Problems

4.1. *Metal–p-Type-Semiconductor Junction.* (a) Draw band diagrams for the junction between a metal and a p-type semiconductor. (b) If ϕ_m is the metal work function and ϕ_s is the semiconductor work function, verify that the junction is a rectifying contact if ϕ_s is larger than ϕ_m, and an ohmic contact if ϕ_s is smaller than ϕ_m. (Neglect surface states in your treatment.)

4.2. *Inversion Layer in MIS Structure.* Using band diagrams, discuss the formation of an n-type inversion layer when a positive potential is applied to the metal in an MIS structure in which the semiconductor is p type. Use the "ideal" MIS structure in which the work functions of the metal and the semiconductor are equal, and neglect surface states at the I–S interface.

4.3. *Nonideal MIS Structure.* Consult the literature (e.g., Reference 17, pages 306–308) and discuss the attainment of equilibrium via charge transfer in a nonideal MIS structure, for which the metal and semiconductor work functions are not equal.

References and Comments

1. H. K. Henisch, *Rectifying Semiconductor Contacts*, Oxford University Press, London (1957), Section 7.3.
2. J. P. McKelvey, *Solid State and Semiconductor Physics*, Harper and Row, New York (1966), Chapter 16.
3. A. van der Ziel, *Solid State Physical Electronics*, Second Edition, Prentice-Hall, New York (1968), Sections 5.4 and 14.1.
4. R. S. Muller and T. I. Kamins, *Device Electronics for Integrated Circuits*, John Wiley, New York (1977), pages 69–83.
5. See, for example, J. I. Pankove, *Optical Processes in Semiconductors*, Prentice-Hall, New York (1971), page 298, for values of the work function and of the electron affinity for semiconductors.
6. See, for example, A. J. Dekker, *Solid State Physics*, Prentice-Hall, New York (1957), page 223, for representative values of the work function for metals.
7. See, for example, A. van der Ziel, Reference 3, Section 14.1a.
8. S. M. Sze, *Physics of Semiconductor Devices*, John Wiley, New York (1969), page 378 and pages 390–393.
9. J. P. McKelvey, Reference 2, Section 16.2, pages 482–485.
10. A. Y. C. Yu, "The Metal–Semiconductor Contact: An Old Device with a New Future," *IEEE Spectrum*, **7**, 83–89 (March 1970), gives several examples.
11. S. M. Sze, Reference 8, pages 378–390; A van der Ziel, Reference 3, pages 266–274.
12. A. van der Ziel, Reference 3, pages 100–101.
13. A. van der Ziel, Reference 3, pages 102–103.
14. A. van der Ziel, Reference 3, page 101; J. P. McKelvey, Reference 2, Section 16.3, pages 485–489.
15. A. Goetzberger and S. M. Sze, "Metal–Insulator–Semiconductor (MIS) Physics," in *Applied Solid State Science*, R. Wolfe (editor), Academic Press, New York (1969), Volume 1, pages 154–238.
16. S. M. Sze, Reference 8, pages 426–444.
17. R. S. Muller and T. I. Kamins, Reference 4, pages 304–314.
18. A. S. Grove, *Physics and Technology of Semiconductor Devices*, John Wiley, New York (1967), pages 317–327.
19. See A. S. Grove, Reference 18, pages 278–285, for a discussion of the effect of deviations from the "ideal" MIS model. See also S. M. Sze, Reference 8, pages 467–471, and A. Goetzberger and S. M. Sze, Reference 15, pages 205–210, for similar discussions.
20. See Reference 17 for a discussion of the important nonideal Al–SiO$_2$–Si structure.
21. S. M. Sze, Reference 8, page 429.
22. W. S. Boyle and G. E. Smith, "Charge-Coupled Devices—A New Approach to MIS Device Structures," *IEEE Spectrum*, **8**, 18–27 (1971), especially pages 20–21 and Figure 5.
23. See, for example, S. M. Sze, Reference 8, page 434, Figure 6(b) for this plot, but drawn for the case of an *n*-type inversion layer in a *p*-type semiconductor.

Suggested Reading '

H. K. HENISCH, *Rectifying Semiconductor Contacts*, Oxford University Press, Oxford (1957). This is an advanced monograph on the physics of metal–semiconductor contacts and surface properties. Chapter 7 is pertinent to our Chapter 4.

A. G. MILNES and D. L. FEUCHT, *Heterojunctions and Metal–Semiconductor Junctions*, Academic Press, New York (1972). Chapters 6 and 7 of this monograph discuss metal–semiconductor junctions and a number of their applications.

A. VAN DER ZIEL, *Solid State Physical Electronics*, Second Edition, Prentice-Hall, New York (1968). This electrical engineering text discusses metal–semiconductor junctions and contacts in a terse and mathematical manner.

S. M. SZE, *Physics of Semiconductor Devices*, John Wiley, New York (1969). This advanced treatise discusses both metal–semiconductor and metal–insulator–semiconductor junctions in Chapters 8 and 9.

R. S. MULLER and T. I. KAMINS, *Device Electronics for Integrated Circuits*, John Wiley, New York (1977). Chapters 2 and 7 of this electrical engineering text discuss metal–semiconductor and metal–oxide–semiconductor structures, including the technologically important metal–SiO_2–silicon system.

5

Metal–Semiconductor and
Metal–Insulator–Semiconductor Devices

Introduction

This short chapter discusses the basic ideas of some of the applications of metal–semiconductor and metal–insulator–semiconductor structures. These include the Schottky metal–semiconductor diode, various insulated-gate field-effect transistors, and charge-coupled devices. The myriad applications of these devices, especially the field-effect transistor, are left to the literature for lack of space, time, and knowledge on the author's part. However, the basic ideas are discussed, some applications are mentioned, and a number of references to this important and changing field are given.

Metal–Semiconductor (Schottky) Diodes

As pointed out in Chapter 4, the current–voltage characteristic of a metal–semiconductor junction is of the form

$$J = J_0[\exp(eV_a/k_BT) - 1] \qquad (5.1)$$

where V_a is the applied voltage and J_0 is a reverse saturation current. The asymmetric current–voltage characteristic is similar to that of a semiconductor p–n junction. For this reason, the Schottky diode can be used as a rectifier.

As pointed out in Chapter 2, minority carrier storage limits the speed with which a semiconductor p–n junction can be switched from a condition of forward bias to one of reverse bias. The metal–semiconductor junction, as noted in Chapter 4, is essentially a majority carrier device. For this reason, minority carrier storage is not a problem in the Schottky diode, and an important application[1-4] of such diodes is their use at high (e.g., microwave) frequencies in mixers, etc.

The Insulated-Gate Field-Effect Transistor (IGFET)

In Chapter 3, we discussed the junction field-effect transistor (JFET). A second type of field-effect transistor has a metal gate electrode insulated from the semiconductor and is usually called an insulated-gate field-effect transistor[5-7] or IGFET. Since the semiconductor involved is generally silicon, and the insulator is then silicon dioxide, this type of FET is often called a metal–oxide–semiconductor field-effect transistor, or MOSFET. Finally, this general class of devices is often called metal–insulator–semiconductor, or MIS, devices. The great importance of these devices is that they can be made very small in integrated circuits used in a wide variety of applications, especially digital circuits.

There are several kinds of insulated-gate field-effect transistors. We consider first the diffused-channel MOSFET, which is shown schematically in Figure 5.1. The source S and drain D are n^+ regions fabricated by diffusion into the p-type semiconductor substrate. A channel of n-type material between source and drain is also made by diffusion. An insulating layer (usually SiO_2) separates the channel from the metal gate. There are also diffused-channel MOSFET devices with p-type channels. We will use the n-channel device as our example.

A potential of either polarity may be applied to the gate; the appropriate source and drain polarities for an n-channel device are shown in Figure 5.1. We consider first a negative potential applied to the gate and look upon the metal gate and n-type semiconductor of the channel as forming a parallel plate capacitor in which the oxide layer is the dielectric

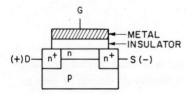

Figure 5.1. Schematic drawing of a diffused n-channel MOSFET or IGFET. The drain, source, and gate are labeled D, S, and G, respectively.

Figure 5.2. Schematic view of the metal gate (*G*), insulator, and *n*-type channel MOSFET with a negative potential applied to the gate *G*; the electric field in the insulator is $\mathscr{E}$. The undepleted part of the channel is shown shaded.

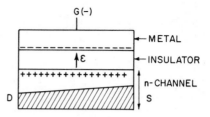

between the plates. This is shown in Figure 5.2. An electric field $\mathscr{E}$ will exist in the insulator, maintained between the metal and the semiconductor of the *n* channel. The applied negative potential will repel electrons from the *n* channel, leaving behind ionized donors bearing a positive charge. These positive charges (shown in Figure 5.2) and the induced negative charges at the metal–insulator interface maintain the electric field $\mathscr{E}$ in the insulator. The effect of the applied negative potential on the gate *G* is to deplete part of the *n* channel of mobile electrons. In Figure 5.2 the *un*depleted portion of the channel is shown as shaded. Since the source to drain current in the *n*-channel MOSFET is composed of majority electrons in the *n* channel, we see that the negative potential applied to the gate depletes the channel of current carriers, thereby reducing the effective cross-section area of the channel in a manner analogous to that taking place in the JFET. The resistance of the *n* channel is thus increased by the negative potential applied to the gate, and the electron current I_{DS} from source to drain is decreased. Just as in the JFET, the depletion region extends further into the channel near the drain, as indicated schematically in Figure 5.2. The overall effect of the negative potential applied to the gate is to deplete the *n* channel of carriers, so we speak of this *n*-channel MOSFET operating in the depletion mode.

If a positive potential is applied to the gate, the opposite happens. Electrons enter the channel from the source, increasing the conductivity of the *n* channel. This is an *n*-channel MOSFET operating in the enhancement mode, in which the conductivity of the *n* channel is increased by the application of a positive voltage to the gate.

We see then that the magnitude and sign of the potential V_{GS} applied to the gate (relative to the source) will determine the conductivity of the *n* channel and hence will determine the current I_{DS} flowing between source and drain for a given value of the voltage V_{DS} between source and drain. To find the current–voltage characteristic of the *n*-channel MOSFET, we know that the current I_{DS} will be larger (for a given value of the voltage V_{DS}) in the enhancement mode than it will in the depletion mode. We thus expect that the current I_{DS} will vary with the potential V_{GS} (between gate

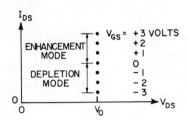

Figure 5.3. Schematic variation of current I_{DS} in an n-channel MOSFET for a particular value V_0 of the source-to-drain voltage V_{DS}. The enhancement and depletion modes of operation are indicated.

and source), for a fixed value V_0 of V_{DS}, approximately as shown in Figure 5.3. In that figure, we see that I_{DS} increases with increasingly positive values of the gate voltage V_{GS} for a fixed value of V_{DS}.

We consider next how I_{DS} will vary with the source to drain voltage V_{DS} for a given value of the gate voltage V_{GS}. We expect that (as for the JFET) the relatively lightly doped n channel will act as a distributed resistance and the voltage V_{DS} across the channel will, as it did for the JFET, produce a higher potential at a point in the channel near the drain than near the source. In the depletion mode, we expect that the channel resistance (dV_{DS}/dI_{DS}) will increase with increasing V_{DS} for a given value of the gate voltage V_{GS}. We thus expect that the I_{DS}–V_{DS} characteristic of an n-channel MOSFET operating in the depletion mode will resemble the JFET characteristic. This is shown in Figure 5.4.[8] In the enhancement mode, the shape of the I_{DS}–V_{DS} curves is quite similar[9] to that of the curves in the depletion mode. The channel resistance increases with increasing magnitude of V_{DS} for any value of the gate voltage.

Comparing Figure 5.4 for the MOSFET with Figure 3.22 for the JFET, we see that the MOSFET is also a voltage-controlled amplifier in that the gate voltage V_{GS} controls the device current I_{DS}, just as in the JFET. We note also that the device current I_{DS} in the MOSFET is composed of majority

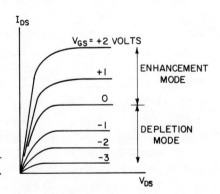

Figure 5.4. Current–voltage (I_{DS}–V_{DS}) characteristics for a diffused n-channel MOSFET. (After Streetman.[8])

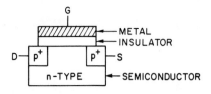

Figure 5.5. Schematic view of an MIS induced-channel MOSFET. The symbols *S*, *D*, and *G* refer to source, drain, and gate, respectively.

carriers (in this case, electrons in the *n* channel) only. The same is true of the JFET.

While we have discussed the *n*-channel MOSFET, all of the ideas obtained pertain also to the *p*-channel MOSFET on making the necessary sign changes in the device currents and voltages.

The diffused-channel JGFET or MOSFET may therefore have either an *n* or a *p* channel, and may be operated in either enhancement or depletion modes. The input resistance (to the gate) is very high (typically many megohms) because of the insulator layer under the gate.

The Induced-Channel MOSFET

The induced-channel (or enhancement mode) MOSFET[10–13] is an insulated-gate field-effect transistor which does not, at equilibrium, have a diffused channel between source and drain. It is a member of the general class of metal–insulator–semiconductor (MIS) devices whose band diagram was considered in Chapter 4. We consider the MIS structure shown schematically in Figure 5.5, which is the same as Figure 4.11 with a source *S*, drain *D*, and gate *G* added. The metal gate in Figure 5.5 is separated from the *n*-type semiconductor by an insulator layer.

The device in Figure 5.5 functions as a field-effect transistor in the following manner. We saw in Chapter 4, Figure 4.16, that the application of a negative potential to the gate *G* of the metal–insulator–*n*-type-semiconductor device can produce a *p*-type inversion layer. This layer, at the insulator–semiconductor interface, is in addition to a depletion layer of

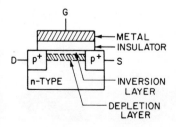

Figure 5.6. Schematic view of the formation of an induced *p*-type channel in the MOSFET of Figure 5.5 by the application of a negative potential to the gate.

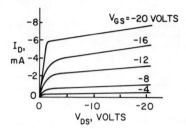

Figure 5.7. Current–voltage ($I_{DS}-V_{DS}$) characteristics of an induced p-channel MOSFET. (After Millman and Halkias.[12])

positive fixed space charge extending into the n-type semiconductor. These results are shown schematically[†] in Figure 5.6. The p-type inversion layer in the n-type semiconductor acts as a channel between the source and drain of the device, so one speaks of an induced-channel[‡] MOSFET.

As we saw in Chapter 4, the increasing magnitude of the negative potential applied to the gate increases the conductivity of the induced channel. The drain-current–drain-voltage ($I_{DS}-V_{DS}$) characteristic[12,13,15] of the induced-channel MOSFET will be similar to the characteristics of the other field-effect transistors we have discussed, including the saturation of the device current with increasing drain voltage. A representative characteristic[12] is shown in Figure 5.7, from which we see that the induced-channel MOSFET also functions as a voltage-controlled amplifier like the other field-effect transistors discussed.

Summary of the Physics of Field-Effect Transistors

All types of FET are voltage controlled amplifiers because an applied electric field controls the flow of majority carriers in a channel which is either diffused or induced. All types of FET have a high input impedence. In the JFET the control voltage is applied to a reverse biased p^+n junction. In the IGFET, the control voltage is applied between gate and source separated by a high-resistance insulator. All types of FET operate by controlling the conductivity of a channel (diffused or induced) with an applied electric field, hence the name field-effect transistor. The IGFET is particularly ap-

---

[†] The shapes of the inversion layer and the depletion layer are shown as uniform for simplicity in Figure 5.6. This will be true only for small drain voltages; the actual shapes for various drain voltages are given by Muller and Kamins.[14]

[‡] The name "enhancement-mode" MOSFET is more commonly used in electrical engineering, but the term "induced-channel" will be retained here because of its physical descriptiveness.

propriate for fabrication in small size, making the principal use of the MOSFET in integrated circuits. For this reason, the IGFET is more important than the JFET.

Applications of the MOSFET

While MOSFET transistors are used in various amplification configurations,[16,17] the most common application of the MOSFET is in digital electronics as an electronically controlled switch.[18] Figure 5.8 shows[19] a partial set of current–voltage (I_{DS}–V_{DS}) curves, for two values of the gate voltage V_{GS}, for a p-channel MOSFET typical of those used in a digital integrated circuit. We can see from Figure 5.8 that the resistance dV_{DS}/dI_{DS} can be changed from its lower value for $V_{GS} = -9$ V to its higher value for $V_{GS} = -4$ V (for a fixed value of V_{DS}). The device can therefore be switched from an "on" state of lower resistance dV_{DS}/dI_{DS} to an "off" state of higher resistance by changing the magnitude of the control voltage V_{GS}. The device is therefore bistable and can be used in a variety of digital circuits.[20]

The primary use of the MOSFET is in integrated circuits for digital applications. An integrated circuit (as opposed to an assembly of discrete components) is a single crystal chip of silicon on which are fabricated both active and passive electronic components and their interconnections. These will not be discussed here for two reasons. First, this vast technology is in a more or less constant state of change. Second, there are excellent expositions[10] of the device physics relevant to integrated circuit technology. Especially recommended for students of physics is the September 1977 issue of *Scientific American* on microelectronics, particularly the article[21] on microelectronic circuit elements. This article discusses, among other things, a comparison of MOS devices of various types. The use of the MOSFET in semiconductor computer memories[22–24] is treated in another article in the same issue.

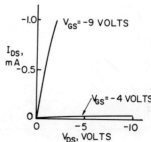

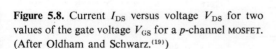

Figure 5.8. Current I_{DS} versus voltage V_{DS} for two values of the gate voltage V_{GS} for a p-channel MOSFET. (After Oldham and Schwarz.[19])

Charge-Coupled Devices

We consider in this section a type of device, based on the MIS structure, whose function could not be duplicated[21] using discrete components, but which is instead inherently based on the concept of the integrated circuit. This is the charge-coupled device (abbreviated CCD), which is also called a charge transfer device.

We return to the result, illustrated in Figure 4.16, that the application of a negative potential $-V_a$ to the gate of a metal–insulator–n-type-semiconductor structure can produce a p-type inversion layer at the insulator–semiconductor interface. Figure 4.16 shows the upward bending of the electron energy bands and the accumulation of holes producing the inversion layer. Since Figure 4.16 is a plot of electron energy as a function of distance, we may (mentally) convert it to a plot of hole energy as a function of distance by multiplying by -1. If we do so to the valence band contour of Figure 4.16, we obtain the curve shown in Figure 5.9. This figure shows a schematic plot of hole energy as a function of distance in the n-type semiconductor of an MIS structure with a negative potential $-V_a$ applied to the metal; the zero of distance is at the insulator–semiconductor interface. Figure 5.9 shows a potential energy well, of depth D, for holes. Further, we saw from the discussion of Figure 4.16 that the larger the magnitude of V_a, the greater the band bending in the semiconductor and the greater the depth D of the potential energy well. Our conclusion is that a negative bias $-V_a$ applied to the gate of the MIS structure produces an energy well for holes at the I–S interface, and, the larger the magnitude of V_a, the deeper the energy well.

Given the existence of this energy well for holes in the biased MIS structure, we see that minority holes in the n-type semiconductor will accumulate there until a condition is reached in which the diffusion of holes away from the surface is just balanced by the drift of holes toward the surface. This steady state is referred to as the condition of saturation.

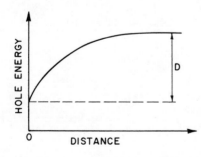

Figure 5.9. Energy well for holes, of depth D, in a metal–insulator–n-type-semiconductor structure with a negative potential applied to the metal gate.

Figure 5.10. Schematic view of a charge-coupled device. The metal gates are shaded. The applied potentials are such that $V_2 > V_1 = V_3 = V_4$. Holes are shown in the deeper energy well under gate 2. The dashed line indicates (approximately) the potential energy contour for holes as a function of distance.

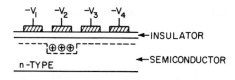

(If the minority hole density corresponding to saturation is exceeded, the excess holes flow back into the bulk semiconductor where they vanish by recombination with majority electrons.) The accumulation and storage of minority carriers in an energy well at the semiconductor surface in a biased MIS device is the basis of the charge-coupled device.[25-27]

Given the possibility of accumulation and storage[28] of carriers in potential energy wells, we may consider charge-coupled devices based on the transfer of minority carriers between different potential wells at different parts of the device. The MIS structure in Figure 5.10 has a series of closely spaced metal gates on a common insulating layer on an n-type semiconductor substrate. The potential $-V_a$ applied to the four electrodes has the values $-V_1, -V_2, -V_3$, and $-V_4$, where V_2 is larger than V_1 and $V_1 = V_3 = V_4$. The dashed line in Figure 5.10 represents the variation of the hole potential energy with distance along the structure; the larger the magnitude of the applied potential, the deeper the energy well for holes. The deeper well is under the gate 2 and is the one in which any available (or introduced) minority holes will accumulate. The regions under gates 1, 3, and 4 are shown free of minority holes, a situation which can occur because silicon can be made with a low density of surface and bulk generation centers. While not really a steady state situation, these wells can remain empty for times of the order of seconds[29] before thermally generated minority carriers accumulate.

In Figure 5.10, we have a situation in which holes have been introduced (in some manner) into the potential energy well under gate 2 and are stored there. Suppose next that V_3 is increased so V_3 is larger than V_2. Then the potential energy contour will change to that shown in Figure 5.11, in which the energy well under gate 3 is now deeper than that under gate 2. Holes will be transferred into the deeper well, as indicated schematically in Figure

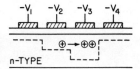

Figure 5.11. Schematic view of transfer of holes between energy wells in a charge-coupled device. The applied potentials are such that $V_3 > V_2 > V_1 = V_4$, so the well under gate 3 is deeper than the well under gate 2.

5.11. The key conclusion is that, in this type of MIS structure, minority carriers (holes in this case) can be transferred between potential energy wells by appropriate manipulation of the bias voltages applied to the metal electrodes. Since the presence or absence of charge carriers in a particular well constitutes information, this device allows the transfer of information along the semiconductor surface. This is the basic idea of the charge transfer or charge-coupled device.

Various methods[27,30] of introducing the minority carriers are in use. These include injection from a *p–n* junction and optical generation by incident photons. Among the methods[27,30] of detecting the minority carriers is, for example, injecting the holes into the *n* region of a reverse-biased diode, thereby producing a current pulse in the external circuit. The literature[25–27] also gives information on "clocking" sequences of voltages applied to the device electrodes in order to transfer the minority carriers along the series of potential energy wells.

The relative simplicity and small size of the charge-coupled device have made it attractive for a number of applications.[27] In particular, the presence or absence of charge in a given potential well provides two stable states, so the charge-coupled device is used as the basis of semiconductor computer memories.[31–33] Another application is the use of charge-coupled devices in imaging,[34,35] i.e., the conversion of a pattern of light and dark areas into an electrical signal. In this application, minority carriers are introduced into the device by optical absorption. These carriers then accumulate in the potential wells of the device, and the resulting charges are transferred and detected. The result is an output whose amplitude varies with the light intensity at different points of the original light pattern.

Problems

5.1. *Area Imaging with a* CCD. Using the literature, discuss area imaging with a charge-coupled device. In particular, since the fabrication of a CCD avoids the diffusion steps necessary in making a bipolar or FET device, consider the use of "exotic" semiconductors (i.e., anything other than silicon) for a CCD imaging device. Choose a particular "exotic" semiconductor for imaging outside the visible, and suggest an application.

5.2. *Speed of* MOS *Devices.* Read the article by Meindl,[21] considering especially the fact that electron mobilities are generally larger than hole mobilities. This means that MOS devices with *n*-type channels tend to be faster than *p*-channel devices. From this point of view, discuss the possibility of using a semiconductor (e.g., InSb) with a very high electron mobility for a MOSFET. Consider as many semiconductor parameters as you can in your discussion of the feasibility and desirability of such a device.

References and Comments

1. R. S. Muller and T. I., Kamins, *Device Electronics for Integrated Circuits*, John Wiley, New York (1977), pages 95–99.
2. J. Millman and C. C. Halkias, *Integrated Electronics*, McGraw-Hill, New York (1972), pages 228–230.
3. D. Cooper, B. Bixby, and L. Carver, "Power Schottky Diodes," *Electronics*, February 5, 1976, pages 85–89.
4. A. Y. C. Yu, "The Metal–Semiconductor Contact: An Old Device with a New Future," *IEEE Spectrum*, **7**, 83–89 (March 1970).
5. B. G. Streetman, *Solid State Electronic Devices*, Prentice-Hall, New York (1972), pages 293–301.
6. S. M. Sze, *Physics of Semiconductor Devices*, John Wiley, New York (1969), Chapter 10.
7. A. S. Grove, *Physics and Technology of Semiconductor Devices*, John Wiley, New York (1967), Chapter 11.
8. B. G. Streetman, Reference 5, page 295, Figure 8-15(a).
9. J. Millman and C. C. Halkias, Reference 2, page 326, Figure 10-13(a).
10. R. S. Muller and T. I. Kamins, Reference 1, Chapters 7 and 8.
11. B. G. Streetman, Reference 5, pages 293–301.
12. J. Millman and C. C. Halkias, Reference 2, pages 322–324.
13. W. G. Oldham and S. E. Schwarz, *An Introduction to Electronics*, Holt, Rinehart, Winston, New York (1972), pages 547–559.
14. R. S. Muller and T. I. Kamins, Reference 1, Figure 8.3, page 351, and Figure 8.5, page 355.
15. R. S. Muller and T. I. Kamins, Reference 1, pages 349–360.
16. J. Brophy, *Basic Electronics for Scientists*, Third Edition, McGraw-Hill, New York (1977), pages 143–150 and 156–169.
17. J. Millman and C. C. Halkias, Reference 2, Chapter 10.
18. W. G. Oldham and S. E. Schwarz, Reference 13, pages 559–562.
19. W. G. Oldham and S. E. Schwarz, Reference 13, Figure 13.9, pages 557–558.
20. W. G. Oldham and S. E. Schwarz, Reference 13, pages 559–584.
21. J. D. Meindl, "Microelectronic Circuit Elements," in *Scientific American*, **237**, 70–81 (September 1977).
22. D. A. Hoges, "Microelectronic Memories," in *Scientific American*, **237**, 130–145 (September 1977).
23. R. S. Muller and T. I. Kamins, Reference 1, pages 370–372.
24. S. Middlehoek, P. K. George, and P. Dekker, *Physics of Computer Memory Devices*, Academic Press, New York (1976), pages 234–268.
25. W. S. Boyle and G. E. Smith, "Charge-Coupled Devices—A New Approach to MIS Device Structures," *IEEE Spectrum*, **8**, 18–27 (July 1971).
26. R. S. Muller and T. I. Kamins, Reference 1, pages 335–341.
27. C. H. Séquin and M. F. Tompsett, *Charge Transfer Devices*, Academic Press, New York (1975).
28. See References 25 and 26 for a discussion of the problems of, and factors affecting, the times of storage and transfer of minority carriers.
29. W. S. Boyle and G. E. Smith, Reference 25, page 21.

30. W. S. Boyle and G. E. Smith, Reference 25, Figures 12 and 13. Note that the captions of these figures are reversed; Figure 12 shows detection methods and Figure 13 shows input schemes.
31. S. Middelhoek, P. K. George, and P. Dekker, Reference 24, pages 269–281.
32. C. H. Séquin and M. F. Tompsett, Reference 27, pages 236–260.
33. L. Altman, "Charge Coupled Devices Move in on Memories and Analog Signal Processing," *Electronics*, **47**, 91–98 (August 8, 1974).
34. C. H. Séquin and M. F. Tompsett, Reference 27, Chapter 5.
35. M. F. Tompsett, W. J. Bertram, and D. A. Sealer, and C. H. Séquin, "Charge Coupling Improves its Image," *Electronics*, **46**, 162–169 (January 18, 1973).

Suggested Reading

J. MILLMAN and C. C. HALKIAS, *Integrated Electronics*, McGraw-Hill, New York (1972). This electrical engineering text discusses the uses of Schottky diodes, field-effect transistors, and integrated circuits.

Scientific American, September 1977: Special Issue on Microelectronics. This issue of the magazine contains a number of articles on the physics and applications of MOSFET transistors. The articles are introductory and physical, and further references are provided.

R. S. MULLER and T. I. KAMINS, *Device Electronics for Integrated Circuits*, John Wiley, New York (1977). This electrical engineering text stresses the physics of various devices from the point of view of their use in integrated circuits. Chapters 7 and 8 discuss the silicon IGFET.

C. H. SÉQUIN and M. F. TOMPSETT, *Charge Transfer Devices*, Academic Press, New York (1975). This monograph covers the physics and applications of charge-coupled devices in detail.

6

Other Semiconductor Devices

Introduction

In this chapter, we discuss some additional topics on semiconductor physics and their applications. After a brief introduction to surface states on semiconductors, some aspects of the band structure at the surface are considered. These concepts are then applied to discussions of metal–semiconductor contacts and photoemission from semiconductors, including the idea of devices with a negative electron affinity. The next topic is the physics of the transferred electron, or Gunn, effect, discussed in terms of band structure using GaAs as the example. One mode (domain formation) of exploitation of the Gunn effect in producing microwave electrical oscillations is considered. Finally, some aspects of the electronic structure of amorphous semiconductors are introduced as the basis for a brief discussion of memory and switching devices.

Semiconductor Surface States

Usually, when we consider the electronic states in a crystal, we neglect the existence of a surface on the crystal. An example would be the Kronig–Penney model[1] of the periodic potential in a perfect crystal. However, surfaces are very important when considering semiconductor devices. In particular, the emission of electrons from semiconductors by incident photons (the photoelectric effect) or electrons (secondary electron emission) is a process that involves the semiconductor surface directly. We now con-

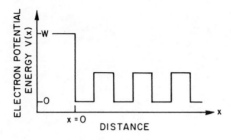

Figure 6.1. Terminated Kronig–Penney model of a one-dimensional crystal, with an energy barrier of height W located at the surface $x = 0$.

sider some basic ideas of the physics of semiconductor surfaces,[2-5] with special reference to those aspects[6] important in electron emission processes.

Consider a variant of the Kronig–Penney model in which[7] there is a surface barrier, i.e., a potential energy step of height W at the surface. Figure 6.1 shows both the surface barrier and the variation of the electron potential energy $V(x)$ as a function of distance x. The surface of the one-dimensional crystal is at $x = 0$. Calculation[7] of the energy band structure using the potential energy function of Figure 6.1 yields results different from those of the usual Kronig–Penney model of an infinite one-dimensional crystal. The familiar Kronig–Penney result is that the electronic structure of the crystal is composed of allowed bands of energies with forbidden gaps between them. When the crystal surface is added to the problem by including the surface potential barrier shown in Figure 6.1, the result is the introduction of energy levels in the forbidden gap. These states are due to the presence of the surface itself, and are called[†] surface states. The wave function for such a surface state is localized in space near the surface. There is generally one such surface state per surface atom, resulting in a surface state density[3] of the order of 10^{15} cm^{-2}. The surface states associated with the semiconductor surface itself are usually called *fast* surface states. (The term "fast" comes from the fact[8] that the transition times between such states and the semiconductor bulk are short, of the order of 10^{-6} sec or less.) In high enough densities, the surface state wave functions can overlap to form bands. The surface properties of a particular crystal will therefore depend on the density and energy distribution of the surface states.

A second kind of state, called a *slow* surface state, is found on semiconductor surfaces. These slow states are associated with the oxide layer, or other adsorbed species, generally found on real (as opposed to ideal) semiconductor surfaces. For these states, the transition times with the bulk

---

[†] We ignore the different types[7] (Tamm and Shockley) of surface states since, for our purposes, all we will need is the existence of surface states.

semiconductor are long (of the order of seconds). The density of slow surface states is believed[5,9] to be larger than that of the fast states, and they are believed also to carry electric charge in a manner similar to donor and acceptor states in a bulk semiconductor. An example is an acceptor surface state due to an oxygen atom on a silicon surface. The oxygen atom can accept an electron from the silicon, creating a hole in the valence band of the bulk silicon crystal. We will consider the effects of both donor and acceptor states on a semiconductor surface, without inquiring as to their detailed identity and nature.[4,10]

Band Structure at the Semiconductor Surface

We discuss the interaction of surface states with the electron energy levels of a bulk semiconductor to investigate how the band structure at the surface[6,11] is modified relative to the bulk band structure. Consider an n-type semiconductor, at 0 K, with a number of acceptor surface states. These acceptor surface states are neutral when empty and are negatively charged when filled. We will assume, for the present, that the number of acceptor surface states is such that, after the rearrangement of charge to be described, all of the acceptor states are occupied by electrons. We assume also that the donor levels in the bulk n-type semiconductor are, initially, all occupied by electrons at the low temperature we are considering. The initial situation is shown in Figure 6.2.

Electrons from bulk donor states flow into the empty acceptor surface states, producing a negative surface charge and leaving behind uncompensated positively charged ionized donors. The result is the creation of a region of fixed positive space charge in the bulk of the semiconductor.

Figure 6.2. Band diagram of an n-type semiconductor with acceptor surface states. The temperature is 0 K, E_{VAC} is the vacuum level, E_F is the Fermi energy, CB and VB refer, respectively, to the conduction and valence bands, and the symbol $(-\bullet-)$ indicates an occupied donor level. The situation pictured is that before any transfer of electrons between bulk states and surface states.

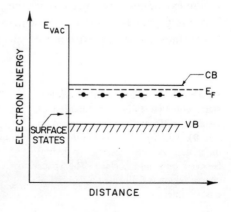

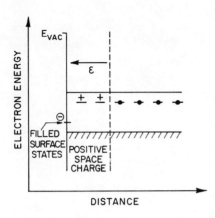

Figure 6.3. Band diagram (schematic) of the situation in Figure 6.2 after transfer of electrons from some donor states to acceptor surface states. Positively charged empty donors are indicated by $\pm$ and $\mathscr{E}$ is the electric field in the positive space charge region. The Fermi level is not shown.

(If the temperature is above 0 K, so some of the donors are ionized, the result is still the same. In this case, electrons from conduction band states near the surface fill the acceptor levels. A region of positive space charge in the bulk, and a negative surface charge, still result.) There is, at equilibrium, an electric field $\mathscr{E}$ in the space charge region, as shown in the band diagram in Figure 6.3. The electric field $\mathscr{E}$ repels further electrons from the surface, and equilibrium is attained.

The electric field $\mathscr{E}$ corresponds to a gradient of electrostatic potential $V(x)$ in the space charge region. Since $\mathscr{E}$ is directed toward the surface, $V(x)$ must increase in the direction away from the surface, as shown schematically in Figure 6.4. The electrostatic potential V is thus higher in the bulk of the semiconductor than it is at the surface, by an amount denoted by Φ_s. The quantity Φ_s is called the surface potential. Since the electrostatic potential is higher by Φ_s in the bulk semiconductor, the electron energy in the bulk is lower by an amount $e\Phi_s$ than it is at the surface. Figure 6.5 gives the band diagram, which shows the positive space charge region extending a distance d from the surface into the bulk semiconductor. The energy bands in the n-type semiconductor are thus "bent" upward at the surface due to the existence of surface acceptor states which have been

Figure 6.4. Schematic plot of electrostatic potential $V(x)$ as a function of distance x (from the surface taken as $x = 0$). The surface potential Φ_s is the difference in electrostatic potential between bulk and surface. The space charge region is delineated by a vertical dashed line.

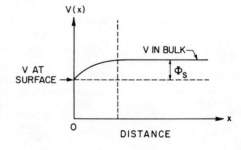

Figure 6.5. Band diagram of the situation in Figure 6.3 after the attainment of equilibrium. The positive space charge region extends a distance d into the semiconductor from the surface. The energy bands are "bent" upward at the surface by an amount $e\Phi_s$, where Φ_s is the surface potential. All surface acceptor states are filled with electrons.

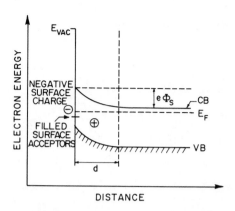

filled with electrons. We note that we have been considering the case in which all of the surface acceptors are full. In that event, as shown in Figure 6.5, the Fermi level at the surface is above the energy of the surface acceptor states since the Fermi function is equal to unity at energies less than the Fermi energy E_F. We summarize the results above by saying that the filled surface acceptors on the n-type semiconductor produce a positive space charge region, a negative surface charge, and upward bending of the energy bands at the surface by an amount $e\Phi_s$, where Φ_s is the surface potential.

In a similar manner, we may examine the surface band structure[6,11] of a p-type semiconductor with donor surface states. Again we assume that, at equilibrium, all of the surface donors are ionized, having given up their electrons to acceptors in the bulk. The result is a positive surface charge and a region of negative space charge in the bulk, the latter being due to fixed ionized acceptors. Because of the space charge, the electrostatic potential is lower in the bulk than at the surface by Φ_s, the surface potential. This in turn means that the electron energy in the bulk is higher by an amount $e\Phi_s$ than it is at the surface. The net result is that the energy bands are "bent" downward at the surface, relative to the bulk, as shown in Figure 6.6. In this figure, we note that the surface donor energy is above the Fermi level at the surface because, in this case, all of the surface donors are empty.

In the cases considered above, both n and p type, the Fermi level position at the surface is determined by the Fermi level position in the bulk because we have assumed a situation in which all of the surface states are ionized. The amount $e\Phi_s$ of band bending (to be discussed later) may be thought of as merely "raising" or "lowering" the conduction and valence band edges at the surface relative to the fixed or "flat" Fermi level.

We consider next the case[12] of a semiconductor in which the number

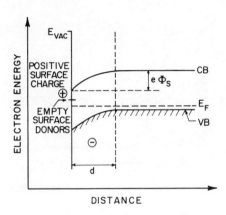

Figure 6.6. Band diagram, at equilibrium, of a *p*-type semiconductor with donor surface states, all of which are ionized. The negative space charge region extends a distance *d* into the semiconductor from the surface. The energy bands are "bent" downward at the surface by an amount $e\Phi_s$, where Φ_s is the surface potential.

of surface states is very large, so that *not* all of them are ionized. For concreteness, we consider donor surface states on a *p*-type semiconductor, all of which lie at roughly the same energy Δ below the conduction band edge. Figure 6.7 shows this situation before the attainment of equilibrium. Electrons flow from the surface donors into the *p*-type semiconductor bulk *only* until the Fermi level at the surface coincides with the surface donor energy. This means that some of the surface donors will remain un-ionized and still occupied by electrons. This fact *fixes* the position of the Fermi level at the surface at the donor energy, i.e., at an energy Δ below the conduction band edge, as is shown in Figure 6.8. On comparing Figure 6.8 with Figure 6.7, we see that, as equilibrium is attained by the flow of electrons into the bulk, the overall effect is to "raise" the Fermi level in the bulk semiconductor until it coincides with the surface donor energy. The Fermi level position at the surface is determined only by the surface donor energy Δ, and is independent of the doping in the *p*-type bulk. One says in this

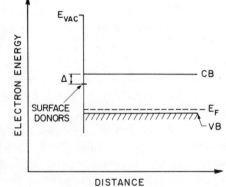

Figure 6.7. Band diagram, before the attainment of equilibrium, of a *p*-type semiconductor with surface donor states an energy Δ below the conduction band edge.

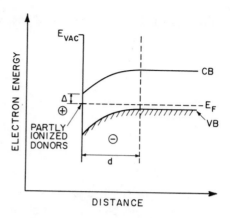

Figure 6.8. Band diagram, at equilibrium, of a p-type semiconductor with a sufficiently large number of surface donors, so that the donors are not completely ionized. The surface Fermi level is pinned at the surface donor energy. The negative space charge region extends a distance d into the bulk, and there is a positive surface charge.

case that the surface Fermi level is "stabilized" or "pinned"[13,14] at the energy Δ, relative to the conduction band edge.

Calculation of the Amount $e\Phi_s$ of Band Bending

We may obtain some simple quantitative results on the amount $(e\Phi_s)$ of band bending, using the situation shown in Figure 6.5. This is the case of an n-type semiconductor with acceptor surface states, all of which are ionized, leading to a region of positive fixed space charge extending a distance d into the bulk. We will call the bulk donor density N_d, and the density of surface acceptors is n_s. Then the space charge density ϱ in the bulk is given by

$$\varrho = +eN_d \tag{6.1}$$

and Poisson's equation for the electrostatic potential $V(x)$ is

$$\frac{d^2V}{dx^2} = \frac{-4\pi\varrho}{\varepsilon} = \frac{-4\pi eN_d}{\varepsilon} \tag{6.2}$$

where ε is the dielectric constant of the semiconductor and $0 \leq x \leq d$. The boundary conditions on the problem are as follows. First, there are n_s filled (i.e., ionized) surface acceptors per unit area, so the negative surface charge density σ is

$$\sigma = -en_s \tag{6.3}$$

If we regard the space charge region in the bulk semiconductor as a capacitor filled with a dielectric of dielectric constant ε, then the electric field $\mathscr{E}$ at the semiconductor surface is of magnitude

$$\mathscr{E} = \frac{4\pi\sigma}{\varepsilon} = \frac{-4\pi e n_s}{\varepsilon} \tag{6.4}$$

Since $\mathscr{E} = -dV/dx$, equation (6.4) gives the first boundary condition on the electrostatic potential as

$$\frac{dV}{dx} = \frac{4\pi e n_s}{\varepsilon} \tag{6.5}$$

at the surface (located at $x = 0$) of the semiconductor.

The second boundary condition stems from the assumption that the electric field $\mathscr{E}$ is confined to the space charge region, so $\mathscr{E} = 0$ at $x = d$, the boundary of the space charge layer in the bulk. Thus we have the second condition that, at $x = d$,

$$\frac{dV}{dx} = 0 \tag{6.6}$$

Integrating Poisson's equation (6.2) gives

$$\frac{dV}{dx} = \frac{-4\pi e N_d x}{\varepsilon} + C \tag{6.7}$$

which, combined with the boundary condition (6.6) at $x = d$, leads to the value $4\pi e N_d d/\varepsilon$ for the constant C. Equation (6.7) becomes

$$\frac{dV}{dx} = \frac{4\pi e N_d}{\varepsilon}(d - x) \tag{6.8}$$

Applying the condition expressed by equation (6.5) at $x = 0$ gives the result that

$$d = n_s/N_d \tag{6.9}$$

for the width d of the positive space charge region.

We integrate equation (6.8) subject to the following boundary conditions on the electrostatic potential. First, we choose $V = 0$ at the surface $x = 0$. Second, this choice of the zero of potential means that V must equal the surface potential Φ_s in the bulk of the semiconductor, as seen from Figure 6.4. We have then that

$$V(x) = \Phi_s$$

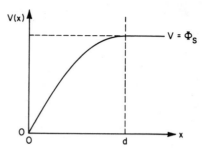

Figure 6.9. Electrostatic potential $V(x)$ as a function of distance x, as given by equation (6.10) for the space charge region $0 \leq x \leq d$; Φ_s is the surface potential. (Arbitrary units are used for V and x.)

for $x = d$. Integrating (6.8) subject to the second of these conditions gives

$$V(x) = (-4\pi e N_d/2\varepsilon)(d - x)^2 + \Phi_s \qquad (6.10)$$

for the spatial variation of the electrostatic potential. The condition that $V = 0$ at $x = 0$ leads to the expression

$$\Phi_s = 2\pi e N_d\, d^2/\varepsilon \qquad (6.11)$$

for the surface potential. Combining (6.11) with (6.9) gives the alternate form

$$\Phi_s = 2\pi e n_s^2/\varepsilon N_d \qquad (6.12)$$

If we plot $V(x)$ as a function of x using equation (6.10), the curve shown in Figure 6.9 is obtained. A representative value[15] for Φ_s is approximately 0.9 V in silicon (300 K) containing 10^{16} acceptors per cm³. We note also that we could also have chosen $V = 0$ at the point $x = d$, a choice that leads to $V = -\Phi_s$ at the surface $x = 0$. This choice would perhaps be in keeping with the name "surface potential" for Φ_s.

Finally, we multiply (6.11) and (6.12) by e to obtain $e\Phi_s$, the amount of band bending, as

$$e\Phi_s = 2\pi e^2 N_d\, d^2/\varepsilon = 2\pi e^2 n_s^2/\varepsilon N_d \qquad (6.13)$$

We will return to equation (6.13) when we discuss photoemission from semiconductors.

Effect of Surface States on Metal–Semiconductor Contacts

In Chapter 4, we discussed metal contacts to semiconductors in terms of the work functions ϕ_m and ϕ_s of the metal and semiconductor, respectively. It was shown that a metal contact to an n-type semiconductor would

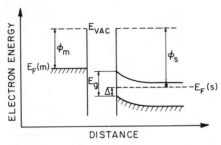

Figure 6.10. Band diagram of a metal (work function ϕ_m) and an n-type semiconductor (work function ϕ_s), where ϕ_m is less than ϕ_s, before equilibrium is established. The surface Fermi level of the semiconductor is pinned at an energy Δ above the surface valence band edge. The semiconductor energy gap is E_g, and the metal and semiconductor Fermi levels are denoted by $E_F(m)$ and $E_F(s)$, respectively. [The position of $E_F(s)$ is exaggerated for clarity.]

be rectifying if ϕ_m were larger than ϕ_s and ohmic if ϕ_m were smaller than ϕ_s; the opposite was true for contacts to a p-type semiconductor. The predictions of these "rules" do not generally work out in practice because the treatment leading to these conclusions neglected the existence of surface states on the semiconductor surface.

Experimentally, it has been found[16] that the height of the energy barrier at the surface of a metal contact to a given semiconductor is approximately independent of the work function ϕ_m of the metal. This is true for a number of different semiconductors, and the explanation[17–20] lies in the "pinning" of the Fermi level at the semiconductor surface. Suppose, in an n-type semiconductor, the Fermi level is pinned at the surface at an energy Δ above the valence band edge at the surface. This situation is (by analogy with Figure 6.8) shown in Figure 6.10, which also shows a metal whose work function ϕ_m is smaller than the work function ϕ_s of the semiconductor, whose energy gap is E_g. Since ϕ_m is smaller than ϕ_s, for an n-type semiconductor one would expect, based on the simpler ideas above, that the energy barrier at the surface would be small (see Figure 4.6), and the contact would be ohmic. However, because the Fermi level in the semiconductor is pinned, then, as equilibrium is established, the metal Fermi level $E_F(m)$ must equalize with the fixed position of the semiconductor Fermi level $E_F(s)$ at the surface. This is shown in Figure 6.11, in which a

Figure 6.11. Band diagram of Figure 6.10, with a small gap between metal and semiconductor, as equilibrium is attained, and the Fermi levels are equalized.

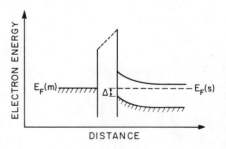

Figure 6.12. Band diagram of the metal–
n-type-semiconductor contact (when
the gap between them has vanished)
at equilibrium. The height of the energy
barrier at the surface is approximately
$E_g - \Delta$. (The symbols are those used
in Figure 6.10.)

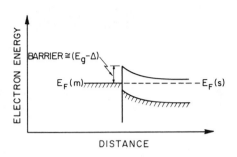

gap still exists between the metal and semiconductor as they come together. The equilibrium band diagram is shown in Figure 6.12, and its key feature, when compared with Figure 4.6, is the existence of an energy barrier, whose height is approximately $E_g - \Delta$, between metal and semiconductor. We would expect the energy band structure in Figure 6.12 to be rectifying, even though it is composed of a metal and n-type semiconductor for which ϕ_m is less than ϕ_s. As seen from these band diagrams, we also expect the height $E_g - \Delta$ of the surface energy barrier to be the same for any value of the metal work function ϕ_m. The barrier height is determined[18] only by the doping and surface properties (i.e., Δ) of the semiconductor and so is independent of the work function of the metal involved. Analogous results may be obtained for metal contacts to p-type semiconductors.

Photoemission from Semiconductors

We now use the results we have obtained on semiconductor surface physics to discuss the photoemission[6,21] of electrons from semiconductors, with some mention of relatively recent developments.

We review in Figure 6.13 the terminology which was introduced in Chapter 4; in the energy band diagram in Figure 6.13, the influence of the surface is, for the moment, neglected. From Chapter 4, we recall the definitions

$$\chi \equiv E_{\text{VAC}} - E_C \tag{6.14}$$

$$\phi_s \equiv E_{\text{VAC}} - E_F \tag{6.15}$$

of the electron affinity χ and the work function[†] ϕ_s of a semiconductor.

[†] It is assumed that the use of ϕ_s for the work function and Φ_s for the surface potential will cause no difficulty.

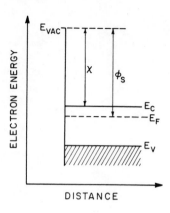

Figure 6.13. Band diagram showing the electron affinity χ and work function ϕ_s of a semiconductor; E_{VAC} is the vacuum level, E_F is the Fermi level, E_C is the energy of the conduction band edge, and E_V is the valence band edge. (Occupied electron states are shaded.)

In equations (6.14) and (6.15), E_{VAC} is the vacuum energy, E_C is the energy of the conduction band edge, and E_F is the energy of the Fermi level.

We may use Figure 6.13 to discuss the threshold energy E_t for photo-emission. The minimum photon energy $E_t = \hbar\omega_t$ that will excite an electron from the semiconductor into the vacuum is, from Figure 6.13,

$$E_t = \hbar\omega_t = E_g + \chi \tag{6.16}$$

where $E_g \equiv E_C - E_V$ is the energy gap of the semiconductor. For the moment, we consider only direct transitions by electrons at the valence band edge (at energy E_V), so no phonon energies are involved in the expression for the threshold energy. We note also that the threshold, or minimum energy, transition of an electron into vacuum will have as its initial state the highest occupied electron energy level. If, as shown in Figures 6.14 and 6.15, the semiconductor is degenerate n or p type, the expression

Figure 6.14. Band diagram illustrating photo-emission of electrons from a degenerate n-type semiconductor. The energy threshold $E_t = \chi - \Delta_n$, where Δ_n is the energy, above the band edge energy E_C, to which the conduction band is filled. (Occupied electron states are shaded.)

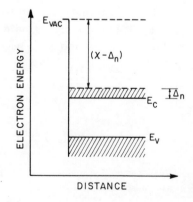

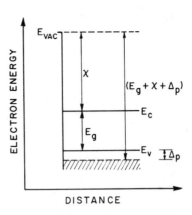

Figure 6.15. Band diagram illustrating photo-emission of electrons from a degenerate p-type semiconductor. The energy threshold $E_t = E_g + \chi + \Delta_p$, where E_g is the energy gap and Δ_p is the depth, below the band edge energy E_V, of the highest occupied electron level. (Occupied electron states are shaded.)

(6.16) for the threshold energy is modified. Equation (6.17),

$$E_t = \chi - \Delta_n \tag{6.17}$$

gives the threshold energy E_t for photoemission of an electron when the conduction band is filled with electrons to an energy Δ_n above the band edge energy E_C. Equation (6.18),

$$E_t = E_g + \chi + \Delta_p \tag{6.18}$$

gives the threshold energy E_t for photoemission of an electron from the valence band when the highest filled electron state is an energy Δ_p below the valence band edge energy E_V. The result is that the threshold energy for photoemission of electrons from a semiconductor varies with the impurity content if the doping is high enough to produce degeneracy.

Effect of the Surface on Photoemission

We now want to discuss the effect of the surface on the photoemission of electrons from a semiconductor. We consider, in Figure 6.16, a p-type semiconductor with donor surface states, in which, as shown in Figure 6.6, the bands are "bent downward" at the surface by an energy $e\Phi_s$, where Φ_s is the surface potential. The width of the space charge layer at the surface is denoted by d, and distance x into the crystal is measured from $x = 0$ at the surface. From the band diagram in Figure 6.16, we see that the electron affinity χ varies with distance x into the crystal, so $\chi = \chi(x)$. If we denote the electron affinity at the surface by χ_s and the electron affinity

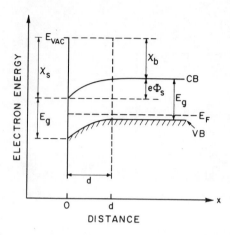

Figure 6.16. Band diagram for a p-type semiconductor showing an amount $e\Phi_s$ of band bending. The electron affinity at the surface and in the bulk are, respectively, χ_s and χ_b, and d is the width of the space charge layer.

in the bulk $(x > d)$ of the crystal by χ_b, then

$$\chi_s = \chi_b + e\Phi_s \qquad (6.19)$$

relates χ_s and χ_b. Equation (6.19) says that the electron affinity at the surface is larger than the electron affinity in the bulk by an energy equal to the band bending.

The threshold energy E_t for photoemission of an electron from the valence band edge into the vacuum is, from equation (6.16), given by

$$E_t = E_g + \chi \qquad (6.16)$$

Since, from (6.19), χ is larger at the surface than in the bulk, the threshold energy for electron emission is smaller in the bulk than at the surface by an amount $e\Phi_s$. This is seen most clearly on rewriting (6.19) as

$$\chi_b = \chi_s - e\Phi_s \qquad (6.20)$$

which states that, if the surface electron affinity χ_s is unchanged,[21] χ_b is decreased. Then the threshold energy E_t for photoemission of electrons from the bulk is given by

$$E_t(\text{bulk}) = E_g + \chi_b = E_g + \chi_s - e\Phi_s \qquad (6.21)$$

Equation (6.21) tells us that, all other things being equal, it is desirable to have a large value of $e\Phi_s$ because the bulk electron affinity and the bulk threshold energy for emission are both reduced.

However, all other things are not equal in this case because an electron photoexcited in the bulk by a photon of energy greater than $E_t(\text{bulk})$

must still physically cross the region of the crystal between $x = d$ and $x = 0$ to reach the surface in order to escape into the vacuum. The photo-excited electron can lose energy by phonon emission during this crossing, and, if d is large, may arrive at the surface with an energy less than E_{VAC}; in such a case the electron will not be emitted into the vacuum. It is therefore desirable that d be small so that the escaping photoelectron makes few energy-losing collisions on its way to the surface. If we write the analog of equation (6.13) for a p-type semiconductor, we obtain

$$e\Phi_s = 2\pi e^2 N_a\, d^2/\varepsilon \qquad (6.22)$$

where N_a is the acceptor impurity concentration, so we have

$$d^2 = \varepsilon(e\Phi_s)/2\pi e^2 N_a \qquad (6.23)$$

Equation (6.23) tells us that, for a given amount $e\Phi_s$ of band bending (determined by the physical state of the semiconductor surface[22,23]) d will be smaller for larger values of the acceptor impurity density N_a. A representative value[22] of d is about 100 Å for acceptor densities in the range 10^{18}–10^{19} cm^{-3}.

As a specific example, consider p-type GaP, whose band diagram is shown in Figure 6.17. The acceptor density is sufficiently high that the Fermi level E_F is very close to the valence band edge; the energy gap $E_g \cong 2.2$ eV. The amount of band bending shown is arbitrary; ϕ is the work function $E_{VAC} - E_F$ at the surface, and χ_b is the electron affinity in the bulk. Since the acceptor density is high, we may assume that E_F is sufficiently close to the valence band that we may write

$$\phi \cong E_g + \chi_b \qquad (6.24)$$

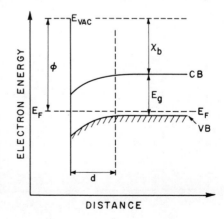

Figure 6.17. Band diagram of a heavily doped p-type semiconductor, with energy gap E_g, surface work function ϕ, and bulk electron affinity χ_b. Since the Fermi level E_F is close to the valence band in this semiconductor, it is approximately true that $\phi \cong E_g + \chi_b$.

Taking a value of $\phi \cong 5$ eV as typical[24] of the III–V semiconductors, we obtain a value of $\chi_b \simeq 2.8$ eV as the energy barrier to an electron to be emitted into the vacuum.

Negative Electron Affinity in Semiconductors

It is clear from Figure 6.17 and equation (6.24) that, if the surface work function ϕ were decreased, the electron affinity χ_b in the bulk would be decreased, resulting in a smaller energy barrier to photoexcited bulk electrons emitted into the vacuum. Given this result, the following possibility now presents itself. Suppose a metal with a very small work function ϕ_m could be applied as a thin layer (i.e., thinner than the electron mean free path). Such a metal layer might be expected to reduce the work function ϕ in equation (6.24) to the smaller value ϕ_m, as shown in Figure 6.18. The thickness of the metal is t, and that of the space charge layer in the semiconductor is d. For an electron at the conduction band edge in the bulk of the semiconductor in Figure 6.18, the energy barrier to emission in the vacuum is χ_b, where

$$\phi_m \cong E_g + \chi_b \tag{6.25}$$

$$\chi_b \cong \phi_m - E_g \tag{6.26}$$

Again, the height of the Fermi level above the valence band has been neglected in the heavily doped p-type semiconductor. From equation (6.26), we see that the smaller the quantity $\phi_m - E_g$, the smaller the bulk electron affinity χ_b, and the lower the energy barrier between a photoexcited electron and emission into the vacuum. The foregoing discussion presupposes that

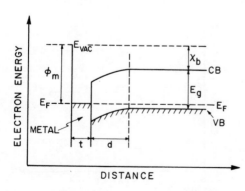

Figure 6.18. Band diagram of a heavily doped p-type semiconductor with a metal layer, of work function ϕ_m, on the surface. The relation $\phi_m \cong E_g + \chi_b$ is approximately true because the Fermi level is close to the valence band.

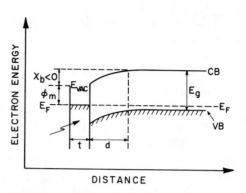

Figure 6.19. Band diagram of a heavily doped p-type semiconductor with a metal of work function ϕ_m on the surface. Since ϕ_m is smaller than the band gap E_g of the semiconductor, χ_b is negative, resulting in a condition of negative electron affinity.

the electron can travel the distance $d + t$ and still arrive at the surface with an energy greater than E_{VAC}, sufficient to escape into the vacuum.

It can also be seen from Figure 6.18 that the threshold energy E_t for photoemission of an electron from the valence band edge into vacuum is

$$E_t = E_g + \chi_b \cong \phi_m \qquad (6.27)$$

so the smaller the value of ϕ_m, the lower the value of the threshold energy E_t.

Examination of equation (6.26) shows the interesting result that, if the energy gap E_g is *larger* than the metal work function ϕ_m, then χ_b is *negative*. The physical meaning of a negative value of χ_b is that, as seen from Figure 6.19, the conduction band edge in the bulk is at a higher energy than the vacuum level at the surface. This means that the energy barrier between a conduction band electron in the bulk and the vacuum level is negative. This situation is referred to as negative electron affinity, and has two main effects. First, the negative energy barrier to emission results in an increase in the number of electrons emitted per absorbed photon. Second, as seen from Figure 6.19, the threshold energy E_t for electron emission is now the band gap energy E_g, since an electron excited only to the bottom of the conduction band now has enough energy to be emitted into vacuum.

Negative electron affinity has been achieved in several semiconductors,[25,26] generally using cesium (or a mixture of cesium and oxygen) as the metal of low work function ϕ_m on the surface of the semiconductor. The procedure for preparing the cesiated surface is generally proprietary, and the Cs–O layer is usually an empirically optimized surface less than an electron mean free path (about 25 Å) in thickness. On p-type GaP, the cesium layer on the surface has a work function[27] $\phi_m \cong 1.2$ eV, resulting, from equation (6.26), in a value of the electron affinity $\chi_b \cong (1.2 - 2.2)$ eV

$= -1.0$ eV. The threshold photon energy for photoemission in GaP with negative electron affinity is[28] at about 2 eV, close to the expected threshold at the band gap energy.

The attainment of negative electron affinity has resulted in improved photoemissive devices, such as photomultiplier tubes, of much higher efficiency. For example, the quantum efficiency[28] (the number of electrons emitted per absorbed photon) is as high as 0.40 for cesiated GaP; this figure is about an order of magnitude higher than that of earlier conventional photoemitters, such as the semiconducting compound K_3Sb. Negative electron affinity has been obtained in other semiconductors, and these have been utilized in various photoemissive devices.[29,30]

We now briefly consider secondary electron emission.[31] When a high-energy electron is incident on a solid semiconductor (or metal), it may excite more than one electron into the vacuum. The electrons so produced are termed secondary electrons. If these secondary electrons are in turn incident on a semiconductor, further "secondaries" can be produced, resulting in a cascade or multiplication of the original electron if there are many stages of secondary electron production. This is the physical idea behind the photomultiplier tube illustrated in Figure 6.20. Photomultiplier tubes produce a cascade of electrons for each photoelectron, and are thus sensitive photon detectors. Recent photomultipliers use photoemissive surfaces (photocathodes) treated to achieve negative electron affinity and thus have high quantum efficiency. If an appropriate semiconductor with a low surface work function or, indeed, with negative electron affinity, is used as the first multiplier stage or dynode, a more efficient secondary emission[30,32] process takes place, resulting in a higher electron multiplication ratio at a lower incident photoelectron energy. Such recent photomultiplier tubes exhibit higher gain and lower noise.

Another interesting device is the "cold cathode,"[33,34] which is a p–n junction in which the surface of the p region is treated to achieve a condition of negative electron affinity. Consider the p^+n junction of GaAs shown in Figure 6.21, in which the p^+ region is very thin. Suppose the p^+ GaAs surface is heated with cesium and oxygen, resulting in negative electron affinity as shown in the band diagram in Figure 6.21. If the p^+n junction is

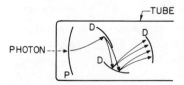

Figure 6.20. Schematic view of a photomultiplier tube with the photoemissive surface P and several secondary emission multiplier stages (dynodes) marked D. The arrows indicate electron paths, showing multiplication.

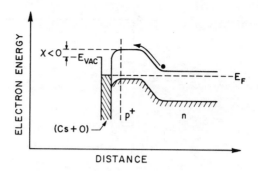

Figure 6.21. Band diagram of a p^+n junction "cold cathode." The vacuum level E_{VAC} at the surface is below the conduction band edge in the p^+ region, leading to a negative electron affinity χ. The injection of an electron from n to p^+ under forward bias is indicated schematically.

forward biased, electrons will be injected into the thin p^+ region. Those electrons that diffuse to the surface will encounter negative electron affinity, and hence an energy barrier of negative height for emission into the vacuum. The result is the production (at high forward currents), of significant electron emission for use as a cathode operating at room temperature, as contrasted with conventional[35] heated cathodes.

Physics of the Transferred Electron (Gunn) Effect

This section will discuss some aspects of the physics of the transferred electron, or Gunn, effect in semiconductors. This effect is essentially the observation of negative differential conductivity in a bulk semiconductor (in contrast to the negative differential conductivity of a degenerate p–n junction). For this reason, one speaks of bulk negative differential conductivity (often abbreviated BNDC). Experimentally, it is found that the electron drift velocity v decreases with increasing applied electric field $\mathscr{E}$ in some n-type semiconductors. The curve in Figure 6.22 shows experimental results[36] for n-type GaAs at room temperature. The electron velocity increases with increasing electric field intensity, until the velocity reaches a maximum at an electric field of approximately 3200 V cm^{-1}. At higher values of the electric field, the electron velocity *decreases* with increasing electric field. Recalling from equation (1.12) that the magnitude J of the electron current density is given by

$$J = nev \qquad (6.28)$$

where n is the electron density, we can write

$$\sigma = dJ/d\mathscr{E} = ne(dv/d\mathscr{E}) \qquad (6.29)$$

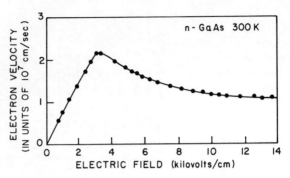

Figure 6.22. Electron velocity as a function of electric field in *n*-type GaAs at room temperature. (After Ruch and Kino.[36])

where σ is the differential electrical conductivity. From Equation (6.29), we see that a negative value of the slope $dv/d\mathscr{E}$, as seen in Figure 6.22 for electric fields larger than 3200 V cm^{-1}, leads to a negative value of the differential conductivity σ.

For this reason, the decrease in electron velocity, with increasing electric field, in a bulk semiconductor is called bulk negative differential conductivity. The value $\mathscr{E}_C$ of the electric field above which $dv/d\mathscr{E}$ is negative is sometimes called the critical value; in Figure 6.22, $\mathscr{E}_C$ is about 3200 V cm^{-1}.

We discuss next the transferred-electron mechanism believed responsible for bulk negative differential conductivity in *n*-type GaAs. We then consider how differential conductivity leads to the Gunn effect, in which the application of a DC electric field of several thousand volts to a bulk sample of GaAs produces electrical oscillations in the microwave frequency range.

To discuss the transferred-electron mechanism leading to negative differential conductivity, we examine the band structure[37,38] of GaAs at room temperature shown in Figure 6.23. The lowest conduction band minimum, at Γ, is 1.42 eV above the valence band edge, which is also at Γ; the energy gap is thus direct at Γ. There are also subsidiary conduction band minima[37] at X, and at L, which are, respectively, 0.48 eV and 0.28 eV above the conduction band minimum at Γ. We will be interested in electrons (in *n*-type GaAs) in the minima at Γ and at L. The effective masses of electrons in the two minima are different. The values,[39] at room temperature, are $m^*(\Gamma) = 0.063m_0$ for electrons in the conduction band at Γ, and $m^*(L) = 0.55m_0$ for electrons in the minimum at L, where m_0 is the free electron mass. Normally, in the absence of an applied electric field, the conduction electrons in GaAs are essentially all in the minimum at Γ, and the electron mobility will be determined by the value of the electron effective mass $m^*(\Gamma)$. However, when an electric field larger than $\mathscr{E}_C$ is applied to the semiconductor, electrons in the conduction band minimum at Γ gain enough energy from the electric field to be transferred (i.e., scattered) into the four higher energy minima at L.

We note next that the combined density of states in the four equivalent minima at L is larger than the density of states in the Γ minimum. This is due to two factors. First, we expect the density of states to be proportional to the electron mass,[40] and $m^*(L)$ is larger than $m^*(\Gamma)$. Second, there are four equivalent minima at L. Qualitatively speaking, we expect that it is relatively unlikely that electrons will be scattered from L (with a higher density of states) to Γ (with a lower density of states). We conclude that the L minima will remain populated with electrons while the applied electric field remains above the critical value.

The transfer of electrons from Γ to L by the electric field means that the electron mobility in the semiconductor will be (at least partially) due to electrons at L. Since electrons at L have a higher effective mass than electrons at Γ, the transfer of electrons from Γ to L results in a decrease in the electron mobility. This is the reason that the electron drift velocity (or, equivalently, the mobility) decreases with increasing electric field in GaAs, as shown in Figure 6.22. The transferred-electron effect thus results, for large enough electric fields, in a bulk negative differential conductivity in GaAs.

It is clear that the existence of the transferred-electron mechanism in GaAs depends directly on its band structure. The band structure requirements[41] in an n-type semiconductor can be seen to be as follows. The electron effective mass in the lowest conduction band minimum should be low relative to that in the higher conduction band minimum. This will

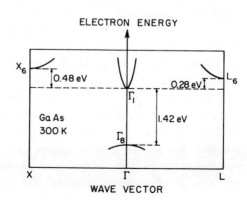

Figure 6.23. Band structure[37] of GaAs at room temperature.[38] The valence band structure is not shown.

lead to electrons having a lower mobility, and to the density of states being larger, in the upper minimum. The energy separation ΔE between the upper and lower minima should be larger than about $4k_BT$ so that the upper minimum is not occupied at low values of the applied electric field. Further, ΔE should be less than half the energy gap so that the critical field $\mathscr{E}_C$ will be small enough that impact ionization (and its consequent increase in conductivity) will be negligible. Finally, the band gap should be larger than about 1 eV, at room temperature, so that conductivity due to intrinsic carriers will not be a problem. From the band structure parameters for GaAs given above, it can be seen that these criteria are satisfied. Other semiconductors in which the transferred-electron effect and bulk negative differential conductivity have been observed are[41,42] InP, CdTe, and ZnSe. These band structure requirements form a necessary, but not sufficient, condition for the existence of the transferred-electron effect. However, other considerations[41] may prevent the exploitation of the effect.

Given the existence of bulk negative differential conductivity in a semiconductor, we next need to discuss how the application of a sufficiently large DC electric field results in electrical oscillations. This effect, originally observed by Gunn, is only one of a number of ways[42-45] in which negative differential conductivity in a semiconductor may be exploited to produce oscillations at microwave frequencies. The mode to be discussed is not particularly efficient or important technologically, but its physics is quite interesting and worth discussing.

Consider a semiconductor, say GaAs, exhibiting bulk negative differential conductivtiy, and which is under an applied DC electric field. We will show that such a situation leads[44] to the production of space charge instabilities in the semiconductor. To see this, we will discuss first an electrically neutral sample of a normal semiconductor, in which the differential conductivity is positive, under an applied DC electric field. From Gauss's law, we have

$$\text{div } \mathscr{E} = 4\pi\varrho/\varepsilon \qquad (6.30)$$

where $\mathscr{E}$ is the electric field in the semiconductor, ϱ is the space charge density, and ε the dielectric constant. Suppose the electric field $\mathscr{E}$ is nonuniform at some point of the semiconductor, possibly due[44] to imperfect doping, or to a statistical fluctuation. Then equation (6.30) says that, if $\mathscr{E}$ is nonuniform, ϱ will be nonzero because div $\mathscr{E}$ is nonzero. Gauss's law thus tells us that a space charge density ϱ will be associated with any nonuniformity of the electric field in the semiconductor. We write the equation of continuity, assuming a constant value of σ in the semiconductor,

$$\partial\varrho/\partial t + \text{div } \mathbf{J} = 0 \qquad (6.31)$$

where $\mathbf{J}$ is the current density, and there are no sources or sinks, in the form

$$\partial\varrho/\partial t + \sigma \text{ div } \mathscr{E} = 0 \qquad (6.32)$$

by using Ohm's law $\mathbf{J} = \sigma\mathscr{E}$. Combining (6.32) and (6.30) gives

$$\partial\varrho/\partial t + 4\pi(\sigma/\varepsilon)\varrho = 0 \qquad (6.33)$$

as the equation governing the time dependence of the space charge density $\varrho(t)$. The solution of (6.33) is

$$\varrho(t) = \varrho_0 \exp(-t/\tau_d) \qquad (6.34)$$

where $\tau_d \equiv \varepsilon/4\pi\sigma$ is the dielectric relaxation time,[46] and ϱ_0 is the initial value of the space charge density at time $t = 0$. If, as is usual in a semiconductor, the conductivity σ is positive, the space charge density ϱ will decrease with time as $\exp[-(4\pi\sigma/\varepsilon)t]$. A typical value of τ_d for a usual semiconductor[46] (germanium with a conductivity of $1\,\Omega$ cm) is about 10^{-12} sec. We see then that a localized space charge density ϱ, due say to one of the nonuniformities in the electric field mentioned above, will die out rapidly in a semiconductor with a positive value of the conductivity.

We may see what is happening physically during this process in the following way. Suppose we have a sample of a semiconductor for which σ is positive. Then, from equation (6.29), $dv/d\mathscr{E}$ is also positive, where v is the electron drift velocity, and $\mathscr{E}$ is the magnitude of the electric field. Consider the situation, shown in Figure 6.24, in which the semiconductor sample is under a DC bias such that $\mathscr{E}$ is constant in the semiconductor. Suppose that a random fluctuation produces a local deviation from charge

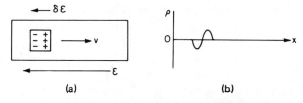

(a)   (b)

Figure 6.24. (a) Schematic view of a dipole region or domain in a semiconductor with a negative value of $dv/d\mathscr{E}$, where $\mathbf{v}$ is the electron drift velocity and $\mathscr{E}$ is the electric field in the semiconductor; (b) schematic variation of the space charge density ϱ as a function of distance x.

neutrality, so a net space charge ϱ in the form of a separation of positive and negative charge is produced. The dipole is shown schematically in Figure 6.24a, and the charge density ϱ as a function of distance is shown, also schematically, in Figure 6.24b. The electric field $\mathscr{E}$ acts on the electrons, which have a drift velocity $\mathbf{v}$ opposite to the direction of $\mathscr{E}$. Because of the space charge in the dipole region, the electric field inside the dipole region will be increased by $\delta\mathscr{E}$, which is also shown in Figure 6.24a. Since σ and $dv/d\mathscr{E}$ are also positive, the electron drift velocity inside the dipole region will be increased by an amount $\delta\mathbf{v}$, where

$$\delta\mathbf{v} = (dv/d\mathscr{E})\, \delta\mathscr{E} \qquad (6.35)$$

The result of the increased electron drift velocity is that electrons inside the dipole region move faster than electrons outside that region, so that the former electrons "overtake" the latter. The effect of this "overtaking" is that the space charge region, or dipole, gradually dies out, restoring space charge neutrality. This is the physical meaning of equation (6.34) with a positive conductivity and hence a positive value of the dielectric relaxation time τ_d.

Now consider the same situation in a semiconductor, like GaAs, in which σ, and hence τ_d, are negative. Formally, this means that the exponent $-t/\tau_d$ is positive in equation (6.34), so that the charge density ϱ *increases* with time. The result is, physically, that a space charge density, produced by, say, a fluctuation, grows with time instead of dying out. We may examine the physical situation shown in Figure 6.24a for the case of negative conductivity and hence a negative value of $dv/d\mathscr{E}$. When the dipole forms, the electric field inside the dipole region increases by $\delta\mathscr{E}$, as before. Now, however, $\delta\mathbf{v}$ is negative because $dv/d\mathscr{E}$ is negative, so electrons inside the dipole region have a drift velocity that is smaller than those outside by an amount again given by equation (6.35). As the electrons move in the direction opposite to $\mathscr{E}$, those inside the dipole region "lag behind" those outside. The result of this "lag" is that the negative and positive space charge regions of the dipole increase, and the space charge density increases instead of dying out. The increased value of ϱ in turn increases the increment $\delta\mathscr{E}$ in the electric field, so the total electric field $\mathscr{E} + \delta\mathscr{E}$ inside the dipole region increases still further. The increased electric field in the dipole decreases the electron drift velocity, thereby accelerating the growth of the dipole region. The dipole region, with its space charge density, is called a domain. The domain moves down the length of the semiconductor sample under the influence of the applied electric field.

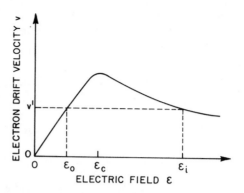

Figure 6.25. Schematic graph of electron drift velocity v as a function of electric field $\mathscr{E}$. The electric fields $\mathscr{E}_i$ and $\mathscr{E}_o$ are, respectively, inside and outside the domain and are shown for the electron velocity $v = v'$, at which the domain is stable. The critical threshold electric field is $\mathscr{E}_C$.

The growth of the domain is a process which might appear capable of continuing indefinitely. What stops the growth of the domain? As the electric field $\mathscr{E}_i$ inside the domain increases, the field $\mathscr{E}_o$ outside the domain decreases in magnitude. This continues until, as shown schematically in Figure 6.25, $\mathscr{E}_o$ falls to a value less than the critical threshold field $\mathscr{E}_C$. At some pair of values of $\mathscr{E}_i$ and $\mathscr{E}_o$ (where $\mathscr{E}_o < \mathscr{E}_C$), the electron drift velocity inside the domain will be equal to the drift velocity outside the domain. This is shown in Figure 6.25; the common value of the electron drift velocity inside and outside the domain is denoted by v', and is approximately 10^7 cm sec^{-1} in GaAs. All of the electrons in the sample now have the same drift velocity, and the stable domain moves down the sample with the velocity v' and without further growth.

When the domain reaches the anode at the end of the sample, it produces a pulse of current in the external circuit. These current pulses are spikes[47] appearing every L/v' seconds, where L is the length of the sample. The transit time τ_t is defined as L/v', so the current pulses have a frequency v'/L. Since v' is of the order of 10^7 cm sec^{-1}, then (for $L = 10^{-1}$ cm), τ_t is of the order of 10^{-8} sec, and the frequency is of the order of 100 MHz. In this way, high-frequency oscillations are produced.

There is a condition which must be fulfilled by the transit time $\tau_t \equiv L/v'$ and the dielectric relaxation time $\tau_d \equiv \varepsilon/4\pi\sigma$ so that the domains will grow and reach a stable size. The transit time must be long compared to the magnitude of the (negative) dielectric relaxation time. Then the domain growth process, described by equation (6.34) with τ_d negative, will have time to grow by many factors of e before the domain reaches the end of the sample. The necessary condition is therefore

$$\tau_t > |\tau_d| \tag{6.36}$$

which can be rewritten as

$$L/v' > \varepsilon/4\pi\sigma = \varepsilon/4\pi ne\mu^* \qquad (6.37)$$

on using the relation $\sigma = ne\mu^*$. In equation (6.37), n is the electron density, and μ^* is an average magnitude of the negative electron mobility given by

$$\mu^* = | (dv/d\mathscr{E}) |_{av} \qquad (6.38)$$

Using (6.37), the condition (6.36) can be written as

$$nL > \varepsilon v'/4\pi e\mu^* \qquad (6.39)$$

For GaAs, the right-hand side of equation (6.39) has a value[48] of about 10^{12} cm^{-2}. When the requirement that nL be larger than 10^{12} cm^{-2} is not met, the formation of domains is not to be expected,[48] and the device is referred to as subcritical.

The mode described above, in which negative differential conductivity leads to the formation and drift of space charge domains, is only one method of exploiting the transferred-electron, or Gunn, effect. Other modes, which are actually more important and efficient, are described in the literature.[42–45]

Physics of Amorphous Semiconductors

It is not yet known if electronic devices made from amorphous semiconductors will be technologically important. However, the physics of amorphous solids is an interesting and active area of research, so some discussion of this field is appropriate. This section will present some of the basic picture of the physics of amorphous semiconductors.

An amorphous solid is one that is not crystalline, in that the amorphous solid lacks the long-range order of the periodic crystal lattice. The amorphous solid exhibits short-range order, extending over distances no greater than a few atomic radii. A familiar example of an amorphous solid is quartz (SiO_2), which exists in both amorphous and crystalline forms. Other examples are tellurium, selenium, germanium, and silicon. Of particular interest from the point of view of applications are the "chalcogenide glasses." These are solid mixtures, in amorphous form, of sulfur, selenium, and/or tellurium. Our interest will be in these amorphous covalent semiconductors.

We wish now to present some of the basic ideas[49-52] of the electronic structure of disordered materials in general, and amorphous semiconductors in particular. We begin by considering a basic model of the electronic structure of a disordered solid, and then apply the results to an idealized picture of an amorphous semiconductor.

The basic model[53] of a disordered solid begins with a perfect crystal and examines the nature of the electronic states in a single isolated band. For a perfect crystal, the states in the band are extended, in that an electron may be found with equal probability in any unit cell. An electron in an extended state may therefore move through the crystal and contribute to the conductivity. To discuss a disordered solid, we consider the introduction (by unspecified means) of randomness into the previously perfect crystal. The disordered material now exhibits fluctuations in atomic configuration; these fluctuations are, of course, not present in the perfect crystal. The effect of the fluctuations is, when strong enough, to change the nature of the wave functions in some parts of the band. Figure 6.26 shows[53] the density of electron states $N(E)$ as a function of electron energy E in the single band of the disordered solid. For electron energies greater than a value E_C', and less than a second value E_C, the character of the wave function is no longer extended, but is now localized in space. In this context, a localized wave function is one whose amplitude is nonzero only in a finite region. For values of the energy between E_C and E_C', the band states are still extended. Now, however, due to the fluctuations in atomic configuration in the disordered solid, there are "tails" of localized states at the upper and lower edges of the band. These tails are shown as shaded in Figure 6.26.

We now apply the results of this basic picture to a model of an amorphous semiconductor. The model is that of an ideal covalent glass, defined[54] as a covalent network of atoms in which the valence of each atom is satisfied. The ideal glass thus contains no impurities or other defects (e.g., "dangling bonds"); real covalent amorphous solids will contain im-

Figure 6.26. Density of states $N(E)$ as a function of electron energy E in an isolated band in a disordered solid. There are tails of localized states (shown shaded) at electron energies less than E_C and greater than E_C'. The band states are extended for energies between E_C and E_C'.

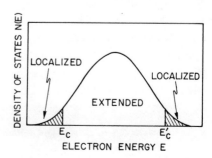

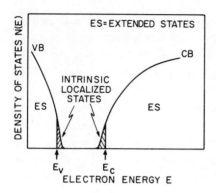

Figure 6.27. Band model of an ideal amorphous semiconductor showing density of states $N(E)$ as a function of electron energy E. There are tails of "intrinsic" localized states restricted to narrow (less than 0.1 eV) energy ranges near the band edges. The valence band (VB) states are localized for $E > E_V$; the conduction band (CB) states are localized for $E < E_C$. (After Tauc.[56])

purities and defects. It is an experimental fact that the DC conductivity σ of amorphous semiconductors has the form

$$\sigma = \sigma_0 \exp(-\varDelta E / k_B T) \tag{6.40}$$

where $\varDelta E$ is an activation energy comparable[55] to those in crystals. The form of equation (6.40), along with optical absorption spectra similar to those of intrinsic crystalline semiconductors, suggest a band model for an amorphous semiconductor which includes valence and conduction bands separated by an energy gap. Such a band model for an ideal covalent amorphous semiconductor is shown in Figure 6.27, which, again, is a plot of density of states $N(E)$ as a function of electron energy E. The valence band states are localized for energies greater than the value E_V; the conduction band states are localized for energies less than E_C. These localized states, shown shaded in Figure 6.27, are those due to the fluctuations inherent in the disordered solid and may be called "intrinsic" localized states. The intrinsic localized states are[56] restricted to narrow (less than 0.1 eV) energy intervals near the band edges.

However, a real amorphous semiconductor will contain impurities and defects such as chain ends, vacancies, dangling bonds, etc. These impurities and defects give rise[†] to a nonzero density of localized electronic states in the gap between the valence and conduction bands shown in Figure 6.27. The result, shown in Figure 6.28, is that, except for the intrinsic localized states very close to the band edges, the states in the gap are due to impurities or defects. The states between the energies E_V and E_C are thus all

---

[†] This is analogous to defects and impurities in crystalline semiconductors giving rise to discrete states in the energy gap.

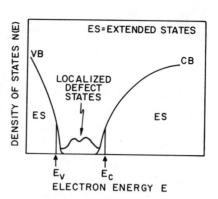

Figure 6.28. Band model of an amorphous semiconductor containing impurities and defects. A nonzero density of localized electron states is present in the gap because of the defects and impurities. The electron states between the energies E_V and E_C are thus localized in nature. The conduction band (CB) and valence band (VB) are labeled. (After Tauc.[56])

localized, while those at energies below E_V and above E_C are extended in nature. In the extended states, the carrier mobility μ is expected to be relatively high; the conduction process in these states is analogous to conduction in crystalline semiconductors. In the localized states between E_V and E_C, on the other hand, the mobility is due[57] to thermally assisted tunneling between localized states. The mobility of carriers, with energies between E_V and E_C, in localized states is thus much lower than that of carriers in the extended states. The carrier mobility $\mu(E)$, plotted as a function of energy E in Figure 6.29, therefore shows sharp decreases at the energies E_V and E_C; these decreases are called mobility edges. For nonzero temperatures, $\mu(E)$ is small but not zero between E_V and E_C; the energy region between E_V and E_C is called the mobility gap. The activation energy ΔE in equation (6.40) thus relates to excitation of carriers from extended states below E_V, across the mobility gap, into extended states above E_C. The activation energy ΔE in an amorphous semiconductor is thus connected with the mobility gap, rather than with the energy gap in the density of states, as it is in a crystalline semiconductor. The mobility gap plays the same role in amorphous semiconductors that the energy gap in the density of states

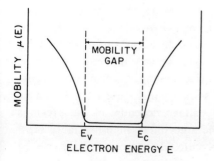

Figure 6.29. Schematic variation of mobility $\mu(E)$ as a function of electron energy E in a covalent amorphous semiconductor at a nonzero temperature.

does in crystalline semiconductors. The band model described above is called the Mott–CFO model after its originators, Mott, Cohen, Fritzsche, and Ovshinsky.

From the point of view of the applications to be described, the key result of the Mott–CFO band model is that the conductivity of an amorphous semiconductor will be low. This is because of the existence of the mobility gap shown in Figure 6.29. The current carriers will be found at energies roughly between E_V and E_C, i.e., in states of low mobility. Representative values[58] of the Hall mobility μ_H in chalcogenide glasses are of the order of 0.1 cm² V^{-1} sec^{-1}. A theoretical estimate[58] of the conductivity mobility μ is $\mu \approx 100\,\mu_H$, leading to a value of μ of the order of 10 cm² V^{-1} sec^{-1}. This value is low compared to mobilities observed in crystalline semiconductors. The conductivity σ varies as shown in equation (6.40), with σ_0 having a value[59] of the order of 1000 Ω^{-1} cm^{-1} for many amorphous chalcogenide semiconductors. The activation energy[59] ΔE is generally between 0.5 and 1.0 eV. The resulting values of the conductivity range from $10^{-3}\,\Omega^{-1}$ cm^{-1} to $10^{-13}\,\Omega^{-1}$ cm^{-1} (at 300 K), depending on the composition and method of preparation of the amorphous semiconductor. These values indicate that the DC conductivity of an amorphous chalcogenide semiconductor at room temperature will be quite low compared to the values we are used to for crystalline semiconductors.

Amorphous Semiconductor Devices

There are two main classes of devices[60,61] based on amorphous chalcogenide semiconductors. These are memory devices and switching devices. While the technological importance[62,63] of these devices is not yet established, it is interesting to discuss them.

We consider first the amorphous chalcogenide semiconductor memory switch. This device is based on the fact[64,65] that there exists in some of these materials a reversible structure change between the amorphous state of high resistance and a crystalline state of low resistance. It is believed that heating[64,66] of the amorphous state can induce crystallization. Further, the small-grain crystalline region so produced can be returned to the amorphous state on heating followed by rapid cooling. A schematic current-voltage plot of such a memory device is shown in Figure 6.30. It can be seen that the resistance dV/dI at zero voltage depends on the previous history of the device. Switching from the amorphous high-resistance OFF state to the low-resistance crystalline ON state takes place when the applied

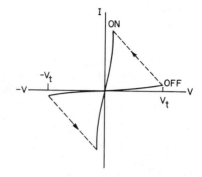

Figure 6.30. Schematic plot of current I versus voltage V for an amorphous semiconductor memory device. The threshold voltage for switching from the high-resistance amorphous OFF state to the low-resistance crystalline ON state is V_t.

voltage exceeds the threshold value V_t; this is indicated by the dotted lines in Figure 6.30. The conducting state is retained even after removal of the applied voltage. The high-resistance amorphous OFF state can be reestablished by heating (with a short pulse of current), which redissolves the crystalline material into a disordered phase. The memory device thus has two stable states.

The second type of amorphous semiconductor application is a switching device.[52,60,61,67] A schematic current–voltage plot for such a device is shown in Figure 6.31. There is a high-resistance OFF state and a low-resistance ON state. If the threshold voltage V_t is exceeded, the device switches from the OFF state to the ON state; if the current is decreased below the holding current I_H, the device switches to its original OFF state. The switching time[68] is less than a nanosecond, and the resistance ratio[68] between the conducting and nonconducting states is of the order of 10^6. It appears[60,67] that the switching mechanism is electronic, not thermal, in nature. Since the details are complex, and not completely settled, the reader is referred to the literature[67] for details.

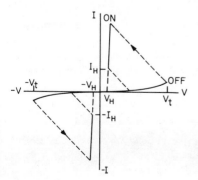

Figure 6.31. Schematic plot of current I versus voltage V for an amorphous semiconductor threshold switching device. The voltage V_t is the threshold for the transition from the OFF state to the ON state; the current I_H is the holding current below which the device reverts to the OFF state. (After Adler et al.[67])

Problems

6.1. *Effect of Surface States on Metal–p-Type-Semiconductor Contacts.* Using drawings analogous to Figures 6.10–6.12, discuss the contact between a metal (of work function ϕ_m) and a p-type semiconductor (of work function ϕ_s), in which the Fermi level is pinned at the surface. Let Δ be the energy of the Fermi level above the valence band edge at the surface, and let E_g be the energy gap of the semiconductor. Consider the case for which ϕ_m is smaller than ϕ_s, and show that one may obtain an ohmic contact in this case, even though the earlier "rule" predicts a rectifying contact.

6.2. *Photoemission in the Infrared.* What semiconductors might be usable as infrared photoemitters? Assume that the Fermi level at the surface of the semiconductor is pinned at an energy $\Delta = E_g/3$ above the valence band edge, where E_g is the band gap. Draw a band diagram, using a value of 1.2 eV for the work function of a Cs + O layer on the semiconductor surface. What is the minimum value of the electron affinity χ? What is the photon energy threshold for electron emission? (For information on actual infrared photoemitters, see R. L. Bell, Reference 13, pages 76–82.)

6.3. *Calculation of Band Bending in GaP.* Suppose the width of the space charge layer of a certain sample of cesiated p-type GaP is found to be 300 Å. If the acceptor density is 10^{19} cm^{-3} and the dielectric constant is 8.4, calculate the energy through which the bands are bent. Express your result as a fraction f of the energy gap 2.2 eV of GaP, and compare your value of f with that given in the literature.[22]

References and Comments

1. See, for example, C. Kittel, *Introduction to Solid State Physics*, Fifth Edition, John Wiley, New York (1976), pages 191–192.
2. J. P. McKelvey, *Solid State and Semiconductor Physics*, Harper and Row, New York (1966), pages 485–489.
3. A. S. Grove, *Physics and Technology of Semiconductor Devices*, John Wiley, New York (1967), pages 144–145, 282–285, 334–337.
4. A. Many, Y. Goldstein, and N. B. Grover, *Semiconductor Surfaces*, North-Holland, Amsterdam (1965), Chapters 5 and 9.
5. S. Wang, *Solid State Electronics*, McGraw-Hill, New York (1966), pages 294–300.
6. J. Van Laar and J. J. Scheer, "Photoemission of Semiconductors," *Philips Technical Review*, **29**, 54–66 (1968), especially pages 56–58.
7. A. Many *et al.*, Reference 4, pages 166–174.
8. A. Many *et al.*, Reference 4, page 348.
9. A. Many *et al.*, Reference 4, pages 357–358.
10. For a review, see S. G. Davison and J. D. Levine, in *Solid State Physics*, H. Ehren-reich, F. Seitz, and D. Turnbull (editors), Academic Press, New York (1970), Volume 25, pages 1–149, especially pages 88–148.

11. H. K. Henisch, *Rectifying Semiconductor Contacts*, Oxford University Press (1957), pages 173–179.

12. J. Van Laar and J. J. Scheer, Reference 6, pages 56–57.

13. R. L. Bell, *Negative Electron Affinity Devices*, Oxford University Press (1973), pages 17–19.

14. S. G. Davison and J. D. Levine, Reference 10, page 135, gives information on Fermi level pinning in a number of semiconductors.

15. A. S. Grove, Reference 3, pages 267–268, Figure 9.4.

16. See S. M. Sze, *Physics of Semiconductor Devices*, John Wiley, New York (1969), pages 397–399, Table 8.4, which gives measured barrier heights for a number of metal–semiconductor contacts.

17. J. Bardeen, "Surface States and Rectification at a Metal–Semiconductor Contact," *Physical Review*, **71**, 717–727 (1947).

18. A. M. Cowley and S. M. Sze, "Surface States and Barrier Height of Metal–Semiconductor Systems," *Journal of Applied Physics*, **36**, 3212–3220 (1965).

19. R. S. Muller and T. I. Kamins, *Device Electronics for Integrated Circuits*, John Wiley, New York (1977), pages 93–95.

20. A. Van der Ziel, *Solid State Physical Electronics*, Second Edition, Prentice-Hall, New York (1968), page 101.

21. J. I. Pankove, *Optical Processes in Semiconductors*, Prentice-Hall, New York (1971), Chapter 13.

22. R. U. Martinelli and D. G. Fisher, "The Application of Semiconductors with Negative Electron Affinity Surfaces to Electron Emission Devices," *Proceedings IEEE*, **62**, 1339–1360 (1974), page 1342.

23. R. L. Bell, Reference 13, Section 6.4, page 74.

24. J. I. Pankove, Reference 21, Table 13-2, page 298.

25. R. U. Martinelli and D. G. Fisher, Reference 22, Table 1, page 1340.

26. R. L. Bell, Reference 13, Chapter 6.

27. J. I. Pankove, Reference 21, page 296.

28. J. I. Pankove, Reference 21, Figure 13-11 (b), page 297.

29. R. U. Martinelli and D. G. Fisher, Reference 22, pages 1345–1352.

30. R. E. Simon, "A Solid State Boost for Electron Emission Devices," *IEEE Spectrum*, **9**, 74–78 (December 1972).

31. A. Van der Ziel, Reference 20, Chapter 10.

32. R. U. Martinelli and D. G. Fisher, Reference 22, pages 1352–1356.

33. R. U. Martinelli and D. G. Fisher, Reference 22, pages 1356–1357.

34. R. L. Bell, Reference 13, Chapter 8, pages 96–102.

35. A. van der Ziel, Reference 20, pages 157–162.

36. J. G. Ruch and G. S. Kino, "Transport Properties of GaAs," *Physical Review*, **174**, 921–931 (1968), Figure 5, page 926.

37. D. E. Aspnes, "Lower Conduction Band Structure of GaAs," in *Gallium Arsenide and Related Compounds (St. Louis), 1976*, L. F. Eastman (editor), Proceedings of the Sixth International Symposium, Conference Series Number 33b, The Institute of Physics, London (1977), pages 110–119.

38. The room-temperature energy separations shown in Figure 6.23 were calculated by the author from results given in Reference 37, Table 1.

39. D. E. Aspnes, Reference 37, Table 2, page 116.

40. C. Kittel, *Introduction to Solid State Physics*, Fifth Edition, John Wiley, New York (1976), page 162.

41. M. E. Jones and R. T. Bate, "Electronic and Optical Phenomena in Semiconductors," *Annual Review of Materials Science*, **1**, 347–372, (1971), pages 351–356.

42. J. A. Copeland and S. Knight, in *Semiconductors and Semimetals*, R. K. Willardson and A. C. Beer (editors), Academic Press, New York (1971), Volume 7, Part A, pages 3–72.

43. See, for example, B. G. Streetman, *Solid State Electronic Devices*, Prentice Hall, New York (1972), pages 429–433, for a summary.

44. H. Kroemer, "Negative Conductance in Semiconductors," *IEEE Spectrum*, **5**, 47–56 (January 1968).

45. H. Kroemer, "The Gunn Effect—Bulk Instabilities," in *Topics in Solid State and Quantum Electronics*, W. D. Hershberger (editor), John Wiley, New York (1972), pages 20–98.

46. S. Wang, Reference 5, pages 274–275.

47. For examples, see H. Kroemer, Reference 45, page 21; J. A. Copeland and S. Knight, Reference 42, page 33.

48. H. Kroemer, Reference 44, page 53.

49. N. F. Mott and E. A. Davis, *Electronic Processes in Non-Crystalline Materials*, Oxford University Press, New York (1971); Second Edition (1978).

50. M. H. Cohen, "Theory of Amorphous Semiconductors," *Physics Today*, **24**, 26–32 (May 1971).

51. E. N. Economou, M. H. Cohen, K. F. Freed, and E. S. Kirkpatrick, "Electronic Structure of Disordered Materials," in *Amorphous and Liquid Semiconductors*, J. Tauc (editor), Plenum Press, New York (1974), pages 101–158.

52. J. Tauc, "Amorphous Semiconductors," *Physics Today*, **29**, 23–31 (October 1976).

53. E. N. Economou *et al.*, Reference 51, pages 106–108.

54. E. N. Economou *et al.*, Reference 51, page 108.

55. M. H. Cohen, Reference 50, page 26.

56. J. Tauc, Reference 52, page 24.

57. H. Fritzsche, "Electronic Properties of Amorphous Semiconductors," in *Amorphous and Liquid Semiconductors*, J. Tauc (Editor), Plenum Press, New York (1974), pages 221–312; page 229.

58. H. Fritzsche, Reference 57, pages 258–262.

59. H. Fritzsche, Reference 57, pages 235–236.

60. D. Adler, "Amorphous Semiconductor Devices," *Scientific American*, **236**, 36–48 (May 1977).

61. H. Fritzsche, "Switching and Memory in Amorphous Semiconductors," in *Amorphous and Liquid Semiconductors*, J. Tauc (editor), Plenum Press, New York (1974), pages 313–359.

62. G. Lapidus, "Amorphous Devices: At the Threshold," *IEEE Spectrum*, **10**, 44–52 (January 1973).

63. R. Allan, "Amorphous Semiconductors Revisited," *IEEE Spectrum*, **14**, 41–45 (May 1977).

64. H. Fritzsche, Reference 61, page 348.

65. D. Adler, Reference 60, pages 45–46.

66. H. Ehrenreich *et al.*, *Fundamentals of Amorphous Semiconductors*, National Academy of Sciences, Washington, D.C. (1972), Chapter 6.

67. D. Adler, H. K. Henisch, and N. Mott, "The Mechanism of Threshold Switching in Amorphous Alloys," *Reviews of Modern Physics*, **50**, 209–220 (1978).
68. D. Adler, Reference 60, page 44.

Suggested Reading

A. MANY, Y. GOLDSTEIN, and N. B. GROVER, *Semiconductor Surfaces*, North-Holland, Amsterdam (1965). This treatise discusses its subject at the advanced level. Chapters 5, on surface states, and 9, on the electronic structure of the surface, are particularly pertinent to our device-related discussion.

J. VAN LAAR and J. J. SCHEER, "Photoemission of Semiconductors," *Philips Technical Review*, **29**, 54–66 (1968). This introductory article discusses photoemission and surface effects, including negative electron affinity, in semiconductors.

H. KROEMER, "Negative Conductance in Semiconductors," *IEEE Spectrum*, **5**, 47–56 (January 1968). A good tutorial review article which stresses basic physics. Some of the details of the GaAs band structure are now superseded by new data, but the application to the Gunn effect is well presented.

J. TAUC (editor), *Amorphous and Liquid Semiconductors*, Plenum Press, New York (1974). A set of articles by several authors discussing, at the advanced level, the physics and applications of amorphous semiconductors.

D. ADLER, "Amorphous Semiconductor Devices," *Scientific American*, **236**, 36–48 (May 1977). An article which emphasizes the physics of amorphous semiconductors and devices.

7

Detectors and Generators
of Electromagnetic Radiation

Introduction

This chapter discusses a number of solid state (mostly semiconductor) detectors and generators of electromagnetic radiation. The absorption of photons by both intrinsic and extrinsic semiconductors is considered first and leads to a discussion of photoconductive and photovoltaic devices. Several important applications of intrinsic photoconductivity (e.g., photography) are treated briefly. Spontaneous photon emission in semiconductors is considered as the basis for p–n junction luminescence in light-emitting diodes. A general discussion of photon amplification by stimulated emission is provided as background for a description of three- and four-level lasers (e.g., ruby). Finally, stimulated emission in semiconductor junction lasers concludes the chapter.

Intrinsic Photon Absorption in Semiconductors

We begin with a brief discussion of processes[1,2] in which the absorption of a photon by a semiconductor results in the production of free charge carriers, either electrons or holes, or both.

Consider first the absorption of a photon by an intrinsic semiconductor[3] with the band structure, at 0 K, shown in Figure 7.1, so the valence band is completely full and the conduction band is completely empty. The band structure in Figure 7.1 is one in which the minimum energy gap E_g is direct. A photon of circular frequency ω_g, where

$$\hbar\omega_g = E_g \tag{7.1}$$

will have just enough energy to excite an electron from the top of the valence band to the bottom of the conduction band. A photon of energy less than E_g will not have sufficient energy to so excite an electron. Excitation of an electron across the energy gap is often called intrinsic excitation and this absorption process in a semiconductor is called intrinsic absorption. The intrinsic absorption process, indicated schematically in Figure 7.1, produces a free electron in the conduction band, and a free hole in the valence band. Intrinsic absorption of a photon thus produces an electron–hole pair in the semiconductor.

Since a photon whose energy is $\hbar\omega_g$ is the lowest-energy photon that can be absorbed by the intrinsic excitation process, there is a threshold for optical absorption by the semiconductor at the photon energy $\hbar\omega_g$. Figure 7.2 shows the optical absorption coefficient α, defined by the relation

$$\alpha \equiv (\omega/c)\varkappa$$

where $\varkappa$ is the imaginary part of the complex refractive index,[4] plotted as a function of photon energy $\hbar\omega$. In Figure 7.2, we note that, above the threshold, the absorption coefficient increases rapidly with photon energy. The functional form $\alpha(\hbar\omega)$ of the dependence of α on $\hbar\omega$ will depend[5] on the details of the band structure of the semiconductors and on the type of transition. The transition of the electron from the valence band to conduction band shown in Figure 7.1 is called a direct transition. This is because

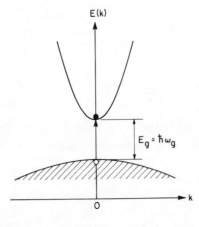

Figure 7.1. Absorption of a photon of energy $\hbar\omega_g = E_g$ by a semiconductor with a minimum direct energy gap equal to E_g located at the zone center. The electron energy $E(k)$ is plotted as a function of wave vector k; filled electron states are shown as shaded. An electron (●) in the conduction band and a hole (○) in the valence band are produced by the absorption of the photon.

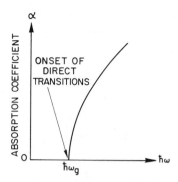

Figure 7.2. Schematic variation of absorption coefficient α as a function of photon energy $\hbar\omega$ for direct transitions. The threshold is located at $\hbar\omega_g = E_g$, where E_g is the minimum direct gap of the semiconductor.

(assuming the photon wave vector to be negligible[†]) the wave vector of the electron does not change in the threshold transition shown, so the transition occurs vertically on the band diagram. Such a direct transition will be the case for a semiconductor (e.g., InSb) in which the minimum energy gap is direct.

We discuss next the more complicated case of intrinsic photon absorption in a semiconductor with an indirect band gap in which the valence band maximum and the conduction band minimum are at different points of the Brillouin zone. Consider such a semiconductor band structure, shown schematically in Figure 7.3. The diagonal arrow in Figure 7.3 shows the excitation of an electron from the valence band maximum at the Γ point of the zone to the conduction band minimum at wave vector k_0 near the zone edge point X, producing an electron–hole pair. Such a transition, involving excitation of an electron across an indirect energy gap, is called an indirect transition and is a transition in which the electron wave vector does change. We can calculate this change because conservation of wave vector requires, for the electron, that

$$\hbar k_i + \hbar k' = \hbar k_f \qquad (7.2)$$

where in equation (7.2) k' is the increase in electron wave vector, k_i and k_f are the initial and final electron wave vectors, and we are again neglecting the wave vector of the photon involved. In Figure 7.3, $k_i = 0$, so $k' = k_f = k_0$.

Where does the "additional" wave vector $k' = k_f - k_i$ come from? It is provided by either the absorption of a phonon of wave vector k', or the emission of a phonon of wave vector $-k'$. Consider first the absorption

---

[†] See Problem 1 of this chapter for an illustrative calculation and example.

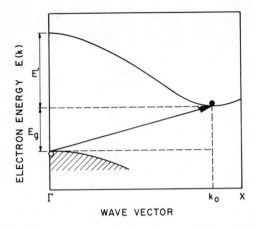

Figure 7.3. Absorption of a photon by a semiconductor with an indirect energy gap, producing a free electron (●) in the conduction band and a free hole (○) in the valence band. The indirect transition is between the valence band maximum at the zone center (Γ) and the conduction band minimum at wave vector k_0 near the zone edge (X).

of a phonon of wave vector k' and energy $\hbar\Omega$. Then the simultaneous conservation of energy and of wave vector requires that

$$E_i + \hbar\Omega + \hbar\omega = E_f \tag{7.3}$$

$$k_i + k' = k_f \tag{7.4}$$

where E_i and E_f are the initial and final electron energies, $\hbar\omega$ is the photon energy, and the photon wave vector has been neglected in equation (7.4). If we rewrite the energy relation (7.3) as

$$E_f - E_i = \hbar\omega + \hbar\Omega \tag{7.5}$$

and note that the threshold electron energy difference $E_f - E_i$ is equal to the minimum energy gap E_g, we see that the threshold photon energy $\hbar\omega_t$ is given by

$$\hbar\omega_t = E_g - \hbar\Omega \tag{7.6}$$

Equation (7.6) tells us that an indirect transition involving the absorption of a phonon of energy $\hbar\Omega$ has its absorption threshold at a photon energy equal to $E_g - \hbar\Omega$, so the threshold in this case is at a photon energy smaller than that of the energy gap. Figure 7.4 shows, schematically, a plot of absorption coefficient α as a function of photon energy $\hbar\omega$ for indirect transitions involving absorption of a phonon of energy $\hbar\Omega$. For a representative semiconductor, $E_g = 1$ eV and $\hbar\Omega$ is about 0.05–0.10 eV. Figure 7.4 is drawn for a hypothetical semiconductor with an indirect band structure (like Figure 7.3) for which $E_g = 0.7$ eV, $\hbar\Omega = 0.1$ eV, and the energy E'

$= 0.15$ eV. Figure 7.4 shows the onset of indirect transitions at a photon energy equal to $E_g - \hbar\Omega$. The figure also shows the onset of direct (vertical) transitions at a photon energy equal to $E_g + E'$; this is seen experimentally as a "knee" in the absorption curve of the semiconductor.[4]

If the indirect transition involves the emission of a phonon of energy $\hbar\Omega$ and wave vector $-k'$, then the threshold photon energy $\hbar\omega_t$ is given by

$$\hbar\omega_t = E_g + \hbar\Omega \qquad (7.7)$$

so the threshold for indirect transitions involving phonon emission is at a photon energy larger than the energy gap.

Finally, it should be remarked that indirect transitions involving the simultaneous participation of more than one phonon have been observed in several semiconductors.[7]

From the point of view of applications, the important results of these considerations are as follows. First, intrinsic excitation in a semiconductor occurs at a photon energy approximately equal to the minimum energy gap. Second, intrinsic excitation produces one free electron–hole pair for each photon absorbed. Third, the direct transition is a two-body process (photon, electron) while the indirect transition is a three-body process (photon, electron, phonon). We therefore expect a lower probability, all other factors being equal, for the indirect transition compared to the direct transition. This is reflected in lower absorption coefficients for indirect transition absorption. Fourth, the photogenerated intrinsic free carriers (electron–hole pairs) are utilized in a variety of applications which we will discuss later.

Figure 7.4. Optical absorption coefficient α as a function of photon energy $\hbar\omega$ for a hypothetical semiconductor with an indirect band gap as in Figure 7.3. The onset of indirect transition at $\hbar\omega = E_g - \hbar\Omega$ due to absorption of phonons of energy $\hbar\Omega$ is shown, as is the onset of direct transitions at $\hbar\omega = E_g + E'$. (Note that the $\hbar\omega$ axis does not go to zero.)

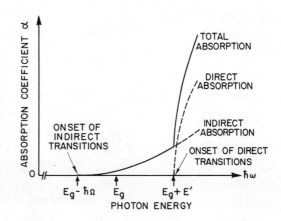

Photon Absorption by Bound States of Impurities in Semiconductors

Impurity conductivity in semiconductors, in which a donor or acceptor impurity is thermally ionized to yield a free electron or hole, was discussed in Chapter 1. At room temperature then, we would expect almost all of the usual donors and acceptors in silicon to be thermally ionized. However, at a low temperature near 0 K, the electron or hole will be bound to the donor or acceptor, as shown schematically in Figures 7.5a and 7.5b. The bound electron or hole may absorb a photon of energy $\hbar\omega$ if $\hbar\omega$ is larger than or equal to E_d for the n-type semiconductor or larger than or equal to E_a for the p-type semiconductor. In the former, the bound electron is excited into the conduction band and, in the latter, the bound hole is excited into the valence band. (Strictly speaking, the process in the p-type case is really the excitation of an electron from a filled state at the top of the valence band into the unoccupied orbital of the acceptor atom, thus creating a hole in the valence band. However, it is usual to speak of the ionization of holes.) In either case a single type of free carrier is created; the processes are indicated symbolically in Figure 7.6. The process of absorption of photons by bound impurities[8] in semiconductors is often called extrinsic absorption or impurity absorption.

The processes described above are excitation of a carrier (electron or hole) from a discrete bound state (donor or acceptor) to a quasicontinuous band of levels (conduction band or valence band). The result is a continuous absorption spectrum with a threshold at the relevant ionization energy. (There may also be observed sharp absorption lines at photon energies less than the ionization energy.[9] These lines are due to transitions between the ground state of the impurity atom and excited, but still bound, states. Such transitions do not produce free carriers because they do not ionize the impurity atom.) Since the density of impurity atoms in a semiconductor (typically in the range 10^{14}–10^{16} cm^{-3} before degeneracy occurs) is quite

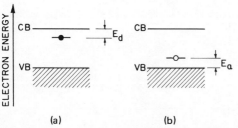

Figure 7.5. Band diagrams showing, schematically (a) an electron (–●–) bound to a donor in an n-type semiconductor, and (b) a hole (–○–) bound to an acceptor in a p-type semiconductor, both at low temperatures. The ionization energy of the donor is E_d, and that of the acceptor is E_a. The shaded areas represent filled electron states, and the ionization energies are exaggerated for clarity.

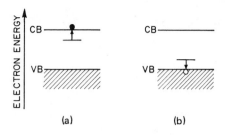

Figure 7.6. Band diagrams showing, schematically, (a) an electron excited from a bound donor state to the conduction band in an n-type semiconductor, and (b) a hole excited from a bound acceptor state to the valence band in a p-type semiconductor.

small compared to the density of intrinsic electrons (10^{22} cm^{-3}), the optical absorption coefficient α for extrinsic absorption is quite small (perhaps 10–100 cm^{-1}) compared to the values for intrinsic direct or indirect absorption (typically 10^3–10^4 cm^{-1}).

Finally, it should be noted that the discussion above dealt tacitly with shallow discrete donors and acceptors, principally in germanium or silicon. There are also deeper impurity levels known in these group IV semiconductors; examples are gold and mercury in germanium. Such deeper levels[10] are usually multiple levels and have larger ionization energies than the shallow levels mentioned above.

From the point of view of applications, extrinsic or impurity absorption of photons produces a single type of free carrier if the photon energy is greater than the ionization energy of the bound impurity state. These photogenerated extrinsic free carriers are useful in a variety of applications to be discussed later.

Threshold Energies for Photon Absorption

One of the important attributes of a detector of electromagnetic radiation is the region of the spectrum in which it absorbs. We will consider some examples of intrinsic and extrinsic absorption in semiconductors which are useful detector materials.

A detector for the visible region of the spectrum is cadmium sulfide, CdS. This semiconductor has a hexagonal crystal structure in its common modification, so its band structure[11] and optical absorption[12] are more complicated than those of the usual cubic semiconductors like germanium. A simplified version[13] of the band structure of CdS is shown in Figure 7.7. The energy gap E_g is direct at the center of the Brillouin zone, and E_g has the values[14] 2.58 eV at 0 K and approximately 2.4 eV at 300 K. This means that the energy threshold for photon absorption by CdS at room

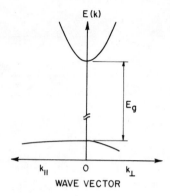

Figure 7.7. Simplified version[13] of the band structure of hexagonal CdS showing electron energy $E(k)$ as a function of wave vectors $k_{\parallel}$ parallel to and $k_{\perp}$ perpendicular to the c axis of the Brillouin zone. The minimum energy gap is direct at the center of the zone.

temperature is at about 2.4 eV. Since a 1-eV photon has a wavelength of 1.240 μm, this absorption threshold in CdS is at a wavelength of approximately 5200 Å, which is in the green region of the visible spectrum. Wavelengths shorter than about 5200 Å are strongly absorbed by CdS, while longer wavelengths (yellow, red) are transmitted. A crystal of CdS thus appears yellowish-orange by transmitted light. Figure 7.8 shows the absorption spectrum[15] of CdS at 300 K, exhibiting the strong absorption at photon energies larger than about 2.4 eV.

There are a number of semiconductors, with energy gaps smaller than about 1 eV, whose absorption thresholds are in the infrared[16] region of the spectrum. Several of these are listed in Table 7.1, which shows the energy gap[14] E_g at 0 K, the threshold wavelength λ_t for absorption, and the type of gap.

Figure 7.8. Absorption spectrum of CdS at 300 K (for radiation polarized with its electric field vector parallel to the c axis. After Gutsche.[15])

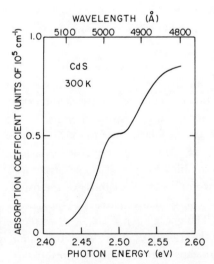

Table 7.1. Semiconductors that Absorb Infrared Photons

Semiconductor	E_g at 0 K (eV)	λ_t (μm)	Type of gap
Si	1.17	1.06	indirect
Ge	0.74	1.68	indirect
InSb	0.24	5.17	direct at Γ
PbSe	0.16	8.27	direct at L

The three lead salt semiconductors PbS, PbSe, and PbTe, all have rather small energy gaps and all are useful as detectors[†] of infrared radiation. Values[17] of E_g at 4 K for these semiconductors are as follows: PbS, 0.286 eV; PbSe, 0.165 eV; PbTe, 0.190 eV. The band structure of each of the three is quite similar, and is shown schematically, in the neighborhood of the minimum energy gap, in Figure 7.9. The gap is direct and is located at the L point of the zone, at which the (111) direction intersects the zone edge. The energy gap in the lead salts is thus located at the edge of the Brillouin zone, rather than at the center or Γ point. (The semiconductor SnTe, for which $E_g = 0.3$ eV at 0 K, has a very similar band structure.) The energy gaps of the lead salts provide thresholds for absorption at wavelengths in the infrared as long as approximately 8 μm.

Solid solutions of SnTe in PbTe are semiconductors[18] whose energy gap E_g displays unusual behavior as a function of the composition of the solid solution. Figure 7.10 shows a plot of the energy gap E_g (at 12 K) of PbTe–SnTe solid solutions as a function of atom per cent SnTe in the crystal. The curve in Figure 7.10 tells us that, if one prepares a series of crystals of PbTe–SnTe solid solutions containing increasing atomic percentages of SnTe, the variation of the energy gap is as follows. The gap decreases from the value for pure PbTe, and becomes zero for a composition of approximately 34 at. % SnTe. For compositions with SnTe concentrations larger than 34 at. %, the energy gap increases until it has the value for pure SnTe. This unusual behavior of the energy gap is useful because it allows one to choose the value of the gap by specifying the tin–lead ratio in the

---

[†] It should be pointed out that the usual polycrystalline lead salt infrared detectors are not exhibiting simple photoconductivity. The physics of these devices is quite complex and the interested reader is referred to the book by Moss *et al.*, Reference 1, pages 380–384.

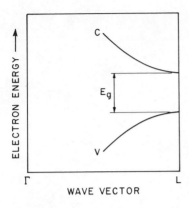

Figure 7.9. Band structure (schematic) of the lead salt semiconductors in the neighborhood of the direct gap E_g at the L point of the Brillouin zone. The conduction and valence bands are labeled C and V, respectively.

PbTe–SnTe solid solution. In particular, it has proved possible to prepare[19] PbTe–SnTe crystals with an energy gap as small as 0.04 eV at 12 K. These semiconductors with very small energy gaps have threshold wavelengths for absorption as long as 30 μm and so are used for detectors in the long-wave infrared region of the spectrum. The same type of variation of energy gap with composition has been observed in semiconducting solid solutions of SnSe in PbSe,[20] and in solid solutions[21] of CdTe (a semiconductor with the zinc blende crystal structure and a value of $E_g = 1.6$ eV at 0 K) and HgTe (a semimetal with the zinc blende crystal structure).

The threshold for intrinsic absorption of photons is determined by the energy gap E_g of the semiconductor. Since the lower limit on realizable values of E_g is a few hundredths of an eV, one turns to photon absorption by bound impurity states to obtain photon absorption at still longer wavelengths. Shallow impurities (e.g., gallium, antimony) in germanium have ionization energies of about 0.01 eV, so their thresholds[22] for extrinsic absorption are in the far infrared, in the neighborhood of 120 μm. Deeper levels[22,23]

Figure 7.10. Energy gap E_g (at 12 K) of solid solutions of SnTe in PbTe plotted as a function of atomic percent SnTe. (Adapted from I. Melngailis and T. C. Harman.[18])

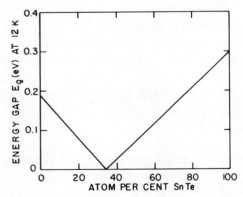

in germanium are obtained by using impurities such as mercury and gold. Mercury produces an acceptor level at 0.087 eV above the valence band and thus mercury-doped germanium absorbs photons at 14 μm and shorter wavelengths. Impurity levels in doped silicon[23a] are also useful, since they have absorption thresholds in the infrared. Shallow impurities (e.g., arsenic, phosphorus) in silicon have ionization energies in the neighborhood of 0.05 eV, so their thresholds are in the 25 μm region of the spectrum.

Photoconductivity in Semiconductors

We have discussed intrinsic and extrinsic absorption of photons in semiconductors and mentioned several examples whose absorption thresholds were determined by the energy gap (intrinsic) or impurity ionization (extrinsic) energy. Intrinsic absorption of a photon produces a photo-generated electron–hole pair, while extrinsic absorption produces free electrons or free holes, respectively, in n- or p-type impurity semiconductors. We turn next to consideration of the use of photogenerated free carriers in various effects used to detect electromagnetic radiation.

The first effect we consider is photoconductivity, which is the change in the conductivity of a solid when irradiated with light. The incident photons produce free charge carriers, electrons or holes, or both, thereby increasing the conductivity of the solid. The conductivity σ of a solid is given by

$$\sigma = q\mu n$$

where q is the electric charge of the carrier, μ is the carrier mobility, n is the density of carriers, and, for simplicity, we assume only one kind of carrier is present. If incident photons produce an additional density Δn of photogenerated carriers, then the increase $\Delta\sigma$ in the conductivity of the solid will be

$$\Delta\sigma = q\mu(\Delta n) \tag{7.8}$$

If, as shown in Figure 7.11, an electric field is placed across the sample, the current I observed with light on the sample will be larger than the current measured with the sample in the dark. Many semiconductors exhibit photo-conductivity,[24] and the phenomenon is the basis of many applications.

We will consider[25] a photoconductor of cross-section area A and length d, as shown in Figure 7.11. A uniform electric field V/d is placed across the sample, which is irradiated uniformly with light such that L photons

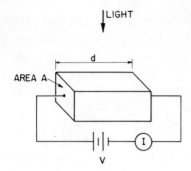

Figure 7.11. Experimental arrangement for observing photoconductivity. The sample is of cross section area A and an electric field V/d is applied across the sample.

are absorbed per second per unit volume of the sample. For simplicity, we consider that these photons produce one kind of photoexcited carrier (say electrons), which then traverse the sample under the influence of the applied electric field. We define the photoconductive gain[26] G of the photoconductor as the ratio

$$G = \frac{\text{number of charges crossing the sample per second}}{\text{number of photons absorbed per second}} \qquad (7.9)$$

The increase $\Delta\sigma$ in the conductivity is given by equation (7.8), so the increase ΔJ in the current density, also called the photocurrent density, is

$$\Delta J = (\Delta\sigma)(V/d) = q\mu(\Delta n)(V/d) \qquad (7.10)$$

The photocurrent $\Delta I = A(\Delta J)$, where A is the cross section-area of the sample. The increase Δn in the carrier density is given by

$$\Delta n = L\tau_e \qquad (7.11)$$

where τ_e is the electron lifetime. Since τ_e is the electron lifetime, L electrons will last τ_e sec. From these results we have that the number of carriers crossing the sample in unit time is

$$(\Delta I/q) = \mu A(\Delta n)(V/d) = \mu A L\tau_e(V/d) \qquad (7.12)$$

and the number of photons absorbed per unit time is LAd because the volume of the sample is Ad. Substituting into the definition (7.9) of the photoconductive gain gives

$$G = \mu\tau_e V/d^2 \qquad (7.13)$$

Since the drift velocity of the electrons is $\mu(V/d)$, the transit time τ_t of the electrons across the sample is

$$\tau_t = \frac{d}{\mu(V/d)} = \frac{d^2}{\mu V} \tag{7.14}$$

and the photoconductive gain G may be expressed as

$$G = \tau_e/\tau_t \tag{7.15}$$

the ratio of the electron lifetime to the transit time across the sample.

We see from equation (7.15) that, if the transit time τ_t is short and the lifetime τ_e is long, many electrons will cross the sample during one lifetime, giving a photoconductive gain greater than unity (since the carrier is replenished at an electrode). A large value of G means a larger current through the photoconductor per absorbed photon, so a large G means a greater photosensitivity for the photoconductor. We can see from equations (7.14) and (7.15) that a high mobility μ and a long carrier lifetime τ_e for a semiconductor favor a high value of the photoconductive gain, which may reach a value[27] of 10^3 in a material in which the carrier lifetime is long.

The sensitivity of a photoconductor is not usually specified in terms of photoconductive gain. Generally, one specifies the incident photon power which will produce a photogenerated signal just equal to the noise of the device under specified conditions.[28,29] This incident power is called the noise equivalent power, abbreviated NEP; the smaller the NEP of a photoconductor, the higher its sensitivity. Another figure of merit that is used is the detectivity D, defined by the relation $D = (\text{NEP})^{-1}$; the larger the detectivity, the higher the sensitivity of the photoconductive device. The responsivity R of a device is the ratio of the signal voltage to the incident power producing the signal.

Figure 7.12 shows an example[30] of photoconductivity in single-crystal solid solutions of SnTe in PbTe. The curves show the responsivity (in volts per watt) as a function of the wavelength of the incident radiation at two temperatures and for two values of the atom percent x of SnTe in the solid solution. The curve at 4.2 K has a threshold at approximately 14 μm in agreement with the energy gap of approximately 0.09 eV of this solid solution with $x = 0.17$. The curves at 77 K show that the threshold for the solid solution with $x = 0.20$ is at a longer wavelength (smaller photon energy) than the threshold for the solid solution with $x = 0.17$. This is as expected since (from Figure 7.10) the solid solution with the larger concentration of SnTe will have the smaller energy gap.

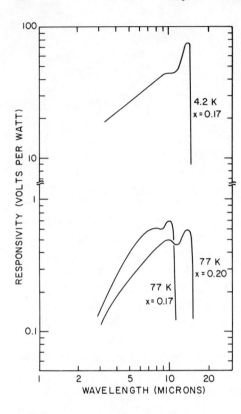

Figure 7.12. Responsivity (in volts per watt) as a function of incident photon wavelength λ (in microns) for solid solutions of SnTe in PbTe; x is the atomic percent of SnTe in the solid solution. (After Melngailis and Harman, Reference 30.)

Photodiodes

Intrinsic photoconductivity in a semiconductor can be exploited also in a nonhomogeneous structure like a semiconductor p–n junction. Consider the photodiode,[31] a reverse-biased p–n junction in which the junction region is illuminated with photons of energy $\hbar\omega$ larger than the band gap of the semiconductor. As shown in Figure 7.13, the absorbed photons generate electron–hole pairs. Those generated within a diffusion length of the depletion layer of the junction (where the photogenerated electrons are on

Figure 7.13. Reverse-biased p–n junction as a photodiode. Electrons (●) and holes (○) are generated on both sides of the junction by photons of energy $\hbar\omega$ and diffuse to the junction. The electric field sweeps the minority carriers down the energy barrier, thereby increasing the reverse saturation current.

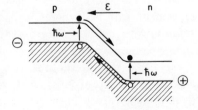

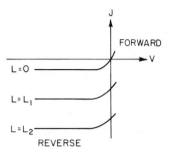

Figure 7.14. Current density J as a function of voltage V shown schematically for an illuminated reverse-biased p–n junction. The illumination intensity L is such that L_2 is larger than L_1.

the p side and photogenerated holes are on the n side) are collected and increase the respective minority carrier densities. These minority carriers are then swept down the energy barrier by the electric field at the junction, thereby increasing the reverse saturation current density. The reverse current through the junction is therefore larger when illuminated than in the dark. This is seen from the current–voltage plot of an illuminated reverse-biased p–n junction, shown schematically in Figure 7.14 for three values $L = 0$, $L = L_1$, and $L = L_2$ of the illumination intensity, where L_2 is greater than L_1. From Figure 7.14, we see that the reverse current through the junction is essentially independent of voltage, but increases with increasing illumination intensity L.

The photodiode is utilizing intrinsic absorption of photons by a semiconductor, and so will have a threshold for photon absorption determined by the band gap of the semiconductor. The reverse-biased p–n junction can also be used to detect energetic particles (e.g., electrons, protons). Such particles, with energies many times the band gap energy, are absorbed over a large distance in the semiconductor, creating many electron–hole pairs. These pairs are collected in the large depletion layer built into these devices and increase the reverse saturation current density.

Photovoltaic Devices

Consider a p–n junction with no external bias applied, as shown in Figure 7.15. The junction is in equilibrium, the energy barrier between n and p sides is eV_0 (where V_0 is the diffusion potential), and the built-in electric field is $\mathscr{E}$. The junction is then illuminated by photons whose energy is greater than the band gap, creating electron–hole pairs. The minority carriers, as also shown in Figure 7.15, are swept down the energy barrier by the built-in electric field. The net result is the separation of the photogenerated electrons and holes, as shown in Figure 7.16, by the p–n junction.

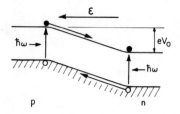

Figure 7.15. Semiconductor p–n junction, without external bias, which is initially at equilibrium, with energy barrier eV_0 (where V_0 is the diffusion potential) and built-in electric field $\mathscr{E}$. Photons of energy $\hbar\omega$ larger than the semiconductor band gap create electrons (●) and holes (○). The minority carriers are swept down the energy barrier.

These separated photogenerated carriers set up an electric field $\mathscr{E}'$ which is opposite in direction to the built-in electric field $\mathscr{E}$. The net electric field across the junction is thus reduced to the value $\mathscr{E} - \mathscr{E}'$. This in turn means that the difference in electrostatic potential between the p and n sides of the junction is reduced from its equilibrium value V_0 to a smaller value $V_0 - V_f$, and the energy barrier is reduced to $e(V_0 - V_f)$, as shown in Figure 7.16. The effect is the same as if a forward-bias voltage V_f were applied to the junction. The overall result is the development of an open-circuit voltage V_f, in the forward direction (the n side negative) across the illuminated junction. The appearance of this voltage across an illuminated p–n junction is called the photovoltaic effect.[32]

We can see also from Figure 7.16 that the maximum value of the open-circuit voltage V_f is the diffusion potential V_0, obtained when the illumination is intense, enough that the photogenerated electric field $\mathscr{E}'$ cancels completely the built-in field $\mathscr{E}$. Then the energy barrier between the p and n sides goes to zero, or, alternatively, the voltage V_f appearing across the illuminated junction is equal to V_0. Further, the maximum value of V_0 in a given semiconductor is E_g/e, where E_g is the band gap. This is because, from equation (2.4) of Chapter 2,

$$eV_0 = E_F(n) - E_F(p) \tag{2.4}$$

where $E_F(n)$ and $E_F(p)$ are, respectively, the Fermi energies on the n and p sides of the junction in equilibrium. Since $E_F(n) - E_F(p)$ will be largest when the Fermi energies are close to the band edges, the maximum value of

Figure 7.16. Semiconductor p–n junction, without external bias, under illumination. Photogenerated electrons (●) and holes (○) produce an electric field $\mathscr{E}'$ opposite to the built-in electric field $\mathscr{E}$. An open-circuit forward (n side negative) voltage V_f appears across the illuminated junction.

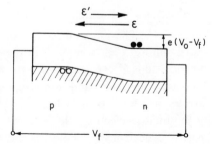

$E_F(n) - E_F(p)$ is the energy gap E_g, so the maximum value of eV_0 is the energy gap. Hence, the larger the energy gap of the semiconductor, the larger the maximum open-circuit voltage in the photovoltaic effect.

The photovoltaic effect in a p–n junction can be used to detect photons in various regions of the spectrum by observing the photogenerated voltage. The old-fashioned exposure meter of photography utilized the photovoltaic effect in the semiconductor hexagonal (grey) selenium, which has an energy gap of about 1.8 eV at 300 K. Junctions fabricated in semiconducting solid solution of SnTe in PbTe have been used as photovoltaic detectors[33] of infrared radiation with wavelengths as long as 30 μm.

If the p–n junction exhibiting the photovoltaic effect is connected across a load resistor, a current will flow in the external circuit when the junction is illuminated. The illuminated junction can therefore deliver power to an external load and the photovoltaic device operating in this manner can convert the energy of the incident photons into electrical energy. Since the junction is now delivering current into an external circuit, the photogenerated voltage appearing across the load will be smaller than the open-circuit photovoltage V_f. The choice of a semiconductor for a specific application of the photovoltaic effect will depend on the distribution in energy of the incident photons. For a particular distribution, there will be an optimum value of the energy gap E_g of the semiconductor. If E_g is too small, some of the energy of the incident photons will be wasted in the production of high-energy hot electron–hole pairs which lose their energy by emitting phonons. If E_g is too large, many photons will pass through the semiconductor without being absorbed.

If the photons whose energy is to be converted into electrical energy come from the sun, then a photovoltaic junction operating as described above is called a solar cell.[34,35] For terrestrial conversion of solar photons,[34] the optimum value[36] of $E_g = 1.4$ eV, suggesting GaAs. However, since many other factors (including economics) influence the choice, silicon ($E_g = 1.1$ eV) and other materials are under intensive development[37] for solar cells for terrestrial use.

Other Applications of Intrinsic Photoconductivity

In this section, we discuss three familiar processes which depend on intrinsic photoconductivity in different semiconductors. These processes are (1) electrophotography, one form of which is the Xerox process; (2) the light-sensitive unit in a television camera tube; (3) photography.

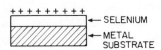

Figure 7.17. First step of the Xerox process, in which a positive charge is built up on the surface of an amorphous selenium film by a corona discharge. (Relative dimensions are not to scale.)

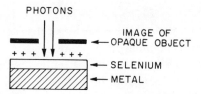

Figure 7.18. Second step of the Xerox process, in which photogenerated carriers in the selenium discharge the illuminated regions, leaving positive surface charge in the dark regions.

We begin with a consideration of electrophotography, a technique whose perhaps most familiar form is the Xerox process[38,39] of copying. The photoconductor involved is amorphous selenium, a semiconductor with an energy gap[40] of approximately 2.1 eV at room temperature, which is vacuum deposited as a thin film on a metal substrate. The selenium has a very high resistivity, of the order[41] of 10^{16} Ω cm, in the dark. The first step in the process is the buildup of a positive charge on the selenium by a high voltage corona discharge, as shown in Figure 7.17. This step is referred to as sensitization. The second step, shown in Figure 7.18, is exposure of portions of the selenium layer to light, generating electron–hole pairs, thereby increasing the conductivity of the illuminated regions. (In the amorphous selenium, the electrons are trapped, and current flow is due to holes alone.) The illuminated regions discharge and the dark regions do not; the resistivity of the semiconductor is so high that charge does not leak off laterally. The result of this step is a positive charge distribution on the surface of the selenium which is a replica of the dark portion of the image to be copied. The third step, shown in Figure 7.19, dusts a dark pigment (the toner) onto the selenium, where it is attracted only to the charged regions which were not illuminated in the second step. The result is a positive image of the material to be copied, in which the dark parts of the original appear dark in the image. The fourth step, shown in Figure 7.20, is the transfer

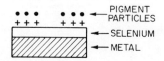

Figure 7.19. Third step of the Xerox process, in which dark pigment particles (●) are attracted to the positively charged regions of the selenium.

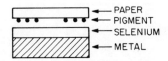

Figure 7.20. Fourth step of Xerox process, in which the dark pigment is transferred to paper and fixed, forming a positive image copy.

of the dark pigment image onto a piece of ordinary paper. The pigment is then fixed onto the paper by heat, forming a permanent positive image which is the final copy.

Our next topic is the light-sensitive unit in a television camera; this unit is often a type of tube called a vidicon.[42,43] The vidicon tube contains a thin layer of a photoconductor sensitive in the visible region of the spectrum. Examples are As_2S_3 and PbO, whose energy gaps[40] are, respectively, approximately 2.5 and 2.3 eV. The photoconductor is prepared in polycrystalline form by evaporation in a poor vacuum, resulting in a photoconductive layer with a very high resistivity of the order of $10^{15}\ \Omega$ cm. Figure 7.21 shows the arrangement of the photoconductive layer in the vidicon tube. Photons are incident on various parts of the photoconductive layer, passing first through a transparent conducting electrode. The photoconductor is scanned from the rear by an electron beam (which moves sequentially over the rear surface), "depositing" negative charges which do not leak off due to the high resistivity of the photoconductive layer. If photons are incident on some part of the front surface of the photoconductive layer, intrinsic electron–hole pairs are generated, increasing the conductivity of the illuminated region. This increased conductivity allows the stored negative charge on the rear surface to be conducted away by an applied positive bias. As a result, when the electron beam returns on its next scan, it "deposits" more electrons in order to return that part of the photoconductive layer to its original condition. The current in the electron beam is thus momentarily increased, giving a pulse of beam current which is essentially information on the light intensity falling on that particular part of the photoconductive layer. In this way, the output signal of the vidicon tube contains information concerning the light falling on the tube. With proper electronic processing, this light pattern information is reproduced on the cathode ray tube of the television receiver.

A newer type[42] of vidicon involves a different type of photosensitive

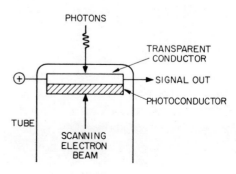

Figure 7.21. Schematic view of a vidicon tube.

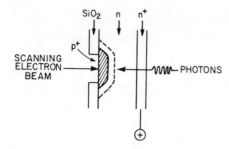

Figure 7.22. Schematic view of one p^+–n junction in a silicon vidicon. The p^+ region is shown shaded, the SiO_2 is an insulator, and the photons pass through the thin n^+ electrode. The dotted line represents the depletion layer of the junction, almost all of which is in the relatively lightly doped n region.

surface. The surface is a wafer of single-crystal silicon on which are fabricated about 10^6 p^+–n junctions using integrated circuit technology. The configuration used in this silicon vidicon is shown in Figure 7.22; photons pass through a thin n^+ layer which serves as an electrode. The scanning electron beam reverse-biases each p^+n junction, extending the space charge region into the more lightly doped n-type layer. If photons are incident on the diode between scans by the electron beams, the depletion region is filled with photogenerated carriers, thus reducing the reverse bias on the junction. On the next scan, the electron beam "deposits" additional electrons, resulting in a pulse of current containing video information. The silicon vidicon exhibits high reliability and response into the infrared because of the relatively small energy gap of silicon.

The third application we will consider is photography. The usual photographic process depends on intrinsic photon absorption in the silver halides AgCl and AgBr. These compounds are ionic semiconductors with thresholds for photon absorption[44] at about 2.5 eV at room temperature. A photographic emulsion[45] consists of silver halide microcrystals in a binding material. A current model[46] of the photographic process in a silver halide AgX (where X is a halogen) is as follows. A photon of sufficient energy is absorbed by an AgX crystal, producing an electron (e^-) and a hole. The electron is then trapped, while the hole can combine with an X^- ion to form a neutral X atom. The trapped electron then combines with a mobile Ag^+ ion,

$$Ag^+ + e^- \rightarrow Ag \qquad (7.16)$$

to form a silver atom. A second photon then produces another electron–hole pair. The hole produces a second X atom, which reacts with the first X atom to produce an X_2 molecule, which escapes. The electron is trapped by the silver atom above,

$$Ag + e^- \rightarrow Ag^- \qquad (7.17)$$

forming an Ag$^-$ ion. An Ag$^+$ ion then combines with the Ag$^-$ ion,

$$Ag^+ + Ag^- \rightarrow Ag_2 \qquad (7.18)$$

to form a stable two-atom molecule of silver. This Ag$_2$ is the basic stable unit of silver, and, on repetition of the process by additional photons, becomes a speck of metallic silver. This silver speck, present in an AgX silver halide microcrystal which has been exposed to light, is called the latent image.

When the latent image is developed, it is exposed to a chemical reagent which converts AgX to silver, in such a way that AgX grains that have been exposed to light are converted at a higher rate[47] than unexposed grains. The result is that the latent image is converted into a metallic silver image, with the silver density highest in the regions that had the highest illumination level.

Summary on Semiconductor Photon Detectors

It is useful to summarize how the physics of the semiconductors involved yields useful photosensitive devices. First, the existence of an energy gap between valence and conduction bands is important from the device standpoint. The photon absorption spectrum for intrinsic excitation has a sharp threshold which can be chosen to match the spectral region of interest. If the energy gap is sufficiently large relative to $k_B T$, the density of thermally excited intrinsic carriers will be small. This allows the photogenerated carriers to change the electrical conductivity appreciably. Second, the appreciable lifetime of carriers in semiconductors allows the maintenance of significant concentrations of photogenerated carriers at usefully low light intensities. Third, the existence of discrete impurity levels, with ionization energies smaller than the band gap energy, yields photoconductivity at low photon energies. Finally, the semiconductor p–n junction permits a variety of useful devices to be made (e.g., photovoltaic detectors) which depend directly on the properties of the junction (e.g., the built-in electric field) for their operation.

Emission of Photons in Semiconductors

We now turn to the use of semiconductors as generators of electromagnetic radiation, discussing some processes whereby an electron and a

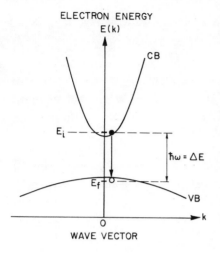

ELECTRON ENERGY
E(k)

Figure 7.23. Band-to-band recombination of an electron (●) and a hole (○) in a direct-gap semiconductor with the emission of a photon of energy ΔE. The initial and final energies of the electron are, respectively, E_i and E_f. The conduction and valence bands are labeled CB and VB, respectively.

hole recombine with the spontaneous emission of a photon. These processes, collectively called luminescence, are the basis of several useful light sources.

We consider first the recombination process in which a free electron in the conduction band makes a transition to an empty (i.e., hole) state in the valence band, emitting a photon. This process is called band-to-band recombination[48,49] of the electron and hole, and can occur in both direct-gap and indirect-gap semiconductors, as shown in Figures 7.23 and 7.24. In these figures, the energy difference ΔE is defined as

$$\Delta E = E_i - E_f \tag{7.19}$$

where E_f and E_i are, respectively, the energies of the final and initial states

Figure 7.24. Band-to-band recombination of an electron (●) and a hole (○) in an indirect-gap semiconductor, with the emission of a photon of energy $\hbar\omega = \Delta E - \hbar\Omega$ and of a phonon of energy $\hbar\Omega$. The quantity $\Delta E = E_i - E_f$, where E_i and E_f are, respectively, the initial and final energies of the electron. The conduction and valence bands are labeled CB and VB, respectively.

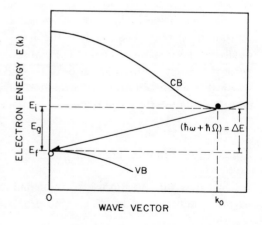

of the electron. Figure 7.23 shows a band-to-band transition of an electron with no change of wave vector (if the photon wave vector is neglected) and the emission of a photon of energy $\hbar\omega = \Delta E$. Since the minimum energy gap is direct, the energy spectrum of the emitted photons will have a low-energy threshold at the band gap energy. The transition shown in Figure 7.23 can take place with emission of a photon of energy ΔE somewhat larger than the energy of the gap if the electron and hole involved have energies above those of their respective band edges.

Figure 7.24 shows a band-to-band transition in an indirect-gap semiconductor. In this process an electron of initial energy E_i and initial wave vector k_i makes a transition to a final state of energy E_f and wave vector k_f with the emission or absorption of a phonon of energy $\hbar\Omega$ and wave vector k'. Conservation of energy requires that

$$E_i = E_f \pm \hbar\Omega + \hbar\omega \qquad (7.20)$$

where $\hbar\omega$ is the energy of the emitted photon. The plus sign refers to the process in which a phonon is emitted and the minus sign to that in which a phonon is absorbed. Conservation of wave vector requires that

$$k_i = k_f \pm k' \qquad (7.21)$$

where the plus and minus signs have the same meaning as in equation (7.20). From equation (7.20), the energy spectrum of emitted photons will have a low-energy threshold when $E_i - E_f = E_g$, the gap energy; the lowest-energy photon emitted will have an energy

$$\hbar\omega = E_g - \hbar\Omega \qquad (7.22)$$

for the transition in which a phonon of energy $\hbar\Omega$ is emitted. Figure 7.24 shows the emission of the minimum-energy photon of energy $E_g - \hbar\Omega$ with the simultaneous emission of a phonon of wave vector k_0. As described earlier for indirect absorption of photons, the indirect band-to-band emission of photons with phonon participation is a three-body process. As such, indirect recombination is less probable than direct electron–hole recombination. An estimate[50] of the electron lifetime with respect to band-to-band recombination in a direct-gap semiconductor is 10^{-6} sec. A similar estimate[50] for an indirect-gap semiconductor is of the order of 0.5 sec, showing that the indirect process is much less probable than the direct process.

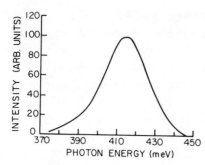

Figure 7.25. Direct band-to-band recombination luminescence in InAs at 77 K. (After Mooradian and Fan.[51])

From the point of view of applications, the main conclusion is that band-to-band recombination of electrons and holes in a semiconductor results in the emission of photons. In both the direct- and indirect-gap cases, there is a continuous spectrum of photon energies above a low-energy threshold at approximately the energy of the band gap. The specific shape and width of the spectrum of emitted photons will depend on the details of the transition probability, band densities of states, impurity content, etc. Figure 7.25 shows the luminescence[51] from n-type InAs (a direct-gap semiconductor), with an electron density of 9×10^{16} cm^{-3}, at 77 K. The photon emission is due to direct radiative recombination of free electrons and free holes. The peak of the emission spectrum at 0.41 eV agress well with the value[52] of the energy gap; the emission at lower photon energies is due to other processes.[53]

Photon emission in semiconductors can also take place by recombination transitions between a band state and an impurity level, and between two impurity levels. The former type of transition is known[54] to occur in both n- and p-type GaAs. Figure 7.26 shows schematically the recombination of an electron in the conduction band of GaAs with a hole bound to an acceptor level about 0.03 eV above the valence band edge. The emitted photon energy $\hbar\omega$ is thus less than the band gap energy by about 0.03 eV. The latter type[55] of transition is also known to occur in various other semiconductors.

Figure 7.26. Schematic recombination of an electron ($\bullet$) in the conduction band (CB) with a hole ($\bigcirc$) bound to acceptor slightly above the valence band (VB). The energy $\hbar\omega$ of the emitted photon is less than the band gap energy E_g. (The acceptor ionization energy is exaggerated for clarity.)

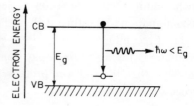

p–n Junction Luminescence

One of the most important applications of luminescence in semi-conductors is in a *p–n* junction, known as a light-emitting diode[56-58] or LED. In our earlier discussion of photon emission in semiconductors we did not discuss the way in which the recombining electrons and holes were produced. In this section, we consider a technologically important device in which the free electrons and holes are injected across a *p–n* junction. Radiative recombination of electrons and holes then produces photons.

Consider the *p–n* junction at equilibrium shown in Figure 7.27, in which both the *p* and *n* sides are degenerate. If a large forward bias is applied to the junction, the energy barrier to the flow of electrons and holes is reduced as shown in Figure 7.28. Electrons are injected from the *n* side to the *p* side, where they recombine with holes, emitting a photon whose energy $\hbar\omega$ is approximately the gap energy of the semiconductor. A similar process takes place with holes injected into the *n* side, also producing photons. In this way, photon emission takes place in the vicinity of the junction. (Actually, the injection of electrons from the *n* side to the *p* side is more complicated than the simplified description given above. At forward bias values smaller than that which produces a significant injection current of holes and electrons, electrons may tunnel from *n* to *p*, either directly or via an intermediate state, with emission of a photon.[59-61])

From the point of view of applications of light-emitting diodes, the main interest is in the emission of photons in the visible region of the spectrum. For this reason, interest has been concentrated on semiconductors whose band gaps are greater than about 1.5 eV; examples[56] are III–V

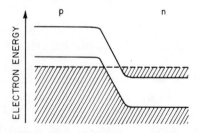

Figure 7.27. Band diagram of a degenerate *p–n* junction at equilibrium. The filled electron states in the conduction band on the *n* side are shown shaded; the empty (hole) states in the valence band on the *p* side are unshaded.

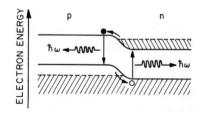

Figure 7.28. Band diagram of a degenerate *p–n* junction under forward bias. Electrons (●) are injected from *n* to *p*, and holes (○) from *p* to *n*, with emission of photons by radiative recombination.

semiconductors and their solid solutions. Since, as discussed earlier, the radiative recombination of electrons and holes is more probable in a direct-gap semiconductor, such materials (e.g., GaAs and various III–V solid solutions) have been the object of considerable research. A figure of merit[58] for a light-emitting diode is the external quantum efficiency η_E, defined as the ratio of the number of photons emitted to the number of electrons flowing across the junction. Representative values[58] (as of 1972) for η_E are in the range 0.01–0.001. Considerable effort[62] has also been made to improve the efficiency of light-emitting diodes with large energy gaps by using semiconductors with indirect gaps. These large gaps correspond to the green region of the spectrum, where the eye is most sensitive.

Light Amplification by Stimulated Emission of Radiation

In our discussion of photon emission from semiconductors, we have been considering only spontaneous emission. We now consider the stimulated emission[63] of photons. Stimulated emission is the transition of an electron from a state of energy E_2 to a state of lower energy E_1 under the influence of an incident photon of energy $\hbar\omega_{21}$, where

$$\hbar\omega_{21} = E_2 - E_1 \qquad (7.23)$$

The incident photon of energy $\hbar\omega_{21}$ stimulates the electronic transition and thereby the emission of a photon whose energy is also $\hbar\omega_{21}$. In certain solid state systems, n_i "input" photons of a particular energy incident on the system will produce the stimulated emission of n_o "output" photons of the same energy. If n_o is larger than n_i, then the process is one of amplification of the number of photons by the process of stimulated emission. The word LASER is an acronym for *l*ight *a*mplification by *s*timulated *e*mission of *r*adiation, that is, the process described above.

We begin by discussing some general ideas[64–67] concerning lasers. Consider a quantum system (as yet unspecified) in which there are three electronic energy levels in a solid, as shown schematically in Figure 7.29. The three electronic energy levels are E_0 (ground state), E_1, and E_2, and various possible electronic transitions in this level system are indicated by arrows. Transition A is from level E_0 to level E_2 and may be excited by the absorption of a photon of energy $E_2 - E_0$ by an electron in the ground state. Transition B from level E_2 to level E_1 is radiationless (phonons are given off) and is characterized by a relaxation time t_{21}. Transition C from

Figure 7.29. Energy level diagram with three electronic levels. The ground state at energy E_0 has an electron population N_0, the intermediate state at energy E_1 has an electron population N_1, and the upper state at energy E_2 has an electron population N_2. Transition A is excited by the absorption of a photon of energy $E_2 - E_0$ by an electron in the ground state. Transition B is radiationless; phonons are given off. Transition C emits photons of energy $E_1 - E_0$.

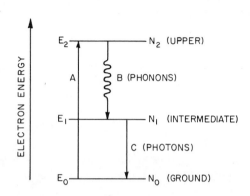

level E_1 to the ground state emits a photon of energy $\hbar\omega_{10} = E_1 - E_0$ and is described by a relaxation time t_{10}. The electron populations of the three levels are denoted by N_0, N_1, and N_2.

The transition that emits photons is C, so we want to consider the possibility of the amplification of photons of energy $\hbar\omega_{10}$ by the use of this transition. In the amplification process an "input" photon of energy $\hbar\omega_{10}$ will stimulate the emission of additional photons of the same energy. Considering the two levels E_1 and E_0, with electron populations, at some instant, N_1 and N_0, an incident photon of energy $\hbar\omega_{10}$ is just as likely to be absorbed in stimulating an $E_0 \to E_1$ transition as it is to stimulate the emission of an additional photon by an $E_1 \to E_0$ transition. For amplification to occur, it is necessary that more electrons in the system be able to emit an $\hbar\omega_{10}$ photon than are able to absorb such a photon. This will be true if there are more electrons in the E_1 level than in the E_0 level. In other words, we want the electron population N_1 of the E_1 level to be larger than the population N_0 of the ground state.

We can see in more detail that this is so as follows.[†] The energy given off, per unit time per unit volume, due to stimulated emission from transitions from E_1 to E_0 is given by

$$\hbar\omega_{10}N_1P_{10} \qquad (7.24)$$

where P_{10} is the transition probability[68] per unit time of the stimulated emission transition from $E_1 \to E_0$. Similarly, the energy absorbed, per unit time per unit volume, due to absorption of photons of energy $\hbar\omega_{10}$ is equal to

$$\hbar\omega_{10}N_0P_{01} \qquad (7.25)$$

[†] In this argument, statistical degeneracies are neglected for simplicity.

where P_{01} is the transition probability per unit time for absorption. The net energy given off, per unit volume per unit time, due to absorption and stimulated emission is just the difference

$$\hbar\omega_{10} P_{10} N_1 - \hbar\omega_{10} P_{01} N_0 \qquad (7.26)$$

Since the transition probabilities P_{10} and P_{01} are equal,[69] the net energy given off is

$$\hbar\omega_{10} P_{10} (N_1 - N_0) \qquad (7.27)$$

If a population inversion exists, N_1 is larger than N_0, net energy is given off due to stimulated emission, and photon amplification can take place. (In the discussion above, spontaneous emission was not included since stimulated emission is the faster process and so dominates spontaneous emission.)

Statistical mechanics tells us that, at equilibrium, the ratio N_1/N_0 is less than unity, and the ground state has the larger electron population. If N_1 should be made, in some way, larger than N_0, then the system would not be in equilibrium, and we would have a situation called a population inversion. We conclude that a population inversion such that N_1 is larger than N_0 is necessary if the $E_1 \rightarrow E_0$ transition is to be used for the amplification of $\hbar\omega_{10}$ photons.

Under certain circumstances, the three-level system in Figure 7.29 allows the achievement of the desired population inversion. Consider the excitation, say optically, of electrons from the ground state to the upper state. Suppose the rate at which electrons in the E_2 (upper) level decay to the E_1 (intermediate) level is very fast compared to the rate at which electrons in the E_1 level decay to the E_0 (ground) state. This is equivalent to the statement that the relaxation times t_{21} and t_{10} are such that

$$t_{10} \gg t_{21} \qquad (7.28)$$

Equation (7.28) is a statement that the electron lifetime in the intermediate state is long compared to the electron lifetime in the upper state. If the condition (7.28) is satisfied, then, as electrons are excited into the upper state, they will decay rapidly into the intermediate state via transition B, and then, much more slowly, into the ground state via transition C. The result will be an accumulation of electrons in the intermediate state relative to the ground state, or, in other words, a population inversion in which N_1 is larger than N_0.

It is clear that the existence of the intermediate state in the three-level system is essential to the attainment of the desired population inversion. If there were only the two levels E_2 and E_0, then incident photons of energy $E_2 - E_0$ could produce, at most, a 50% population of the E_2 level. This is because such a photon is just as likely to stimulate an $E_0 \rightarrow E_2$ absorption transition as it is to stimulate an $E_2 \rightarrow E_0$ emission transition. The existence of the intermediate state, coupled with the difference in lifetime expressed in equation (7.28), allows the achievement of the population inversion between the intermediate and ground states in the three-level system.

Once the population inversion is achieved, and N_1 is larger than N_0, then spontaneously emitted photons of energy $\hbar\omega_{10}$ can cause stimulated emission of additional $\hbar\omega_{10}$ photons with concurrent amplification. This process is what is called laser action or lasing.

In the three-level system in Figure 7.29, we can see that, in order to obtain a population inversion, more than half of the electrons originally in the ground state must be excited to the intermediate state (via the upper state). This means that considerable energy must be put into the exciting transition A in order to achieve the population inversion, and leads to a relatively inefficient[64] laser when, as in the three-level case, the final state of the emission transition is also the ground state. The four-level system[64] shown in Figure 7.30 is one that avoids this disadvantage. Considering such a set of energy levels in a solid, electrons are excited by photons from the ground state E_0 to the upper state E_3 in transition A. A fast radiationless transition B from level E_3 to level E_2 is characterized by a relaxation time t_{32}. The slow transition C from level E_2 to level E_1 emits a photon of energy $\hbar\omega_{21} = E_2 - E_1$ and is described by the relaxation time t_{21}. Finally, a fast radiationless transition D, with relaxation time t_{10}, returns the electron to the ground state. If the relaxation times are such that

$$t_{21} \gg t_{32}, t_{10} \tag{7.29}$$

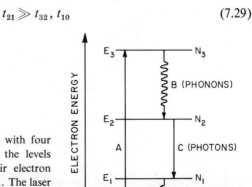

Figure 7.30. Energy level diagram with four electronic levels. The energies of the levels are E_0, E_1, E_2, and E_3, and their electron populations are N_0, N_1, N_2, and N_3. The laser transition C emits photons of energy $\hbar\omega_{21}$ $= E_2 - E_1$.

a population inversion with N_2 larger than N_1 may be achieved. Laser action, with the stimulated emission of photons of energy $\hbar\omega_{21}$, is obtained by transition C, in which the final state E_1 is relatively unpopulated as compared to the ground state E_0 in the three-level scheme.

Solid State Lasers

A very important class of laser systems, which exhibit three- and four-level schemes similar to Figures 7.29 and 7.30, is that of solids containing small amounts of an impurity atom. The host solid may be a crystal (such as Al_2O_3) or a glass, and the impurity atom is generally one in which electronic transitions can take place between states of inner, incompletely filled electron shells. An example of such a three-level scheme is that of the Cr^{+3} ion in Al_2O_3 (ruby); a four-level system is that of the Nd^{+3} ion in yttrium aluminum garnet (YAG).

The energy level scheme[70] of the Cr^{+3} ion in Al_2O_3, the ruby laser, is shown in simplified form in Figure 7.31. At just less than 2 eV above the ground state is a very closely spaced pair of ^2E states, shown as a single level in Figure 7.31. Above ^2E there are two broad bands of states, referred to as "blue" and "green." In the operation of the ruby laser, electrons are optically excited (transitions A) by pumping from the ground state to the blue and green bands by a broad-spectrum photon source like a xenon

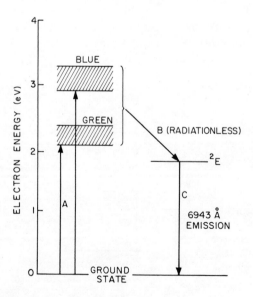

Figure 7.31. Simplified energy level scheme of the Cr^{3+} ion in Al_2O_3 (ruby). The ^2E state is actually a closely spaced (3.6 meV) pair; the shaded regions are broad bands of electron states; arrows represent electronic transitions.

Figure 7.32. Schematic view of cylindrical laser crystal. The mirrored surfaces are shown shaded.

— AXIS

lamp. The lifetime of an electron in these excited states is about 10^{-7} sec, after which they decay via radiationless transition B to the intermediate ^2E state with the emission of phonons. Considering the ^2E states as a single level, the lifetime of an electron in the ^2E state is relatively long, about 5×10^{-3} sec. Electrons accumulate in the intermediate state as pumping continues, and a population inversion of the intermediate state relative to the ground state is produced. Laser action takes place, via transition C in Figure 7.31, with the emission of a photon of wavelength 6943 Å, the familiar red line of the ruby laser.

In practice, single crystals of ruby in a cylindrical shape are constructed so that the two ends are plane and parallel, and silvered[†] on the outside to act as mirrors, as shown schematically in Figure 7.32. Assume that the pump radiation has effected a population inversion via transitions A and B in Figure 7.31. There will eventually (in a small fraction of a second) be spontaneous emission of photons due to transition C in Figure 7.31. Many of these photons will leave the cylinder through the sides, but some will travel along the axis of the rod, being reflected at the ends and causing stimulated emission of photons of wavelength 6943 Å via transition C. As these photons traverse the rod (which is still being pumped), there is an "avalanche" of stimulated emission of photons, and amplification takes place. The photons are extracted from the crystal rod by having one mirror slightly transparent. The fact that the output of the laser is due to stimulated, as opposed to spontaneous, emission of photons results in the coherence[71] of the laser emission.

While ruby is still an important solid state laser because of its visible output, there are several other solid state lasers, generally four-level systems with outputs in the near infrared. Discussions of various solid state lasers are to be found in the literature.[72]

Next, we mention some of the solid state physics factors[73] relevant to laser action. Considering a crystal in which the active impurity resides in a passive host, the host crystal must be transparent to both the exciting pump radiation and the laser emission. This means that the host crystal must be transparent from the visible through the near infrared. Suitable materials are generally ionic insulators with large energy gaps, like Al_2O_3,

---

† In practice, external mirrors are more commonly used.

$CaWO_4$, and yttrium aluminum garnet (YAG). Various glasses form another such group of hosts. (The energy level scheme of the active impurity ion must also be suitable, but that is more a problem in atomic physics than in solid state physics even though, of course, the energy levels of the free ion will be modified by interaction with the host crystal potential.) Last, the phonon spectrum of the host crystal will influence the details of the nonradiative transitions in the energy level scheme used for laser action.

Semiconductor Injection Lasers

A semiconductor injection laser[74-76] also involves an inversion of the electron population between two bands of energy levels. However, in this case, the population inversion is accomplished by the injection of electrons into the p-type region of a p–n junction, rather than with the optical excitation used in, for example, the ruby laser.

Consider a degenerate p–n junction, whose band diagram at equilibrium is shown in Figure 7.33. The energy gap of the semiconductor is E_g. The conduction band on the n side is filled with electrons to an energy Δ_n above the band edge, and the valence band on the p side is filled with holes to an energy Δ_p below the band edge. If a forward-bias voltage V_a is applied to the junction, the electron energies on the n side will be increased an amount eV_a relative to the p side. For an applied voltage V_a approximately equal to E_g/e, the situation shown in Figure 7.34 obtains, with electrons being injected into the p region. In this situation, there is an effective population inversion established in a narrow region near the junction. There are filled electron states in the conduction band at a higher energy than empty states in the valence band. Electrons in the active region containing the inverted population can combine with holes with the emission of photons. As the

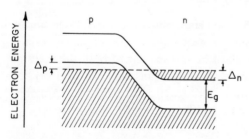

Figure 7.33. Equilibrium band diagram of a degenerate p–n junction. Filled electron states are shown shaded. The energy gap of the semiconductor is E_g. The conduction band on the n side is filled with electrons to an energy Δ_n above the band edge; the valence band on the p side is filled with holes to an energy Δ_p below the band edge. The figure is drawn for Δ_n larger than Δ_p.

Figure 7.34. Band diagram of the degenerate p–n junction under an applied forward bias V_a approximately equal to E_g/e, where E_g is the energy gap of the semiconductor. Injection of electrons into the p side forms a region of population inversion near the junction.

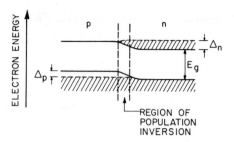

forward current through the junction is increased, a threshold value of the current is reached. Above this threshold, the photon emission is stimulated; below threshold, the emission is spontaneous junction luminescence. The process shown in Figure 7.34 is typical of GaAs junction lasers,[77] in which electrons are injected into the p side, so recombination and photon emission take place in the p region of the junction.

If we approximate the energy width of the filled and empty states in the region of population inversion in Figure 7.34 by Δ_n and Δ_p, respectively, we will have the band structure shown in Figure 7.35. This figure is drawn for a direct-gap semiconductor like GaAs, and shows that the electron population is inverted with respect to the emission of photons with energies as large as $E_g + \Delta_n + \Delta_p$ and as small as E_g, if we consider only the process of band-to-band recombination of electrons and holes. Actually, Figure 7.35 is not correct[78] because the high impurity densities necessary for forming a degenerate junction will introduce a hole impurity band in the p region where the population inversion exists. It is believed[79] that the lasing transition in GaAs junctions is from the electron distribution in the conduction band to an acceptor state, so the energy of the emitted photon may be less than the gap energy. In any case, we conclude that stimulated

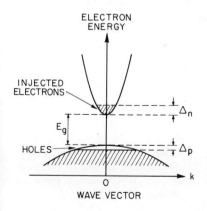

Figure 7.35. Semiconductor band structure within the region of population inversion. The direct energy gap is E_g, and the energy widths of the inverted populations of injected electrons and of holes are Δ_n and Δ_p, respectively.

emission is possible for a range of photon energies near E_g, and will discuss below a process which "selects" one energy for which emission becomes dominant.

A semiconductor junction laser is generally fabricated so the structure forms an optical resonant cavity of length L, such that there are an integral number m of half-wavelengths contained in the cavity length. If λ is the wavelength in the semiconductor and λ_0 is the wavelength in vacuum, then

$$\lambda = \lambda_0/n \qquad (7.30)$$

where n is the refractive index. Then the cavity condition is

$$L = m(\lambda/2) = m(\lambda_0/2n) \qquad (7.31)$$

where m is an integer. The vacuum wavelengths λ_0 satisfying equation (7.31) are given by

$$\lambda_0 = 2nL/m \qquad (7.32)$$

for integral values of m. The stimulated emission spectrum above the threshold current will be at a wavelength or wavelengths which are the modes (7.32) of the optical resonant cavity. The laser emission is thus "selected" by the resonant cavity from the range of possible transitions in the neighborhood of the gap energy E_g.

Junction lasers have been made in a number[80] of direct-gap semiconductors. The emitted photon energy due to a single mode of emission is usually approximately equal to the energy gap when considering the simple and more probable case of direct recombination (i.e., no phonon participation) of electrons and holes. However, since a variety of emission transitions[81] with slightly different energies (due to impurity levels) is known, the emission is only approximately at the gap energy. Junction lasers using narrow gap semiconductors[82] have been operated at emission wavelengths as long as 32 μm.

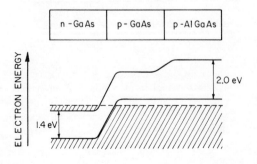

Figure 7.36. Band diagram of a heterojunction structure at equilibrium. Filled electron states are shown as shaded.

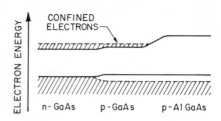

Figure 7.37. Band diagram of a hetero-junction structure at high forward bias, showing confinement of injected electrons. Filled electron states are shown as shaded.

As a final point on semiconductor injection lasers, we consider hetero-junction structures.[83,84] A heterojunction is a junction between two semiconductors with different energy gaps. An example is a heterostructure with a junction between p-type GaAs ($E_g = 1.4$ eV) and a p type solid solution of AlAs and GaAs. The latter has a variable energy gap depending on the composition of the solid solution; in the case shown in Figure 7.36, the energy gap of the AlGaAs is 2.0 eV. The doping of the p-type GaAs and the p-type AlGaAs is the same, so the equilibrium band diagram is as shown in Figure 7.36. Under high forward bias, electrons are injected from the n-type GaAs into the p-type GaAs, giving an inverted population as shown in the band diagram in Figure 7.37. The effect of the p-GaAs–p-AlGaAs heterojunction is to provide an energy barrier which helps prevent the diffusion of the electrons injected into the p-type GaAs. This confinement of the inverted electron population results in the achievement of stimulated emission at lower values of the threshold current.

Summary on Solid State Lasers

It is useful to summarize how some aspects of the physics of the solids considered give them their utility as lasers. First, the energy levels of various ions (e.g., Cr^{+3}, Nd^{+3}) in transparent host crystals or glasses provide the three- or four-level systems which make a population inversion possible. Second, in these three- or four-level schemes, differences in electron lifetime in upper and intermediate states allow the achievement of an inverted population. Third, in semiconductor injection lasers, a population inversion can be achieved because of the following. Injection of electrons produces a large number of electrons on the p side of the junction. The fact that the p side can be doped degenerately gives a large number of vacant final states, thereby contributing to the population inversion.

Problems

7.1. *Direct Transition.* Using the band structure of InSb in Figure 1.6, consider a direct transition at an electron wave vector k (say 0.1×10^7 cm^{-1}) near the zone center. Estimate the photon energy involved, and show that the wave vector of this photon is negligible compared to that of the electron.

7.2. *Indirect Transitions with Phonon Emission.* Using equations for the conservation of energy and wave vector, show that equation (7.7) is true for indirect transitions involving the emission of phonons of energy $\hbar\Omega$ in a semiconductor whose minimum (indirect) energy gap is E_g.

7.3. *Automobile Exhaust Monitor.* It is desired to monitor the CO_2 concentration in auto exhaust gas by using the CO_2 molecular absorption line whose wavelength is 4.25 μm. Discuss the design of a photon source (either a laser or a light-emitting diode) and a detector suitable for this spectral region. Comment on the factors that must be taken into account if the system is to be used "on board" an automobile.

7.4. *Ultraviolet LED.* Discuss possible designs for an ultraviolet light-emitting diode (or injection laser) at a photon energy of about 2.8 eV or greater. Suggest semiconductors and doping impurities; in particular, consider recent results[62] concerning the possible use of indirect-gap semiconductors.

References and Comments

1. T. S. Moss, G. J. Burrell, and B. Ellis, *Semiconductor Opto-Electronics*, John Wiley, New York (1973), Chapter 3.
2. J. I. Pankove, *Optical Processes in Semiconductors*, Prentice-Hall, New York (1971), Chapter 3.
3. C. Kittel, *Introduction to Solid State Physics*, Fifth Edition, John Wiley, New York (1976), pages 209–211.
4. See, for example, J. M. Stone, *Radiation and Optics*, McGraw-Hill, New York (1963), pages 376–383.
5. T. S. Moss *et al.*, Reference 1, pages 55–69.
6. See, for example, the data on germanium of W. C. Dash and R. Newman, *Physical Review*, **99**, 1151 (1955).
7. See, T. S. Moss *et al.*, Reference 1, pages 68–69, for examples.
8. T. S. Moss *et al.*, Reference 1, pages 90–94; J. I. Pankove, Reference 2, pages 62–66.
9. See H. M. Rosenberg, *Low Temperature Solid State Physics*, Oxford University Press, New York (1963), pages 249–252 for examples.
10. For impurity energy levels in germanium and silicon, see P. R. Bratt, "Impurity Germanium and Silicon Infrared Detectors," in *Semiconductors and Semimetals*, R. K. Willardson and A. C. Beer (editors), Academic Press, New York (1977), Volume 12, pages 44–45.
11. D. Long, *Energy Bands in Semiconductors*, John Wiley, New York (1968), pages 123–134.

12. D. L. Greenaway and G. Harbeke, *Optical Properties and Band Structure of Semi-conductors*, Pergamon Press, Oxford (1968), pages 88–95.

13. Adapted from D. Long, Reference 11, page 129, Figure 7.8.

14. D. Long, Reference 11, page 197.

15. E. Gutsche, J. Voight, and E. Ost, in *Proceedings of the Third International Conference on Photoconductivity, 1969*, E. M. Pell (editor), Pergamon Press, Oxford (1971), page 106.

16. H. Levinstein, *Physics Today*, **30**, 23–28 (November 1977).

17. R. Dalven, in *Solid State Physics*, F. Seitz, D. Turnbull, and H. Ehrenreich (editors), Academic Press, New York (1973), Volume 28, pages 179–224.

18. I. Melngailis and T. C. Harman, in *Semiconductors and Semimetals*, R. K. Willardson and A. C. Beer (editors), Academic Press, New York (1970), Volume 5, pages 111–174.

19. T. C. Harman in *The Physics of Semimetals and Narrow-Gap Semiconductors*, D. L. Carter and R. T. Bate (editors), Pergamon Press, Oxford (1971), pages 363–382.

20. T. C. Harman and I. Melngailis, in *Applied Solid State Science*, R. Wolfe (editor), Academic Press, New York (1974), Volume 4, pages 1–94.

21. D. Long and J. L. Schmidt, in Reference 18, pages 175–255.

22. T. S. Moss *et al.*, Reference 1, pages 290–296.

23. A. G. Milnes, *Deep Impurities in Semiconductors*, John Wiley, New York (1973), pages 175–177.

23a. P. R. Bratt, "Impurity Germanium and Silicon Infrared Detectors," in *Semiconductors and Semimetals*, R. K. Willardson and A. C. Beer (editors), Academic Press, New York (1977), Volume 12, pages 39–142, especially pages 108–113.

24. R. H. Bube, in *Photoconductivity and Related Phenomena*, J. Mort and D. M. Pai (editors), Elsevier Scientific Publishing Co., Amsterdam (1976), pages 117–153.

25. C. Kittel, *Introduction to Solid State Physics*, Fourth Edition, John Wiley, New York (1971), pages 628–632.

26. R. H. Bube, *Photoconductivity of Solids*, John Wiley, New York (1960), pages 59–60, 74–77.

27. T. S. Moss *et al.*, Reference 1, pages 191–192.

28. T. S. Moss *et al.*, Reference 1, pages 168–169.

29. P. W. Kruse, "The Photon Detection Process," in *Optical and Infrared Detectors*, R. J. Keyes (editor), Springer-Verlag, New York (1977), pages 42–47.

30. I. Melngailis and T. C. Harman, *Applied Physics Letters*, **13**, 180–183 (1968), Figure 2.

31. B. G. Streetman, *Solid State Electronic Devices*, Prentice Hall, New York (1972), pages 232–239.

32. T. S. Moss *et al.*, Reference 1, pages 153–158; J. I. Pankove, Reference 2, pages 302–312.

33. I. Melngailis and T. C. Harman, Reference 18, page 158.

34. H. J. Hovel, "Solar Cells," in *Semiconductors and Semimetals*, R. K. Willardson and A. C. Beer (editors), Academic Press, New York (1975), Volume 11.

35. T. S. Moss *et al.*, Reference 1, pages 192–197.

36. T. S. Moss *et al.*, Reference 1, page 193.

37. J. duBow and L. Curran, *Electronics*, **49**, 86–99 (November 11, 1976); J. Javetski, *Electronics*, **52**, 105–122 (July 19, 1979).

38. M. D. Tabak, S. W. Ing, and M. E. Scharfe, *IEEE Transactions on Electronic Devices*, **ED-20**, 132–139 (1970).

39. F. W. Schmidlin, in Reference 24, pages 421–478.
40. R. H. Bube, Reference 26, pages 233–234.
41. W. R. Beam, *Electronics of Solids*, McGraw-Hill, New York (1965), page 202.
42. B. G. Streetman, Reference 31, pages 244–246.
43. W. R. Beam, *Electronics of Solids*, McGraw Hill, New York (1965), pages 202–205.
44. C. E. K. Mees and T. H. James, *The Theory of the Photographic Process*, Third Edition, Macmillan, New York (1966), pages 19–30.
45. See, for example, *Color As Seen and Photographed*, Second Edition, Eastman Kodak Co., Rochester (1972), pages 30–39.
46. C. E. K. Mees and T. H. James, Reference 44, pages 103–111.
47. C. E. K. Mees and T. H. James, Reference 44, page 278.
48. J. I. Pankove, Reference 2, pages 124–131.
49. T. S. Moss *et al.*, Reference 1, pages 198–200; 202–206.
50. S. Wang, *Solid State Electronics*, McGraw Hill, New York (1966), page 278.
51. A. Mooradian and H. Y. Fan, *Radiative Recombination in Semiconductors* (Seventh International Conference on the Physics of Semiconductors, Paris, 1964), Academic Press, New York (1965), pages 39–46.
52. D. Long, Reference 11, page 111.
53. J. I. Pankove, Reference 2, pages 125–126.
54. J. I. Pankove, Reference 2, pages 132–136.
55. T. S. Moss *et al.*, Reference 1, pages 210–216.
56. C. J. Neuse, H. Kressel, I. Ladany, *IEEE Spectrum*, **9**, 28–38 (May 1972).
57. B. G. Streetman, Reference 31, pages 239–244.
58. W. V. Smith, in *Topics in Solid State and Quantum Electronics*, W. D. Hershberger (editor), John Wiley, New York (1972), pages 264–269.
59. T. S. Moss *et al.*, Reference 1, pages 224–228.
60. J. I. Pankove, Reference 2, pages 177–193.
61. P. J. Dean, in *Applied Solid State Science*, R. Wolfe (editor), Academic Press, New York (1969), Volume 1, pages 24–29.
62. C. B. Duke and N. Holonyak, Jr., *Physics Today*, **26**, 23–31 (December 1973).
63. L. I. Schiff, *Quantum Mechanics*, Third Edition, McGraw-Hill, New York (1968), pages 403–404.
64. W. Koechner, *Solid State Laser Engineering*, Springer-Verlag, New York (1976), Chapter 1.
65. W. V. Smith, Reference 58, pages 241–249.
66. C. Kittel, Reference 3, pages 525–527.
67. W. V. Smith and P. P. Sorokin, *The Laser*, McGraw-Hill, New York (1966), pages 5–7.
68. L. I. Schiff, Reference 63, pages 404–405 and 414.
69. See, for example, H. Eyring, J. Walter, and G. E. Kimball, *Quantum Chemistry*, John Wiley, New York (1944), pages 113–114.
70. W. Koechner, Reference 64, page 46; C. Kittel, Reference 3, pages 528–529.
71. W. Koechner, Reference 64, page 2.
72. W. Koechner, Reference 64, Chapter 2; W. V. Smith, Reference 58, Table 1, pages 254–255.
73. W. V. Smith, Reference 58, pages 249–253.
74. A. Yariv, *Quantum Electronics*, Second Edition, John Wiley, New York (1975), pages 219–238.

75. W. V. Smith, Reference 58, pages 269–274.
76. B. G. Streetman, Reference 31, pages 262–272.
77. W. V. Smith, Reference 58, page 270.
78. See, for example, B. Lax, *Science*, **141**, 1247–1255 (1963).
79. A. Yariv, Reference 74, pages 230–231.
80. A. Yariv, Reference 74, page 237, gives a table of examples.
81. W. V. Smith, Reference 58, page 267.
82. T. C. Harman and I. Melngailis, in *Applied Solid State Science*, R. Wolfe (editor), Academic Press, New York (1974), Volume 4, pages 1–94, Section 13.
83. T. S. Moss *et al.*, Reference 1, pages 241–242.
84. M. B. Panish and I. Hayashi, in *Applied Solid State Science*, R. Wolfe (editor), Academic Press, New York (1974), Volume 4, pages 235–328.

Suggested Reading

T. S. Moss, G. J. Burrell, and B. Ellis, *Semiconductor Opto-Electronics*, John Wiley, New York (1973). A good book on the physics of the interaction of radiation with semiconductors, covering both basic and applied topics.

J. I. Pankove, *Optical Processes in Semiconductors*, Prentice-Hall, New York (1971). This book is more of a research monograph than the book above, but covers many of the same topics.

W. Koechner, *Solid State Laser Engineering*, Springer-Verlag, New York (1976). Chapter 1 of this monograph offers a brief introduction to optical amplification for the non-specialist, while Chapter 2 describes solid state laser materials (excluding semi-conductors) in some detail.

W. V. Smith, in *Topics in Solid State and Quantum Electronics*, W. D. Hershberger (editor), John Wiley, New York (1972). This collection of articles contains a chapter by Smith discussing both optically pumped solid state lasers and semiconductor injection lasers.

A. Yariv, *Quantum Electronics*, Second Edition, John Wiley, New York (1975). Chapters 9 and 10 of this advanced textbook discuss many aspects of laser physics, including non-solid-state lasers.

8

Superconductive Devices and Materials

Introduction

This chapter discusses some of the applications of superconductivity. After a brief review, the wave function for a condensed phase of Cooper pairs is introduced and used to discuss the Josephson effects. The physics of the DC and AC Josephson effects is developed, and current–voltage plots are described as preparation for a treatment of the effect of electromagnetic radiation on Josephson junctions. Quantization of magnetic flux in a superconducting ring leads to the idea of superconducting quantum interference and devices (DC squids) based thereon. Finally, the chapter concludes with a discussion of superconducting materials, with emphasis on the factors determining the magnitude of the transition temperature.

Review of Some Aspects of Superconductivity

We recall[1-4] a few points concerning superconductivity that will be useful in discussing some superconducting devices utilizing the Josephson effect. (Other aspects connected with superconductive materials will be introduced as needed in later sections.)

We remember that the BCS ground state of a superconductor differs from the ground state of a normal metal. In a normal metal at 0 K, the electrons fill the Fermi sphere up to the Fermi energy E_F. Since this free-electron model of the normal metal includes no interaction between the electrons, the energy of an excited electronic state may be arbitrarily small. In a superconductor, however, there is an attractive electron–electron interaction which leads to the existence of an energy gap $E_g = 2\Delta$ (at the Fermi energy) between the ground state of the superconductor and the lowest excited state. The attractive electron–electron interaction leads to the formation of electron pairs (Cooper pairs), in which the electrons occupy states of opposite wave vector and spin. Since each electron is a fermion, the pair will behave (approximately) as a boson since exchanging both electrons multiplies the wave function by $(-1)^2$. We will consider the Cooper pair as a "particle" that behaves like a boson.

We will focus our attention on pairs because we will want to discuss Josephson effect devices that involve pairs in their operation. Since we are considering pairs as bosons, we expect qualitatively that, if the temperature is low enough, almost all of the pairs will be in their lowest quantum state. This would be a Bose–Einstein condensation, in momentum space, in which all of the pairs are in the same quantum state and have the same wave function. The assembly of pairs, all in the same state, is often referred to as the condensed phase or condensate of pairs.

Wave Function of the Condensed Phase of Pairs

We next discuss the wave function for an assembly of pairs, following the macroscopic approach of Feynman.[5,6] If the wave function for one particle in three dimensions is $\psi(\mathbf{r})$, then, ignoring the time dependence, we recall[7] that the probability $P(\mathbf{r})$, where

$$P(\mathbf{r}) = \psi^*(\mathbf{r})\psi(\mathbf{r}) \tag{8.1}$$

is the probability that the particle will be in a unit volume located at $\mathbf{r}$. There is[7] an equation of continuity for P such that

$$\mathbf{V} \cdot \mathbf{S} + \partial P/\partial t = 0 \tag{8.2}$$

In equation (8.2), $\mathbf{S}$ is the probability current density given[7] by

$$\mathbf{S} = (1/2m)[\psi \hat{p}^*\psi^* + \psi^*\hat{p}\psi] \tag{8.3}$$

where $\hat{p} \equiv -j\hbar\mathbf{V}$ is the momentum operator in the absence of a magnetic field.

We know also, from equation (8.1), that, for one particle of wave function $\psi(\mathbf{r})$, the quantity

$$P(\mathbf{r})\,dV = \psi^*(\mathbf{r})\psi(\mathbf{r})\,dV \tag{8.4}$$

is the probability that the particle will be found in a volume element dV located at $\mathbf{r}$. If we consider a simple case for which $\psi^*\psi$ is constant in space, we have (assuming normalization of ψ) that

$$\int \psi^*\psi \, dV = 1 \tag{8.5}$$

in this case. Since $\psi^*\psi$ is constant, we obtain

$$\psi^*\psi = 1/V \tag{8.6}$$

where V is the total volume. Equation (8.6) suggests that we may, in a crude way, interpret the quantity $\psi^*\psi$ as the number of particles (in this case, one) per unit volume. In this sense, $\psi^*\psi$ is the "particle density" for one particle described by the wave function ψ. Consider next a large number N of identical particles, all in the same quantum state. These N particles all have the same wave function $\psi(\mathbf{r})$, and, as we did for one particle, we interpret

$$\psi^*(\mathbf{r})\psi(\mathbf{r}) = \varrho = N/V \tag{8.7}$$

where ϱ is the particle density. This interpretation is reasonable; the particle density $\varrho = N/V$ should certainly be N times larger for N particles than it is for one particle in the same volume V.

This crude plausibility argument is designed to introduce the idea[5] that, for a large number of particles all in the same state, and thus all with the same wave function $\psi(\mathbf{r})$,

$$\psi^*(\mathbf{r})\psi(\mathbf{r}) = \varrho(\mathbf{r}) \tag{8.8}$$

where $\varrho(\mathbf{r})$ is the particle density in the system at point $\mathbf{r}$. Further, from (8.8), if each particle has an electric charge q,

$$\varrho q = q\psi^*\psi \tag{8.9}$$

is the electric charge density. Since $\psi^*\psi$ is the particle density, the quantity $\mathbf{S}$ given by equation (8.3) is the particle flux, i.e., the number of particles crossing unit area in unit time. Then the electric current density $\mathbf{J}$ is given by

$$\mathbf{J} = q\mathbf{S} = (q/2m)[\psi \hat{p}^*\psi^* + \psi^* \hat{p}\psi] \tag{8.10}$$

Equations (8.9) and (8.10) tell us that, when there is a large number of particles in exactly the same state, the electric charge density and the electric current density can be calculated *directly* from the wave function ψ.

Both of the above are *macroscopic* physical quantities, so the wave function has a macroscopic significance for the case of many particles in the same state.

Equation (8.8) suggests that we write the wave function $\psi(\mathbf{r})$ in the form

$$\psi(\mathbf{r}) = [\varrho(\mathbf{r})]^{1/2} \exp[j\theta(\mathbf{r})] \qquad (8.11)$$

where $\theta(\mathbf{r})$ is the phase of the wave function, and $\varrho(\mathbf{r})$ is the particle density in the system. If we consider one dimension for simplicity, equation (8.11) reduces to

$$\psi(x) = [\varrho(x)]^{1/2} \exp[j\theta(x)] \qquad (8.12)$$

The wave function ψ given by (8.11) or (8.12) involves the particle density ϱ, which certainly has a classical macroscopic physical meaning. We can investigate the meaning of the phase[8,9] $\theta(x)$ of the wave function by calculating the electric current density $\mathbf{J}$ (in one dimension) using equations (8.10) and (8.12). We have

$$\psi\hat{p}^*\psi^* = \hbar\varrho(d\theta/dx) + (1/2)j\hbar(d\varrho/dx) \qquad (8.13)$$

$$\psi^*\hat{p}\psi = \hbar\varrho(d\theta/dx) - (1/2)j\hbar(d\varrho/dx) \qquad (8.14)$$

which, when substituted in equation (8.10), gives

$$J = (\hbar\varrho q/m)(d\theta/dx) \qquad (8.15)$$

Since the magnitude J of the electric current density and the particle density ϱ are both macroscopic variables, $d\theta/dx$, the spatial variation of the phase of the wave function, is also a macroscopic variable or observable. Since

$$J = \varrho q v \qquad (8.16)$$

where v is the velocity of particle flow, we have, on combining (8.16) with (8.15), that

$$mv = \hbar(d\theta/dx) \qquad (8.17)$$

a result which, extended to three dimensions, becomes

$$m\mathbf{v} = \hbar\,\nabla\theta \qquad (8.18)$$

Equation (8.18) tells us that the particle momentum mv is related to the gradient $\nabla\theta$ of the phase $\theta(\mathbf{r})$ of the wave function. The absolute phase is not observable, but, if $\nabla\theta$ is known everywhere, then θ is known except for a constant. If the phase is defined at one point of the system, then that constant is determined, and the phase is known everywhere in the system.

We next apply these results to an assembly of Cooper pairs at a very low temperature. All of the pairs are in the same quantum state and have the same wave function. We write the pair wave function $\psi(\mathbf{r}, t)$ as

$$\psi(\mathbf{r}, t) = |\psi(\mathbf{r}, t)| \exp[j\theta(\mathbf{r}, t)] \tag{8.19}$$

where $\theta(\mathbf{r}, t)$ is the phase of the pair wave function. Since all of the pairs are in the same state, the density $\varrho(\mathbf{r}, t)$ of pairs is given by

$$\varrho(\mathbf{r}, t) = \psi^*(\mathbf{r}, t)\psi(\mathbf{r}, t) \tag{8.20}$$

so, from (8.19)

$$\varrho(\mathbf{r}, t) = |\psi(\mathbf{r}, t)|^2 \tag{8.21}$$

and we may write (8.19) in the form

$$\psi(\mathbf{r}, t) = [\varrho(\mathbf{r}, t)]^{1/2} \exp[j\theta(\mathbf{r}, t)] \tag{8.22}$$

We regard the wave function (8.22) as the wave function of the *entire* assembly of pairs. The phase $\theta(\mathbf{r}, t)$ in (8.22) is[10] the phase of the entire condensate of pairs and is, as discussed above, a physical observable of the system.

Consider next a superconductor in which no magnetic field is present and in which there is no center-of-mass motion of the pairs. In such a case, there is no net velocity of the pairs, so no supercurrent flows. From equation (8.17), we have, in one dimension,

$$d\theta/dx = 0 \tag{8.23}$$

since the pair velocity v vanishes in this case. The conclusion from equation (8.23) is that the phase θ of the condensate of pairs is a constant in space in a superconductor in which no current flows.

We note that the fact that there is no center-of-mass motion in this case is in agreement with the fact that the electrons of the pair have opposite wave vectors and spins, so the net momentum equals zero. If a current does flow in the superconductor, then the pair velocity v is a nonzero constant, so equation (8.17) becomes

$$d\theta/dx = C \tag{8.24}$$

$$\theta = Cx + \theta_0 \tag{8.25}$$

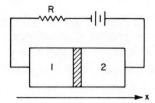

Figure 8.1. Schematic view of a Josephson junction, composed of two superconductors separated by an insulating barrier, which is shown shaded. The barrier, usually a metallic oxide, is of the order of 10 Å thick; its width is exaggerated in the drawing. The direction x is through the junction, as shown.

where C and θ_0 are constants. When a supercurrent flows, the phase θ varies linearly in space for a one-dimensional time-independent situation.

We may summarize the results of this section as follows. The superconductor is regarded as an assembly of Cooper pairs, all in the same state. The wave function ψ of the entire assembly of pairs is given by equation (8.22), where $\psi^*\psi = \varrho$, the pair density. The phase θ of the wave function is a macroscopic observable quantity related, through equation (8.15), to the current density J of pairs in the superconductor. If no current flows, the phase is everywhere constant in the superconductor.

The Josephson Effects

We now apply the ideas developed in the previous sections to the physics of the Josephson effect,[6,11-13] a term applied to the effects arising from the tunneling of pairs† from one superconductor, through an insulating barrier, into a second superconductor. The experimental arrangement is shown schematically in Figure 8.1. There are two Josephson effects. The DC Josephson effect is the fact that, if a DC current is passed through the junction from an external source, no voltage is observed across the junction. The AC Josephson effect is the observation that a DC voltage across the junction causes high-frequency current oscillations across the junction. We will discuss the DC and AC effects in that order.

Physics of the DC Josephson Effect

The approach[11-13] to the DC Josephson effect will be as follows. We consider a Josephson junction in which sides (1) and (2) of Figure 8.1 are the same superconductor. We will set up equations describing the pair wave function ψ on the two sides of the junction, and include the existence

---

† The tunneling of pairs is to be distinguished from single-electron, or Giaver, tunneling.

of tunneling by pairs. We will then consider a DC current passed through the junction and observe its effect, via the phase, on the wave function.

Let ψ_1 and ψ_2 be the wave functions on the sides (1) and (2) of the Josephson junction. The time-dependent Schrödinger equation must hold on both sides, so

$$j\hbar(\partial\psi_1/\partial t) = U_1\psi_1 \tag{8.26}$$

$$j\hbar(\partial\psi_2/\partial t) = U_2\psi_2 \tag{8.27}$$

where U_1 and U_2 are the energies of lowest states of the superconductors on sides (1) and (2). Next, we include tunneling in equations (8.26) and (8.27). We recall[14] that, in one dimension, tunneling through an energy barrier of width d, as shown in Figure 8.2, is the existence of a nonzero wave function in the region of space beyond d. Figure 8.2 shows an incident particle of plane-wave wave function $A\exp(jkx)$ incident on an energy barrier of height V_0 and width d; the energy of the incident particle is less than V_0. Physically, we may think of tunneling as the "leaking" of the wave function $A\exp(jkx)$ through the energy barrier, resulting in a nonzero transmitted wave function $C\exp(jkx)$ on the other side $(x > d)$ of the barrier.

We may incorporate tunneling into equations (8.26) and (8.27) as follows. Assume that the tunneling of pairs from side (2) to side (1) increases the amplitude ψ_1 of the pair wave function on side (1). Further, assume that the time rate of increase of ψ_1 is proportional to ψ_2, the amplitude of the pair wave function on side (2). We write the time rate of change of ψ_1 in the form[11]

$$\hbar T\psi_2 \tag{8.28}$$

where the constant T is characteristic of the junction and is a measure of the transfer of pairs from side (2) to side (1). The dimension of T will be that of a frequency. We add the expression (8.28) to equation (8.26) for $\partial\psi_1/\partial t$ and obtain

$$j\hbar(\partial\psi_1/\partial t) = U_1\psi_1 + \hbar T\psi_2 \tag{8.29}$$

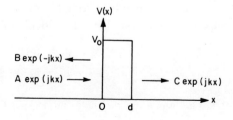

Figure 8.2. Plot of energy $V(x)$ as a function of distance x showing an energy barrier height V_0 and width d. The wave functions of the incident, reflected, and transmitted particles are, respectively, $A\exp(jkx)$, $B\exp(-jkx)$, and $C\exp(jkx)$.

Equation (8.29) neglects[†] any decrease in ψ_1 due to pairs tunneling back from side (1) to side (2). In the same way, we add a term $\hbar T\psi_1$ to equation (8.27) to represent tunneling of pairs from side (1) to side (2), obtaining

$$jh(\partial\psi_2/\partial t) = U_2\psi_2 + \hbar T\psi_1 \qquad (8.30)$$

In obtaining (8.30), we have assumed that the constant T is the same for tunneling in both directions.

Since T is a measure of tunneling through the barrier, $T = 0$ when the barrier is very thick, and there is no coupling via tunneling between the two pieces of superconductor. The two sides act like separate superconductors with different values of the phase θ since the phase θ_1 in side (1) is independent of the phase θ_2 in side (2). If the barrier is very thin, and approaches zero in thickness, the coupling between the two pieces of superconductor is large, and the properties of the system change continuously from those of two isolated superconductors to those of a single superconductor. In particular, as the width d of the barrier goes to zero, the phases θ_1 and θ_2 will become equal, as long as no supercurrent is flowing, in accordance with equation (8.23).

We next consider equations (8.29) and (8.30) for the case of appreciable tunneling. For simplicity, we consider the situation in which the superconductors on sides (1) and (2) of the junction are the same. Then, the lowest-state energies U_1 and U_2 are the same, so we set

$$U_1 = U_2 \equiv U \qquad (8.31)$$

We define the energy U as the zero of energy in the problem, so $U \equiv 0$ and it vanishes from equations (8.29) and (8.30). We obtain

$$jh(\partial\psi_1/\partial t) = \hbar T\psi_2 \qquad (8.32)$$

$$jh(\partial\psi_2/\partial t) = \hbar T\psi_1 \qquad (8.33)$$

Equation (8.32) and (8.33) are the time-dependent Schrödinger equations for the pair wave functions ψ_1 and ψ_2 on the two sides of the junction, including the effect of tunneling. We note that, if there is no tunneling and $T = 0$, then both ψ_1 and ψ_2 are constant in time, as is reasonable.

We now introduce complex wave functions of the form (8.12) for the superconductor on sides (1) and (2) of the junction. We have

$$\psi_1 = [\varrho_1]^{1/2} \exp(j\theta_1) \qquad (8.34)$$

$$\psi_2 = [\varrho_2]^{1/2} \exp(j\theta_2) \qquad (8.35)$$

[†] See Problem 8.1 at the end of this chapter for a possible way to remedy this neglect.

where the subscripts refer to sides (1) and (2), and we are considering the one-dimensional case, so the ψ's, ϱ's, and θ's are all functions of x and of the time. The direction of x is through the junction, as shown in Figure 8.1. Our next step will be to substitute the wave functions (8.34) and (8.35) into the coupled Schrödinger equations (8.32) and (8.33), thereby obtaining equations relating ϱ_1, ϱ_2, θ_1, and θ_2. We then connect the phases θ_2 and θ_1 with the current density J flowing through the junction.

However, before making those substitutions, let us see if we can get a qualitative idea of the result we expect when a current density flows through the junction. Consider a current density J, determined by the battery and resistor, flowing through the junction in Figure 8.1. From equation (8.15), since J is not zero, we expect $d\theta/dx$ to be nonzero, and

$$\frac{d\theta}{dx} = \frac{mJ}{\hbar \varrho q} \tag{8.36}$$

From equation (8.36), we expect that there will be a gradient $d\theta/dx$ in the phase of the wave function between the two sides of the junction when a current is flowing. Further, we expect that the phase gradient will be almost entirely across the oxide barrier. That this is so can be seen from equation (8.36), which says that $d\theta/dx$ is inversely proportional to the pair density ϱ for a given value of J. In the superconductor itself, the density of pairs is large, of the order of 10^{22} cm^{-3}, assuming a very low temperature so pairing is essentially complete.[15] In the oxide barrier layer, the density of tunneling pairs is small, typically of the order[16] of 10^{10} cm^{-3}. Thus, $d\theta/dx$ is small in the superconductor and large in the oxide barrier layer. We will assume that, approximately, the phase θ is constant in space in the two superconducting sides of the junction. This is shown in Figure 8.3, where $\theta = \theta_1$ in side (1), $\theta = \theta_2$ in side (2), and the width of the oxide barrier is Δx. Then $d\theta/dx = 0$ in the superconductors, and the entire gradient of phase

$$d\theta/dx = (\theta_2 - \theta_1)/\Delta x \equiv \Delta\theta/\Delta x \tag{8.37}$$

appears across the oxide barrier. Our central physical result is that the flow

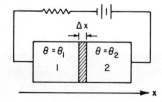

Figure 8.3. Current flowing through a Josephson junction, with a phase gradient $\Delta\theta/\Delta x = (\theta_2-\theta_1)/\Delta x$ across the oxide barrier layer (shaded). The phases on sides (1) and (2) are θ_1 and θ_2, respectively.

of a current through the junction produces a change of phase $\Delta\theta \equiv \theta_2 - \theta_1$ across the oxide barrier layer.

We now substitute the wave functions (8.34) and (8.35) into the coupled Schrödinger equations (8.32) and (8.33). Since we are assuming that the phase is constant in space within the superconductor itself, we have $\theta = \theta(t)$ in the superconductor on both sides of the junction. Substituting the wave functions ψ_1 and ψ_2 into the Schrödinger equations yields

$$ j\hbar \frac{\partial}{\partial t} (\varrho_1^{1/2} e^{j\theta_1}) = \hbar T \varrho_2^{1/2} e^{j\theta_2} \tag{8.38} $$

$$ j\hbar \frac{\partial}{\partial t} (\varrho_2^{1/2} e^{j\theta_2}) = \hbar T \varrho_1^{1/2} e^{j\theta_1} \tag{8.39} $$

Performing the differentiation, and equating[11] real and imaginary parts of the resulting two equations, yields

$$ \dot{\varrho}_1 = 2T(\varrho_1\varrho_2)^{1/2} \sin(\Delta\theta) \tag{8.40} $$

$$ \dot{\varrho}_2 = -2T(\varrho_1\varrho_2)^{1/2} \sin(\Delta\theta) \tag{8.41} $$

$$ \dot{\theta}_1 = -T(\varrho_2/\varrho_1)^{1/2} \cos(\Delta\theta) \tag{8.42} $$

$$ \dot{\theta}_2 = -T(\varrho_1/\varrho_2)^{1/2} \cos(\Delta\theta) \tag{8.43} $$

where $\Delta\theta \equiv \theta_2 - \theta_1$ and $\dot{\varrho}_1 = \partial\varrho_1/\partial t$, $\dot{\theta}_1 = \partial\theta_1/\partial t$, etc. Since the superconductors on sides (1) and (2) of the junction are the same, we set the pair densities equal on both sides of the junction, so

$$ \varrho_1 = \varrho_2 \tag{8.44} $$

The condition (8.44) leads, from equations (8.42) and (8.43), to the result, since $\dot{\theta}_1 = \dot{\theta}_2$, that

$$ \frac{\partial}{\partial t} (\theta_2 - \theta_1) \equiv \frac{\partial}{\partial t} (\Delta\theta) = 0 \tag{8.45} $$

Equation (8.45) says that the phase difference $\Delta\theta$ across the junction is constant in time. The condition (8.44) applied to equations (8.40) and (8.41) yields

$$ \dot{\varrho}_2 = -\dot{\varrho}_1 \tag{8.46} $$

an equation that says the time rate of increase of the pair density ϱ_2 is equal to the time rate of decrease of the pair density ϱ_1. The magnitude J of the current density flowing from side (1) to side (2) is proportional to $\partial\varrho_2/\partial t$, so we conclude, from (8.40) or (8.41), that J is proportional to $\sin(\varDelta\theta)$. We write

$$J = J_0 \sin(\varDelta\theta) \qquad (8.47)$$

where the constant J_0 is a function[17] of the properties of the barrier, including the constant T, and the temperature. Since the magnitude of $\sin(\varDelta\theta)$ varies between zero and one, we see from equation (8.47) that the magnitude J varies from zero to J_0. The quantity J_0 is thus the maximum possible value of the DC supercurrent density through the junction and is called[6] the critical current density. The values[18] of the corresponding critical currents may vary from a fraction of a microampere to tens of milliamperes. The relation given by (8.47) is thus valid for values of J less than the critical current density J_0.

We note especially that no electric field appears in equation (8.47) for the DC supercurrent density J. If an experiment is set up, as shown in Figure 8.3, in which a current density J, smaller than J_0, is passed through the junction, the current flows as a direct supercurrent and *no voltage* appears across the junction. This is the DC Josephson effect. When J is increased until it equals the critical current density J_0, the junction can no longer sustain[+] a supercurrent. While no voltage is developed across the junction for values of the current density below the critical value, a phase difference $\varDelta\theta$, given by

$$\varDelta\theta = \sin^{-1}(J/J_0) \qquad (8.48)$$

is produced across the junction. As the current density J through the junction (determined by the external circuit in Figure 8.3) increases from zero to a value less than J_0, the phase difference $\varDelta\theta$ is developed according to equation (8.48). As a final point, we note that, for the junction, the current density is related to the phase *difference* $\varDelta\theta$ between the two sides of the junction, while from equation (8.15), the current density in a single superconductor is related to the phase *gradient* $d\theta/dx$.

To summarize our results for the DC Josephson effect, the DC supercurrent density J through the junction is described by equation (8.47), where $\varDelta\theta$ is the difference in the phase of the wave function between the two sides of the junction. Equation (8.47) is valid for values of J less than the critical value J_0, the maximum supercurrent density the junction can sustain. For values of J smaller than J_0, no voltage is observed across the junction.

[+] A discussion in terms of the energy difference (the coupling energy[19]) between the two sides of the junction is given by Clarke.[6]

Physics of the AC Josephson Effect

Suppose the current J through a Josephson junction has a value larger than the critical value J_0. In this situation, a new effect, the AC Josephson effect,[6,11-13] is observed, in which a DC voltage is produced across the junction, resulting in high-frequency current oscillations.

If J is larger than J_0, the junction can no longer maintain a DC supercurrent. Some of the current is carried by single normal electrons tunneling through the barrier, and a DC voltage V appears across the junction. If a potential difference V exists between the two sides of the junction,[†] then a particle of electric charge q will change its potential energy by qV on passing through the barrier. For a Cooper pair, $q = -2e$, and the change in potential energy per pair is $-2eV$. This is equivalent to saying that a pair on one side of the junction is at potential energy $-eV$, and a pair on the other side is at potential energy $+eV$.

We now modify equations (8.26) and (8.27) to include these potential energy terms in the Hamiltonian, obtaining

$$j\hbar(\partial\psi_1/\partial t) = U_1\psi_1 + eV\psi_1 \tag{8.49}$$

$$j\hbar(\partial\psi_2/\partial t) = U_2\psi_2 - eV\psi_2 \tag{8.50}$$

Setting, as before, $U_1 = U_2 = U \equiv 0$, and including, again as before, the tunneling terms $\hbar T\psi_2$ and $\hbar T\psi_1$ in (8.49) and (8.50), respectively, yields

$$j\hbar(\partial\psi_1/\partial t) = \hbar T\psi_2 + eV\psi_1 \tag{8.51}$$

$$j\hbar(\partial\psi_2/\partial t) = \hbar T\psi_1 - eV\psi_2 \tag{8.52}$$

as the equations into which we again substitute the wave functions (8.34) and (8.35). From (8.51) we obtain

$$j\hbar(\tfrac{1}{2}\varrho_1^{-1/2}\dot\varrho_1 e^{j\theta_1} + j\varrho_1^{1/2}e^{j\theta_1}\dot\theta_1) = \hbar T\varrho_2^{1/2}e^{j\theta_2} + eV\varrho_1^{1/2}e^{j\theta_1} \tag{8.53}$$

Multiplying by $\varrho_1^{1/2}\exp(-j\theta_1)$ gives, with $\Delta\theta \equiv \theta_2 - \theta_1$, the result

$$j\hbar(\tfrac{1}{2}\dot\varrho_1 + j\varrho_1\dot\theta_1) - eV\varrho_1 = \hbar T(\varrho_1\varrho_2)^{1/2}[\cos(\Delta\theta) + j\sin(\Delta\theta)] \tag{8.54}$$

Equating real and imaginary parts of (8.54) gives

$$\dot\varrho_1 = 2T(\varrho_1\varrho_2)^{1/2}\sin(\Delta\theta) \tag{8.55}$$

$$\dot\theta_1 = (-e/\hbar)V - T(\varrho_2/\varrho_1)^{1/2}\cos(\Delta\theta) \tag{8.56}$$

---

† Strictly speaking, this argument should be given in terms of the chemical potential.[19,20]

Equation (8.55) is the same as (8.40) found for the DC Josephson effect. Equation (8.56), when compared with (8.42), shows that the term $(e/\hbar)V$ has entered the expression for $\dot{\theta}_1$, the time rate of change of the phase θ_1. In a similar manner, from equation (8.52), one obtains

$$\dot{\varrho}_2 = -2T(\varrho_1\varrho_2)^{1/2}\sin(\Delta\theta) \tag{8.57}$$

$$\dot{\theta}_2 = (e/\hbar)V - T(\varrho_1/\varrho_2)^{1/2}\cos(\Delta\theta) \tag{8.58}$$

where equations (8.57) and (8.58) may be compared with (8.41) and (8.43).

We again set, as we did in equation (8.44), the pair densities ϱ_1 and ϱ_2 equal, so (8.56) and (8.58) become

$$\dot{\theta}_1 = (-e/\hbar)V - T\cos(\Delta\theta) \tag{8.59}$$

$$\dot{\theta}_2 = (e/\hbar)V - T\cos(\Delta\theta) \tag{8.60}$$

which, in turn, lead to the result

$$\frac{d}{dt}(\Delta\theta) \equiv \dot{\theta}_2 - \dot{\theta}_1 = 2eV/\hbar \tag{8.61}$$

Since the DC voltage V across the junction is a constant, we may integrate the differential equation (8.61) for the phase difference $\Delta\theta$, obtaining

$$\Delta\theta = (\Delta\theta)_0 + (2eV/\hbar)t \tag{8.62}$$

where $(\Delta\theta)_0$ is the phase difference $\Delta\theta$ at time $t = 0$. Equation (8.62) gives the time dependence of the phase difference across the junction. Equation (8.47) for the supercurrent density J is still valid,[21] but now, with V not zero, the phase difference $\Delta\theta$ changes with time as given by equation (8.62). Substituting expression (8.62) for $\Delta\theta$ into equation (8.47) for J gives

$$J = J_0\sin[(\Delta\theta)_0 + \omega t] \tag{8.63}$$

where we have defined

$$\omega \equiv 2eV/\hbar \tag{8.64}$$

Physically, equation (8.63) tells us that the supercurrent J oscillates with time at the frequency ω given by (8.64), so ω is a function of the nonzero voltage V across the junction. Note that so far we have been dealing with the supercurrent J, given by (8.63), only. To obtain the total current, one must add[22] the current due to the tunneling of single normal electrons.

To summarize, the result of the AC Josephson effect is that a current greater than the critical value produces a voltage V across the junction. This results in an alternating supercurrent J, given by (8.63), of frequency $\omega \equiv 2eV/\hbar$, through the junction. A voltage $V = 10^{-6}$ V corresponds to a frequency of approximately 484 MHz.

Voltage–Current Curves for Josephson Junctions

We now discuss a plot[23,24] of the voltage V across a tunnel junction of the kind indicated in Figure 8.1, as a function of the DC current density J through the junction. We consider the situation in which the impedance R of the current source is large compared to that of the junction. As the current is increased from zero, no voltage appears across the junction for currents less than the critical current J_0. This is the DC Josephson effect and is shown as curve (1) on the voltage–current plot in Figure 8.4. Curve (2) in that figure is the current due to the tunneling[25] of single normal electrons; this begins to increase sharply at a voltage approximately equal to $2\Delta/e$, where 2Δ is the energy gap of the superconductor. The normal electron tunnel current is very small for voltages less than $2\Delta/e$. When the critical current is exceeded, there is a discontinuous jump from zero voltage to a voltage value on the normal electron current curve (2). This is shown in Figure 8.4 as the horizontal dashed line connecting curves (1) and (2). There is now a nonzero voltage across the junction. As the current is increased further, the voltage across the junction follows curve (2), and the DC current through the junction is due to normal electron tunneling. The supercurrent through the junction is now oscillatory in time, as given by equation (8.63) for the AC Josephson effect. If the current is decreased to zero, hysteresis effects[24,26] may be observed. Finally, current sources

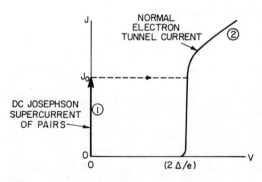

Figure 8.4. Schematic plot of voltage V across a Josephson tunnel junction as a function of the current density J. The dashed line represents the discontinuous increase in V when the critical current J_0 is exceeded; 2Δ is the energy gap of the superconductor. Curve (1) is the DC current due to pair tunneling; curve (2) is the current due to normal single-electron tunneling.

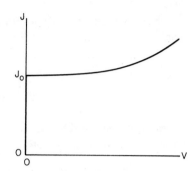

Figure 8.5. Schematic current–voltage characteristic of a point contact Josephson junction, showing the critical current J_0 of the device.

with an impedance less than that of the tunnel junction produce somewhat different current–voltage curves.[24]

It should also be pointed out that there are several other types[24,27,28] of junctions in addition to the oxide tunnel junction shown schematically in Figure 8.1. Among these is the point contact junction in which a superconductor, sharpened to a radius of a few micrometers, is pressed against a block of superconducting material. The current–voltage characteristic of these other types of junctions differs[18,24] from that shown in Figure 8.4, and looks approximately like that in Figure 8.5, which is a schematic curve for a point contact.

Effect of Electromagnetic Radiation on the Junction

We now consider the effect[27–31] of high-frequency (microwave and far infrared) electromagnetic radiation on a Josephson junction. It will be found that changes are observed in the curve of DC current as a function of voltage. These changes enable the junction to be used as a detector[18,28,30,32] of electromagnetic radiation.

Consider a junction across which a constant DC voltage V_0 is produced, resulting in the alternating supercurrent of the AC Josephson effect. The frequency of the alternating supercurrent will depend, from equations (8.63) and (8.64), on the DC voltage. Suppose further that microwave electromagnetic radiation of frequency ω_0 is incident on the junction, inducing an AC voltage

$$V_1 \cos \omega_0 t \qquad (8.65)$$

across the junction. The radiation given by (8.65) is mixed by the junction with the alternating Josephson supercurrent, resulting in frequency modulation of the latter. The experimental result is the alteration, referred to above, of the DC current–voltage characteristic of the junction.

Following Clarke,[29] we may calculate the effect of the applied radiation on the junction on the assumption that both the DC current and the incident radiation have low source impedances. The total voltage V across the junction is the sum of the applied DC and induced AC voltages, so

$$V = V_0 + V_1 \cos \omega_0 t \qquad (8.66)$$

From equation (8.61), the time rate of change $d(\Delta\theta)/dt$ of the phase difference $\Delta\theta$ across the junction is

$$d(\Delta\theta)/dt = 2eV/\hbar = (2e/\hbar)(V_0 + V_1 \cos \omega_0 t) \qquad (8.67)$$

since the total voltage V across the junction is given by (8.66). Integrating (8.67) gives

$$\Delta\theta(t) = (\Delta\theta)_0 + \left(\frac{2eV_0}{\hbar}\right)t + \left(\frac{2eV_1}{\hbar\omega_0}\right) \sin \omega_0 t \qquad (8.68)$$

for the time dependence $\Delta\theta(t)$ of the phase difference, and where V_0, V_1, and ω_0 are constant; $(\Delta\theta)_0$ is the value of the phase difference at time $t = 0$. The supercurrent density J is, as before, obtained by substituting the phase difference (8.68) into equation (8.47) for J, yielding

$$J(t) = J_0 \sin\left[(\Delta\theta)_0 + \left(\frac{2eV_0}{\hbar}\right)t + \left(\frac{2eV_1}{\hbar\omega_0}\right) \sin \omega_0 t\right] \qquad (8.69)$$

showing that the supercurrent is a function of time. Equation (8.69) can be expanded in a Fourier–Bessel series[33] using the relations[34]

$$\sin(X \sin \alpha) = \sum_{n=-\infty}^{\infty} B_n(X) \sin(n\alpha) \qquad (8.70)$$

$$\cos(X \sin \alpha) = \sum_{n=-\infty}^{\infty} B_n(X) \cos(n\alpha) \qquad (8.71)$$

where, to avoid confusion with the current density J, the symbol $B_n(X)$ has been used for the Bessel function of order n of argument X. Using relations (8.70) and (8.71), equation (8.69) may be expanded to give[29]

$$J(t) = J_0 \sum_{n=-\infty}^{\infty} \left\{B_n\left(\frac{2eV_1}{\hbar\omega_0}\right) \sin\left[\left(n\omega_0 + \frac{2eV_0}{\hbar}\right)t + (\Delta\theta)_0\right]\right\} \qquad (8.72)$$

for the time-dependent supercurrent density $J(t)$.

We may examine the properties of the supercurrent through the junction by considering equation (8.72). The equation is a sum of components at infinitely many frequencies

$$n\omega_0 + (2eV_0/\hbar) \tag{8.73}$$

The amplitude of the component at frequency $n\omega_0 + (2eV/\hbar)$ is

$$J_0[B_n(2eV_1/\hbar\omega_0)]\sin(\Delta\theta)_0 \tag{8.74}$$

In particular, when the frequency $n\omega_0 + (2eV_0/\hbar)$ vanishes, so

$$n\hbar\omega_0 = -2eV_0 \tag{8.75}$$

where $n = 0, \pm1, \pm2, \ldots$, there will be a component of the supercurrent density (8.72) at zero frequency. A DC supercurrent will therefore be present in the junction whenever the DC voltage V_0 across the junction has one of the values

$$V_0 = n(\hbar\omega_0/2e) \tag{8.76}$$

where n is any integer. These zero-frequency supercurrents correspond to the appearance of a DC Josephson effect at a nonzero voltage V_0 across the junction when the junction is irradiated with electromagnetic radiation of frequency ω_0. These results predict a series of spikes of DC supercurrent superimposed on the DC current–voltage characteristic of the junction. The spikes occur, from equation (8.76), at voltages $\hbar\omega_0/2e$, $2\hbar\omega_0/2e$, $3\hbar\omega_0/2e$, etc. The voltage separation between adjacent spikes is $\hbar\omega_0/2e$ and is determined by the radiation frequency ω_0; a frequency of approximately 484 MHz will correspond to a spike separation of 1 μV. In practice the spike structure has not been observed.[35] Instead, for the usual experimental arrangement of a high-impedance source of DC current, current steps[29] are produced in the DC junction characteristic. The voltage separation of the steps is also equal to $\hbar\omega_0/2e$. Figure 8.6 shows[36] the steps induced in a Sn–SnO–Sn tunnel junction by 4 GHz electromagnetic radiation.

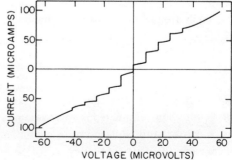

Figure 8.6. Steps induced in the DC current–voltage characteristic of an Sn–SnO–Sn tunnel junction by 4 GHz electromagnetic radiation. The spacing of the steps is approximately 8.5 μV (after Clarke[36]).

From equation (8.74), the amplitude J_{DC}^{n} of step n of the DC super-current, induced by the radiation, is proportional to the Bessel function of order n and argument $2eV_1/\hbar\omega_0$, so

$$J_{DC}^{n} \propto B_n(2eV_1/\hbar\omega_0) \tag{8.77}$$

The step n of amplitude J_{DC}^{n} appears on the DC current–voltage characteristic at the voltage $n\hbar\omega_0/2e$ determined by equation (8.76). If we consider the step for $n = 0$ at zero voltage, the amplitude of the zero-voltage step is

$$J_{DC}^{0} \propto B_0(2eV_1/\hbar\omega_0) \tag{8.78}$$

where, since the voltage $n\hbar\omega_0/2e$ equals zero for $n = 0$, J_{DC}^{0} given by (8.78) is the critical current of the junction. From the properties[37] of the Bessel function of order zero, $B_0(x)$ decreases monotonically with increasing x, for x less than approximately 2.4. The amplitude J_{DC}^{0} of the $n = 0$ step responds to incident radiation (at low values of V_1 where $2eV_1$ is less than about $2\hbar\omega_0$) by decreasing with increasing V_1. The critical current of the junction therefore decreases[18] with increasing microwave power incident on the device.

On the other hand, the amplitude J_{DC}^{1} of the $n = 1$ step, appearing at the voltage $\hbar\omega_0/2e$ on the DC current–voltage characteristic, is

$$J_{DC}^{1} \propto B_1(2eV_1/\hbar\omega_0) \tag{8.79}$$

From the properties[37] of the Bessel function of order unity, $B_1(x)$ increases with increasing x, for x less than about 1.8. The amplitude J_{DC}^{1} of the $n = 1$ step increases with increasing V_1 (for low values of V_1, such that $2eV_1 \lesssim 1.8\hbar\omega_0$).

Finally, we note that equation (8.76) is satisfied, for $n = 0$, by *any* incident frequency ω_0. If broadband radiation, containing many frequencies, is applied to a junction, the critical current will respond to, and be modified by, each frequency component incident. Since,[38] for small values of x, $B_0(x)$ may be expressed as the series

$$B_0(x) = 1 - x^2/4 + \cdots \tag{8.80}$$

equation (8.78) shows that J_{DC}^{0} is proportional to V_1^2/ω_0^2 for small values of V_1/ω_0. This result leads to the conclusion[18] that the junction will act as a broadband square-law detector, and becomes less sensitive with increasing incident frequency ω_0 of the radiation.

Based on their response to incident electromagnetic radiation, different types of Josephson junctions can be used as detectors in a number of different ways. Recent review articles[18,28,39] discuss the field, with particular emphasis on more realistic models of the practical junctions used.

Quantization of Magnetic Flux in a Superconducting Ring

In preparation for a discussion of the physics of superconductive quantum interference devices, we discuss the quantization of magnetic flux in a superconducting ring.[40] First, we return to our discussion of the phase θ of the wave function of a superconductor, and include the effect of a magnetic field $\mathbf{B}$. It is well known[41] that the total momentum $\mathbf{p}$ of a particle, of mass m and electric charge q, moving with velocity $\mathbf{v}$ in field of magnetic induction $\mathbf{B}$ is

$$\mathbf{p} = m\mathbf{v} + (q/c)\mathbf{A} \tag{8.81}$$

where the magnetic vector potential $\mathbf{A}$ is given by $\mathbf{B} = \mathbf{\nabla} \times \mathbf{A}$. The quantity $(q/c)\mathbf{A}$ is called the field momentum and $m\mathbf{v}$ is the kinetic momentum. The relation (8.18),

$$m\mathbf{v} = \hbar\,\mathbf{\nabla}\theta \tag{8.18}$$

between the particle velocity $\mathbf{v}$ and the gradient $\mathbf{\nabla}\theta$ of the phase of the wave function, was obtained for the situation in which no magnetic field was present. In that case, $\mathbf{B} = 0$, so $\mathbf{A} = 0$. To include a nonzero magnetic field, we add the field momentum $(q/c)\mathbf{A}$ to the left-hand side of (8.18), obtaining

$$\mathbf{p} = m\mathbf{v} + (q/c)\mathbf{A} = \hbar\,\mathbf{\nabla}\theta \tag{8.82}$$

which we rewrite as

$$m\mathbf{v} = \hbar\,\mathbf{\nabla}\theta - (q/c)\mathbf{A} \tag{8.83}$$

Next, we derive a relation for the current density J in a magnetic field. From (8.16),

$$\mathbf{J} = \varrho q\mathbf{v} \tag{8.84}$$

is the current density of particles of charge q and density ϱ. Substituting the velocity $\mathbf{v}$ from (8.83) into (8.84) gives us

$$\mathbf{J} = (\varrho q/m)[\hbar\,\mathbf{\nabla}\theta - (q/c)\mathbf{A}] \tag{8.85}$$

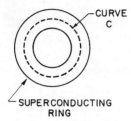

Figure 8.7. Superconducting ring, showing the (dashed) curve C used as the path of integration in the line integral in equation (8.88).

SUPERCONDUCTING
RING

as the current density **J** in a magnetic field described by the vector potential **A**.

We now want to consider **J** in the interior of a superconducting ring, shown schematically in Figure 8.7. From Maxwell's equations[42] under static conditions, when all time derivatives vanish,

$$\mathbf{\nabla} \times \mathbf{B} = (4\pi/c)\mathbf{J} \tag{8.86}$$

From the Meissner effect, we know also that $\mathbf{B} = 0$ inside a superconductor.[†] If $\mathbf{B} = 0$ in equation (8.86), then $\mathbf{J} = 0$, and, from (8.85),

$$\hbar\,\mathbf{\nabla}\theta = (q/c)\mathbf{A} \tag{8.87}$$

inside the superconducting ring.

Consider next the line integral of $\mathbf{\nabla}\theta$ around a path C well inside the superconducting ring, as shown in Figure 8.7. The line integral is

$$\oint_C \mathbf{\nabla}\theta \cdot d\mathbf{l} = \varDelta\theta \tag{8.88}$$

where $\varDelta\theta$ is the change in the phase of the wave function on going around the path C in the ring. Recalling from our earlier discussion that the wave function has the macroscopic significance $\varrho^{1/2}$ from equation (8.8), we require that the wave function be single valued in the ring. From equation (8.11), we see that this means that the change in phase $\varDelta\theta$ must be an integral multiple of 2π since $\exp(j2\pi) = 1$. Our requirement that the wave function be single valued leads to the condition

$$\varDelta\theta = 2\pi s \tag{8.89}$$

where s is an integer, on the change of phase $\varDelta\theta$.

[†] For simplicity, we neglect penetration of the magnetic field into the superconductor.

We examine next the line integral of $\mathbf{A}$ around the curve C in the ring, and, using Stokes' theorem, obtain

$$\oint_C \mathbf{A} \cdot d\mathbf{l} = \int_{S'} (\nabla \times \mathbf{A}) \cdot \mathbf{n} \, dS \qquad (8.90)$$

where $\mathbf{n}$ is the unit normal to the element of area dS of the surface S' bounded by curve C. Since $\mathbf{B} = \nabla \times \mathbf{A}$, equation (8.90) becomes

$$\oint_C \mathbf{A} \cdot d\mathbf{l} = \int_{S'} \mathbf{B} \cdot \mathbf{n} \, dS = \Phi \qquad (8.91)$$

where Φ is the magnetic flux passing through the curve C. Since, from (8.87),

$$\mathbf{A} = (\hbar c/q) \nabla \theta \qquad (8.92)$$

we have also

$$\oint_C \mathbf{A} \cdot d\mathbf{l} = (\hbar c/q) \oint_C \nabla \theta \cdot d\mathbf{l} = (\hbar c/q) \, \Delta\theta = 2\pi s (\hbar c/q) \qquad (8.93)$$

on using (8.88) and (8.89). Comparing (8.93) with (8.91) gives the result

$$\Phi = s(hc/q) \qquad (8.94)$$

where the particle charge $q = |-2e| = 2e$ for Cooper pairs. We rewrite (8.94) as

$$\Phi = s\Phi_0 \qquad (8.95)$$

giving the flux Φ through the superconducting ring as an integral multiple of

$$\Phi_0 \equiv (hc/2e) \qquad (8.96)$$

where Φ_0 is called the flux quantum and has the approximate value 2×10^{-7} G cm^2.

Our conclusion is that the magnetic flux Φ through a superconducting ring is quantized in multiples of the flux quantum or fluxoid Φ_0. The flux Φ is the sum of the flux Φ_e from the external sources and the flux Φ_s from supercurrents flowing in the surface of the ring. Since there is no quantization imposed on Φ_e, the flux Φ_s must adjust itself[40] in order that the total flux Φ take on a quantized value that is a multiple of Φ_0.

Superconducting Quantum Interference

To begin a discussion of superconducting quantum interference,[43-45] we consider the effect of a magnetic field on the DC supercurrent flowing through a Josephson junction. From equations (8.93) and (8.94), we have

$$\Delta\theta = (q/hc)\Phi \qquad (8.97)$$

for a superconducting ring, where Φ is the quantized magnetic flux enclosed by the ring, and $\Delta\theta$ is the change in the phase of the wave function on performing the line integral in equation (8.88). Equation (8.97) shows that there is a connection between the phase θ of the superconducting wave function and the magnetic flux Φ.

We now want to find a relation between Φ and the DC supercurrent J flowing in a Josephson junction. We consider the junction shown schematically in Figure 8.8, where a and b are the end points of the junction. From equation (8.82), the total momentum $\mathbf{p}$ of a particle in a magnetic field is given by

$$\mathbf{p} = m\mathbf{v} + (q/c)\mathbf{A} = \hbar\,\boldsymbol{\nabla}\theta \qquad (8.82)$$

where $\mathbf{v}$ is the velocity, m the mass, q the charge, $\mathbf{A}$ the magnetic vector potential, and $\boldsymbol{\nabla}\theta$ the gradient of the phase. We apply equation (8.82) to the junction in Figure 8.8, obtaining

$$\hbar\,\Delta\theta = \hbar(\theta_b - \theta_a) = \hbar\int_a^b \boldsymbol{\nabla}\theta \cdot d\mathbf{l} = \int_a^b \mathbf{p} \cdot d\mathbf{l} \qquad (8.98)$$

which becomes

$$\hbar\,\Delta\theta = \int_a^b m\mathbf{v} \cdot d\mathbf{l} + (q/c)\int_a^b \mathbf{A} \cdot d\mathbf{l} \qquad (8.99)$$

where $\Delta\theta$ is the phase difference across the junction, and the line integrals are taken between the end points b and a.

We now use equation (8.99) to consider a superconducting loop L containing two Josephson junctions, (1) and (2), as shown in Figure 8.9. No voltage is applied. The phase differences $\Delta\theta_1$ and $\Delta\theta_2$ across the two

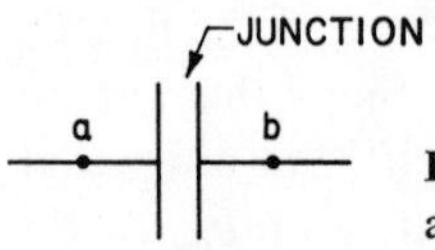

Figure 8.8. Schematic Josephson junction with end points a and b.

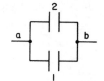

Figure 8.9. Superconducting loop L containing two Josephson junctions (1) and (2), where a and b are points on the loop.

junctions are

$$\hbar \, \Delta\theta_1 = \int_a^b m\mathbf{v} \cdot d\mathbf{l} + (q/c) \int_a^b \mathbf{A} \cdot d\mathbf{l} \tag{8.100}$$

$$\hbar \, \Delta\theta_2 = \int_a^b m\mathbf{v} \cdot d\mathbf{l} + (q/c) \int_a^b \mathbf{A} \cdot d\mathbf{l} \tag{8.101}$$

where $\mathbf{A}$ is the vector potential of the magnetic field through the loop, and $\mathbf{v}$ is the particle (pair) velocity in the supercurrent flowing in the loop. Rewriting equation (8.100) as

$$-\hbar \, \Delta\theta_1 = \int_b^a m\mathbf{v} \cdot d\mathbf{l} + (q/c) \int_b^a \mathbf{A} \cdot d\mathbf{l} \tag{8.102}$$

and adding (8.102) to (8.101) gives

$$\hbar(\Delta\theta_2 - \Delta\theta_1) = \oint_L m\mathbf{v} \cdot d\mathbf{l} + (q/c) \oint_L \mathbf{A} \cdot d\mathbf{l} \tag{8.103}$$

where, in equation (8.103), the line integrals are around a closed curve in the superconducting loop L containing the two Josephson junctions. The first integral in equation (8.103) is[46] usually small, and will be neglected, so equation (8.103) becomes

$$\hbar(\Delta\theta_2 - \Delta\theta_1) = (q/c) \oint_L \mathbf{A} \cdot d\mathbf{l} \tag{8.104}$$

Next, using equation (8.91), we have

$$(q/c) \oint_L \mathbf{A} \cdot d\mathbf{l} = (q/c)\Phi \tag{8.105}$$

where Φ is the total magnetic flux through the loop L. Equations (8.104) and (8.105) give

$$\Delta\theta_2 - \Delta\theta_1 = (2e/\hbar c)\Phi \tag{8.106}$$

on putting $|q| = 2e$ for pairs. Equation (8.106) connects the phase differences $\Delta\theta_2$ and $\Delta\theta_1$ across the two junctions with the total magnetic flux

Φ through the superconducting loop containing the junctions. We note that $\Delta\theta_2$ and $\Delta\theta_1$ are equal if the flux Φ is zero.

Equation (8.106) is satisfied[43] if

$$\Delta\theta_2 = \theta_0 + (e/\hbar c)\Phi \tag{8.107}$$

$$\Delta\theta_1 = \theta_0 - (e/\hbar c)\Phi \tag{8.108}$$

where θ_0 is an introduced constant whose significance will be discussed below. Since (8.107) and (8.108) give the phase differences across the two junctions, we may use equation (8.47) to find the DC supercurrents J_1 and J_2 through junctions (1) and (2). We obtain

$$J_1 = J_{01} \sin[\theta_0 + (e/\hbar c)\Phi] \tag{8.109}$$

$$J_2 = J_{02} \sin[\theta_0 - (e/\hbar c)\Phi] \tag{8.110}$$

where J_{01} and J_{02} are, respectively, the maximum DC supercurrents through junctions (1) and (2). Since the individual supercurrents J_1 and J_2 are, from (8.109) and (8.110), functions only of the flux Φ, the total supercurrent $J_T \equiv J_1 + J_2$ through the entire loop device is given by

$$J_T = J_{01} \sin[\theta_0 + (e/\hbar c)\Phi] + J_{02} \sin[\theta_0 - (e/\hbar c)\Phi] \tag{8.111}$$

If we assume that junctions (1) and (2) are identical and have the same maximum current J_0, where $J_{01} = J_{02} \equiv J_0$, we obtain

$$J_T = J_0\{\sin[\theta_0 + (e/\hbar c)\Phi] + \sin[\theta_0 - (e/\hbar c)\Phi]\} \tag{8.112}$$

which can be written as

$$J_T = 2J_0(\sin\theta_0)\cos(e\Phi/\hbar c) \tag{8.113}$$

showing that the total supercurrent J_T varies harmonically with $e\Phi/\hbar c$, and hence with the total magnetic flux through the loop containing the junctions.

Equation (8.113) represents a total supercurrent J_T due to interference between the supercurrents J_1 and J_2, given by (8.109) and (8.110), through the individual junctions (1) and (2). From equations (8.107) and (8.108), the phase differences $\Delta\theta_1$ and $\Delta\theta_2$ across the two junctions are functions of the total magnetic flux Φ through the loop. As Φ the is varied by changing the applied magnetic field, the currents J_1 and J_2 change also. The result is the interference effect expressed in equation (8.113).

The quantity θ_0 is, from equations (8.107) and (8.108), the value of the phase difference $\Delta\theta_2 = \Delta\theta_1$ across both junctions when the magnetic flux $\Phi = 0$. We may therefore regard[45] $\sin \theta_0$ as a quantity that is free to adjust to the current through the loop. Its maximum value is unity. Further, we have lumped into J_0 the effect[47] (which we did not discuss) of the magnetic flux on the maximum supercurrents of the individual junctions.[†] Equation (8.113) shows that the total supercurrent J_T has maxima when

$$e\Phi/\hbar c = s\pi \qquad (8.114)$$

where s is an integer, or, equivalently, when

$$\Phi = s\Phi_0$$

where $\Phi_0 \equiv (hc/2e)$ is the flux quantum defined in (8.96).

Physically, we expect that a plot of total current J_T as a function of magnetic field will show oscillations. There will be maxima in the total current whenever the value of the magnetic field is such that an integral number of flux quanta pass through the superconducting loop. The effect has been observed experimentally,[44,45] and is termed superconducting quantum interference.

The Superconducting Quantum Interference Device (Squid)

We now discuss a device, the DC squid, based on superconducting quantum interference. We consider, as in the previous section, the two identical junctions on a superconducting ring, as shown schematically in Figure 8.9. The total supercurrent through the device is given by equation (8.113), rewritten as

$$J_T = 2J_0(\sin \theta_0) \cos(\pi\Phi/\Phi_0) \qquad (8.115)$$

as a function of magnetic flux Φ through the loop of the device. Since the maximum value of the quantity $\sin \theta_0$ is unity, the total supercurrent will have a maximum magnitude $|J_{max}|$ given by

$$|J_{max}| = 2J_0 |\cos(\pi\Phi/\Phi_0)| \qquad (8.116)$$

where this current is composed of superconducting electron pairs passing

---

[†] This leads to the long-period diffraction effects[44,45] observed experimentally.

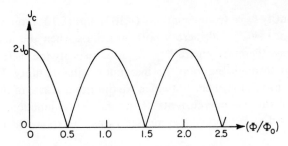

Figure 8.10. Plot of critical current J_c as a function of total flux Φ, as given by equation (8.117).

through the entire device of two junctions on a superconducting loop. Equation (8.116) is correct[48] if Φ is the total flux from both applied external sources and from supercurrents in the device induced by external magnetic fields. Since $|J_{\max}|$ is the maximum current the squid can carry at a given value of the flux Φ, we identify $|J_{\max}|$ with the critical current J_c (the current at which a voltage first appears across the junction). From (8.116), then,

$$J_c = 2J_0 \, | \cos(\pi\Phi/\Phi_0) | \qquad (8.117)$$

where Φ is the total magnetic flux through the loop of the squid. Figure (8.10) is a plot of equation (8.117), showing the periodic variation of J_c with the flux Φ.

One is generally interested in the variation of the squid critical current as a function of the external *applied* magnetic flux Φ_e. The *induced* flux is Φ_i, where

$$\Phi_i \cong LI \qquad (8.118)$$

for a squid loop of self-inductance L carrying a current I. If L is small, such that

$$LI \ll \Phi_0 \qquad (8.119)$$

then the induced flux Φ_i will be small, and the total flux Φ, where

$$\Phi = \Phi_e + \Phi_i \qquad (8.120)$$

will be given by

$$\Phi \cong \Phi_e \qquad (8.121)$$

Then equation (8.117) for the squid critical current becomes

$$J_c = 2J_0 \, | \cos(\pi\Phi_e/\Phi_0) | \qquad (8.122)$$

where Φ_e is the external applied magnetic flux through the loop. The variation of J_c as a function of Φ_e given by equation (8.122) is the same as that shown in Figure 8.10, and is the case for the limit in which the condition (8.119) holds. Equation (8.122) shows that the critical current of the squid is a periodic function of the applied magnetic flux Φ_e, and that the minimum critical current is zero. This is approached if the condition (8.119) holds for the device. However, in practical devices,[49] (8.119) is not usually satisfied, and the self-inductance L is of magnitude

$$L \approx \Phi_0/J_0 \qquad (8.123)$$

where J_0 is the critical current of each junction. Since the inductance L given by (8.123) is larger than that required to satisfy (8.119), the maximum decrease in the critical current will be smaller than the value $2J_0$ given by equation (8.122), and the critical current will not go to zero in real devices. What is observed in a plot of critical current J_c as a function of applied flux Φ_e is shown schematically in Figure 8.11. There are still maxima and minima in J_c, but the minimum critical current is not zero as it was for the idealized case of very small inductance on which equation (8.122) is based.

The two-junction DC squid is thus a device in which the critical current is a function of the magnetic flux passing through the loop. The squid can therefore act as a magnetometer[19,50,51] by measuring the critical current. Since the maxima in the critical current in Figure 8.11 are Φ_0 apart on the Φ_e axis, the device can measure magnetic fields much smaller[19] than 10^{-7} G (assuming a maximum device area of about 1 cm²). Other techniques[19,50,52a] have been used to obtain even higher sensitivities with DC squids. Finally, we mention the superconducting voltmeter,[19,50,51,53] the idea behind which is shown schematically in Figure 8.12. In this device, the voltage V to be measured generates a current $I = V/R$ in a superconducting loop of in-

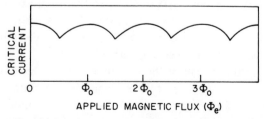

Figure 8.11. Plot of critical current J_c as a function of applied flux Φ_e, in the case of a DC squid for which the self-inductance is not small. The positions of the maxima and minima are the same as in Figure 8.10, but the minimum value of J_c is not zero (after Clarke[19]).

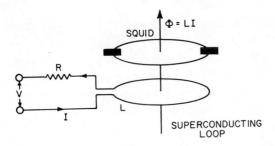

Figure 8.12. Superconducting voltmeter (schematic) (after Clarke[19]).

ductance L, producing a magnetic flux $\Phi = LI$. This flux is then measured[†] with a squid magnetometer, yielding a noise-limited voltage sensitivity of the order of 10^{-15} V. There are other types of squid devices and many other interesting applications based on the physics we have discussed. The interested reader is referred to the literature.[52–56]

Superconducting Materials: The Transition Temperature

We now turn our attention to superconducting materials, examining some of the factors that determine their superconducting properties. We will emphasize the solid state physical parameters affecting the magnitude of the transition temperature T_c.

We start by considering the magnitude of T_c in the BCS microscopic theory.[57–59] The transition temperature of a superconductor is extremely important from the point of view of applications. One would, of course, like to have available superconducting materials with values of T_c as high as possible so that simpler refrigerants (e.g., liquid hydrogen) may be used. Further, one would like to operate at a temperature of about $T_c/2$ in order that the critical field[60] of the superconductor not be too small in applications involving magnetic fields. The search for higher values of T_c goes on, with the immediate objective a T_c of 30 K, so that pumped liquid hydrogen (at about 14 K) could be used as the refrigerant. As of this writing (mid-1979), the maximum observed T_c is about 23 K for the alloy Nb_3Ge,[61] thereby breaking the "hydrogen barrier"[62] of 20.4 K.

To begin our discussion of the magnitude of T_c, we recall that, in the BCS picture, there is an attractive electron–electron interaction because of the interaction of the electrons with the lattice phonons of the solid.

---

† In practice, the flux change is not measured directly. See the paper by Clarke (Reference 53, pages 116–118) for a discussion of practical superconducting voltmeters using squids.

This attractive interaction between electrons is described[63] by a potential energy $V_{k,k'}$ of interaction between two electrons with wave vectors k and k', and with energies E_k and $E_{k'}$. The simplest form of $V_{k,k'}$ is the BCS "one square well" model in which

$$V_{k,k'} = -V \tag{8.124}$$

where V is a positive constant, for electron energies E_k and $E_{k'}$ such that

$$|E_k - E_F| < \hbar\omega_D, \qquad |E_{k'} - E_F| < \hbar\omega_D \tag{8.125}$$

where ω_D is the Debye frequency, and E_F is the Fermi energy. For electron energies E_k and $E_{k'}$ other than those given by (8.125), $V_{k,k'} \equiv 0$. The model described by equations (8.124) and (8.125) is thus one with a net attractive potential energy of interaction, in which the phonon-mediated attraction overcomes the Coulomb repulsion, for electrons with energies within $\hbar\omega_D$ of the Fermi energy. For other electron energies, $V_{k,k'}$ is zero, and there is no attraction. The criterion[64] for an attractive interaction, and hence for superconductivity in this model, is simply that V be positive. Then the net attraction will result in the formation of Cooper pairs.

Using the BCS theory with the model of (8.124) and (8.125), one may obtain[59] the following results. The critical transition temperature T_c is given by

$$T_c = 1.14\Theta_D \exp[-1/N(0)V] \tag{8.126}$$

where $\Theta_D \equiv \hbar\omega_D/k_B$ is the Debye temperature, and $N(0)$ is the density of states (of one spin) at the Fermi energy E_F. Equation (8.126) is true for zero magnetic field, and for values of T_c less than Θ_D. An important restriction on the conditions under which (8.126) holds is that it must also be true that

$$N(0)V \ll 1 \tag{8.127}$$

This condition is called the weak-coupling limit. For most superconductors, $N(0)V$ is less than[59] about 0.3. The magnitude of the superconducting energy gap $E_g(0) \equiv 2\Delta(0)$, at $T = 0$ K, is given by

$$\Delta(0) = 2\hbar\omega_D \exp[-1/N(0)V] \tag{8.128}$$

again in the weak coupling limit (8.127). The quantity $\Delta(0)$ is the energy gap parameter at 0 K; combining (8.126) and (8.128) gives the relation

$$E_g(0) \equiv 2\Delta(0) = 3.52k_B T_c \tag{8.129}$$

Equation (8.129) relates the energy E_g, necessary to break up a Cooper pair, to the transition temperature T_c. A discussion of the comparison of experimental results with the various predictions of the BCS theory may be found in the literature.[65]

From the point of view of applications, we would now like to consider some of the solid state factors influencing, through equation (8.126), the magnitude of T_c. From that equation, we see that, as $N(0)V$ increases, T_c increases, so we expect that large values of the attractive interaction V and/or of the density of states $N(0)$, at the Fermi energy, will favor relatively large values of the transition temperature. Also, large values of Θ_D favor high values of T_c, but the exponential in $-1/N(0)V$ is the dominant factor. The fact that Θ_D is the less important parameter is seen in the experimental values[1] of $T_c = 7.2$ K for lead, for which $\Theta_D = 105$ K, while $T_c = 1.2$ K for aluminum, whose value of Θ_D is 428 K.

We consider first the quantity V in equation (8.126), subject to the weak-coupling limit. Since the attraction between electrons is due to the electron–phonon interaction, we would expect V to be large if the electron–phonon interaction is, in some sense, large. Since it is the scattering of electrons by phonons that is primarily responsible for the resistivity of a metal at room temperature, we would expect[1] metals with a high resistivity at room temperature to be more likely to be superconductors at low temperatures. This idea may be illustrated by writing[66]

$$\varrho = m^*/ne^2\tau \tag{8.130}$$

where ϱ is the resistivity, m^* is the effective mass, n the electron density, e the charge on the proton, and τ an electron–phonon collision time. If the electron–phonon interaction V is large, we might expect, roughly, that τ^{-1} will be large and ϱ will be large. Table 8.1 shows such data for a few metallic elements, with values of τ^{-1} calculated[67] from equation (8.130); the elements are arranged in order of increasing room-temperature resistivity. The values of τ^{-1} increase as the resistivity increases, and these values of τ^{-1} may be compared with values[68] of V which may be obtained from experimental data as follows. The electronic specific heat C_e of a normal metal is given[69] by

$$C_e = \tfrac{1}{3}\pi^2 k_B^2 D(E_F)T \equiv \gamma T \tag{8.131}$$

where $D(E_F)$ is the density of states of both spins at the Fermi energy E_F. Then

$$D(E_F) = 2N(0) \tag{8.132}$$

and we obtain

$$N(0) = (3\gamma/2\pi^2 k_B^2) \tag{8.133}$$

Table 8.1. Resistivity and Superconductivity of Several Elements

	ϱ (295 K, Ω cm)^a	n (cm^{-3})^b	$(m^*/m)^c$	τ^{-1} (sec^{-1})	V (eV cm^3)^d
Ag^e	1.6×10^{-6}	5.86×10^{22}	1.00	2.6×10^{13}	$<1.2^f$
Al	2.74×10^{-6}	18.1×10^{22}	1.48	9.3×10^{13}	0.97
W	5.3×10^{-6}	12.6×10^{22}	1.69*	1.0×10^{14}	0.59
Cd	7.27×10^{-6}	9.27×10^{22}	0.73	2.6×10^{14}	2.5
In	8.75×10^{-6}	11.5×10^{22}	1.37	2.0×10^{14}	2.1
Nb	14.5×10^{-6}	5.56×10^{22}	11.7*	1.9×10^{13}	0.34
Pb	21.0×10^{-6}	13.2×10^{22}	1.97	3.9×10^{14}	1.7

^{a} C. Kittel, Reference 1, page 356, Table 1.
^{b} N. W. Ashcroft and N. D. Mermin, Reference 2, page 5, Table 1.
^{c} Thermal values from C. Kittel, Reference 1, page 167, Table 5; m is the free electron mass. The values marked with an asterisk were calculated by the author from the data in Table 5.
^{d} G. Gladstone, M. A. Jensen, and J. R. Schrieffer, Reference 68, Table VI.
^{e} Not superconductive.^{d}
^{f} Presumably the interaction is repulsive for silver.

relating the density of states $N(0)$ of one spin, at the Fermi energy, to the experimental coefficient γ determined from heat capacity measurements. Combining equation (8.133) for $N(0)$ with the BCS equation (8.126) for T_c allows a calculation[68] of V from experimental values of γ, Θ_D, and T_c. From the values of V in Table 8.1, we see that τ^{-1} generally increases as V increases, agreeing with the intuitive idea that high resistivity suggests a large electron–phonon interaction. Even though V represents only very roughly[68] the net attractive interaction between electrons, these illustrative results indicate the correlation between resistivity and the attractive interaction. The values of τ^{-1} and V in Table 8.1 are plotted in Figure 8.13. Silver is omitted from the figure because it is not superconductive, but is included in the table to exhibit its low value of τ^{-1}.

It is interesting to plot values[68] of V for several different metallic

Figure 8.13. Electron–electron attractive interaction V as a function of τ^{-1}, where τ is the electron–phonon collision time in equation (8.130). Data and calculations of Table 8.1.

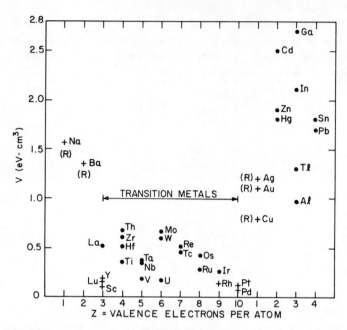

Figure 8.14. Electron–electron attractive interaction V (in eV cm³) as a function of the number Z of valence electrons per atom. Superconductors are indicated by dots; nonsuperconductors (at atmospheric pressure) by crosses, in which cases the value of V is an upper limit. The large values marked (R) are presumably repulsive.

elements as a function of the number Z of valence electrons per atom; such a graph is shown in Figure 8.14. This plot shows a number of interesting qualitative trends. First, V is very small (or repulsive) for monovalent metals and the alkaline earth metals, which (except for Be[70]) are not superconductive at atmospheric pressure. Second, the nontransition metals of valence 2, 3, 4 have relatively high values of V, and they superconduct at relatively high (above 0.5 K) temperatures. Third, almost all of the transition metals[68] superconduct in spite of their relatively low values of V. The value of V is seen to be fairly small and roughly constant for values of Z between 4 and 8, i.e., for most of the transition metals.

We may obtain some additional information by considering the variation of the density of states $N(0)$ as a function of Z. From equation (8.133), one may calculate[68] values[†] of $N(0)$ from the experimental values of γ;

[†] The density of states calculated from γ using equation (8.133) is not the "bare" or unrenormalized density of states $N(0)$. However, since the behavior[71] of γ is generally indicative of the behavior of $N(0)$, in the interest of simplicity we will use the symbol $N(0)$ for the density of states calculated from γ.

these values of the density of states are plotted against Z in Figure 8.15. The values of $N(0)$ are relatively small for the nonsuperconducting monovalent metals and for the alkaline earths. The density of states varies greatly with Z for the transition metals, with peaks in $N(0)$ for $Z = 3$ (nonsuperconducting Sc, Lu, Y, and superconducting La), for $Z = 5$ (superconducting V, Nb, and Ta), and for $Z = 9$ and 10 (nonsuperconducting Rh, Pt, and Pd).

Finally, Figure 8.16 shows the transition temperatures[72] T_c of several superconductors (and upper limits[68] on T_c for nonsuperconductors) plotted as a function of the number Z of valence electrons per atom. The nonsuperconductivity of the alkali metals correlates with their small density of states *and* (presumably) repulsive values of V, as seen from Figures 8.14 and 8.15; the same is true of the alkaline earth and noble metals. The superconductivity of the nontransition metals of valence 2, 3, 4 at relatively high temperatures is in accordance with their large values of the electron–phonon interaction V (Figure 8.14) and is presumably in spite of their small values of the density of states $N(0)$ as seen in Figure 8.15. The values of T_c for the transition metals show complex behavior,[73] with peaks in T_c at $Z = 5$ and $Z = 7$ and a relative minimum (excepting uranium) at $Z = 6$. These results may be correlated with the maximum in $N(0)$ at $Z = 5$ and the minimum in $N(0)$ at $Z = 6$; both effects are seen in Figure 8.15. The maximum in T_c at $Z = 7$ correlates with the moderately large values of $N(0)$ and of V for rhenium and technetium seen in Figures 8.14 and 8.15.

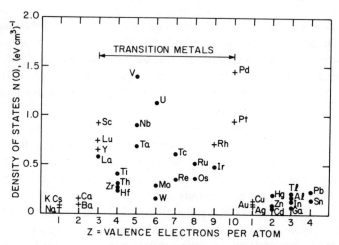

Figure 8.15. Density of states $N(0)$, calculated[68] from equation (8.133), plotted as a function of the number Z of valence electrons per atom. Superconductors are indicated by dots; nonsuperconductors (at atmospheric pressure) by crosses.

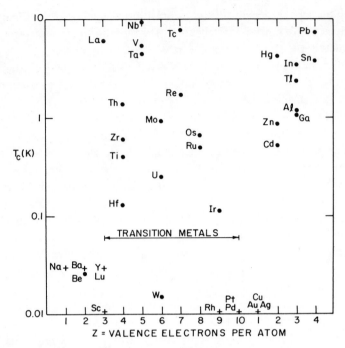

Figure 8.16. Transition temperatures T_c of superconductors[72] (dots) and upper limits[68] on T_c for nonsuperconductors (crosses), plotted as a function of the number Z of valence electrons. Note that T_c is plotted on a logarithmic scale.

These conclusions are in accord with the empirical results of Matthias,[74,75] often referred to as "Matthias' rules." These rules include the observation of maxima in T_c at $Z = 5$ and $Z = 7$, and a minimum in T_c at $Z = 6$, results that correlate[76] well with the description in terms of $N(0)$ and V given above. Further discussion of the systematics of superconductivity may be found in the literature.[77]

We now want to consider superconductors with high transition temperatures in the neighborhood of 20 K. All of these high-temperature superconductors are alloys or compounds, often of nonstochiometric composition. In fact,[78] all known superconductors with values of T_c greater than 17 K are compounds or alloys containing niobium; those with values of T_c above 18 K occur in the A15 crystal structure.[79] In order to discuss superconductors with high values of T_c, it is necessary to drop the weak-coupling requirement that $N(0)V \ll 1$.

The strong-coupling regime is one in which $N(0)V$ is not small compared to unity, so the result (8.126) for T_c is no longer valid. An outline[80] of the results of the application of the strong-coupling regime to the BCS theory is as follows. In the theory, there is a well-known equation, the BCS gap equation, which, in the weak-coupling limit with the attractive interaction V given by (8.124) and (8.125), leads to the well-known results (8.126) and (8.128) for T_c and Δ. In the strong-coupling regime, the BCS gap equation becomes a set of coupled integral equations called the Eliashberg equations. Solution of the Eliashberg equations will give an expression for T_c, in terms of the density of states at the Fermi energy and the electron–phonon interaction, valid in the strong-coupling regime. However, the results of the strong-coupling treatment are generally expressed, not in terms of $N(0)$ and V directly, but as a function of new parameters, which we now introduce.

The quantity V in equation (8.124) is the *net* attractive interaction between electrons due to the electron–phonon interaction, so

$$V = V_{ph} - V_c \tag{8.134}$$

where V_{ph} represents the attractive interaction and V_c the repulsive Coulomb interaction. Consider a quantity λ defined by

$$\lambda \equiv N(0)V_{ph} \tag{8.135}$$

where $N(0)$ is the usual density of electron states of one spin at the Fermi energy. The parameter λ is thus a measure of the attractive electron–phonon interaction; its value[81] ranges from 0.1 to 1.5 in different metals. Introducing a second parameter μ defined by

$$\mu \equiv N(0)V_c \tag{8.136}$$

we see that μ is a measure of the Coulomb repulsion between the electrons forming a pair. Our original variable $N(0)V$ is then given by

$$N(0)V = \lambda - \mu \tag{8.137}$$

However, consideration of the retardation[81] of the phonon-mediated attraction relative to the much more rapid Coulomb repulsion suggests that (8.137) be modified to

$$N(0)V = \lambda - \mu^* \tag{8.138}$$

where the quantity μ^* is related to μ by the relation

$$\mu^* = \mu/[1 + \mu \ln(E_F/k_B\Theta_D)] \tag{8.139}$$

and where E_F is the Fermi energy, and Θ_D is the Debye temperature. In terms of λ and μ^*, the weak-coupling result (8.126) for T_c becomes[82]

$$T_c \propto \Theta_D \exp[-1/(\lambda - \mu^*)] \qquad (8.140)$$

For strong coupling, the electron–phonon interaction is large enough that $N(0)V_{ph}$, and hence λ, are close to unity. Solution of the Eliashberg equations in the strong-coupling regime results in the approximate relation

$$T_c \cong \frac{\Theta_D}{1.45} \exp\left[-\frac{1.04(1+\lambda)}{\lambda - \mu^*(1+0.62\lambda)}\right] \qquad (8.141)$$

Equation (8.141), giving T_c as a function of λ, μ^*, and Θ_D, is usually called the McMillan equation, and values of T_c calculated[83] using it agree quite well with the experimental values, including those for the strong coupling superconductors lead and mercury, for which λ has the values 1.55 and 1.60, respectively.

Our objective will be to use equation (8.141) to discuss some factors influencing the magnitude of T_c. Clearly, as λ increases in (8.141) T_c increases, so an increase in the strength λ of the electron–phonon interaction will tend to increase the transition temperature. The quantity μ^* may be determined[84] from experimental data on the isotope shift coefficient of a superconductor; most values of μ^* lie near 0.13. Regarding μ^* as a constant in the McMillan equation (8.141) allows us to regard T_c as a function of the electron–phonon interaction strength λ (and, of course, of Θ_D). We now consider the solid state factors which influence the magnitude of λ. It can be shown[85] that λ may be expressed as

$$\lambda = \frac{N(0)\langle I^2 \rangle}{M\langle \omega^2 \rangle} \qquad (8.142)$$

where M is the ionic isotopic mass in the superconductor and $\langle I^2 \rangle$ is the average of the square of the matrix element I of the electron–phonon interaction. The quantity $\langle \omega^2 \rangle$ is the average of the square of the phonon frequency ω, taken over the phonon distribution for the superconductor. As before, a large value of the density of states $N(0)$ favors a large λ, as does a large value of the electron–phonon matrix element I. A small atomic mass M also increases λ and favors a relatively high T_c. This may be the reason[70] beryllium alone among the alkaline earth metals is a superconductor at atmospheric pressure. Finally, a decrease in $\langle \omega^2 \rangle$ will increase λ; the resultant expected decrease in $\langle \omega \rangle$, and hence in Θ_D, in (8.141) will not be as important as the increase in λ in the exponential term. The factors in equation (8.142) for λ may be used to discuss the transition temperatures of actual superconducting materials.[86,87] It has been suggested[88] that the origin of large values of λ (of about 2) in superconductors with high values of T_c lies in especially large values of the quantity $N(0)\langle I^2 \rangle$. However, these questions are still not resolved, and are the subject of intensive research, particularly on the A15 compounds which exhibit transition temperatures above 20 K.

As a final point on the magnitude of T_c, it may be noted that the McMillan equation (8.141) is a valid representation of T_c as a function of λ only for values of λ less than about 2. Above $\lambda \cong 2$, recent results[88] have shown that

$$T_c \propto (\langle \omega^2 \rangle \lambda)^{1/2} \tag{8.143}$$

suggesting that T_c increases monotonically with increasing values of λ. This would mean that there is, in principle, no natural[†] maximum[89] superconducting transition temperature. However, the question is still under discussion.[90]

Problems

8.1. *DC Josephson Effect Equations with Reverse Flow.*[‡] In the derivation of equation (8.32), the effect on ψ_1 of "backflow" tunneling from side (1) to side (2) was neglected. The same is true for tunneling from side (2) to side (1) in equation (8.33). To explore this question, add a term $-\hbar T \psi_1$ to equation (8.32) and a term $-\hbar T \psi_2$ to equation (8.33). Proceed with the development and obtain the equivalents of equations (8.40)–(8.43). Comment on the physical meaning and reasonableness of your results.

8.2. *Voltage Measurement with a DC Squid.* Ignoring noise considerations and any capacitance, consider measurement of a voltage V in a circuit whose resistance R is such that the time constant (determined by the inductance L of the squid circuit) is 1 sec. Estimate the voltage sensitivity of measurements with the squid.

8.3. *Magnitude of Coulomb Term in the McMillan Equation.* Using representative values of the solid state parameters in equation (8.139) defining μ^*, show that μ^* is of the order of 0.1 for a typical metal.

[†] In this statement, "natural" refers to a maximum arising from the structure of the equation alone. There may be maxima in transition temperatures due to lattice instabilities, etc.

[‡] The author would like to thank B. Black for pointing out this question and its answer.

References and Comments

1. C. Kittel, *Introduction to Solid State Physics*, Fifth Edition, John Wiley, New York (1976), Chapter 12.
2. N. W. Ashcroft and N. D. Mermin, *Solid State Physics*, Holt, Rinehart, and Winston, New York (1976), Chapter 34.
3. M. Tinkham, *Introduction to Superconductivity*, McGraw-Hill, New York (1975), Chapter 1.
4. E. A. Lynton, *Superconductivity*, Third Edition, Methuen, London (1969).
5. R. P. Feynman, R. B. Leighton, and M. Sands, *The Feynman Lectures on Physics*, Addison-Wesley, Reading, Massachusetts (1975), Volume III, Sections 21-4, 21-5.
6. For an introductory discussion of a microscopic quantum mechanical treatment of the Josephson equations, see J. Clarke, "The Josephson Effect and e/h," *American Journal of Physics*, **38**, 1071–1095 (1970), Section II.1.
7. L. I. Schiff, *Quantum Mechanics*, Third Edition, McGraw-Hill, New York (1968).
8. R. P. Feynman, R. B. Leighton, and M. Sands, Reference 5, Sections 21-3, 21-5.
9. D. Bohm, *Quantum Theory*, Prentice-Hall, New York (1951), Sections 4.9, 6.6, 6.7, 6.8.
10. J. Clarke, Reference 6, page 1073.
11. C. Kittel, Reference 1, pages 390–392.
12. R. P. Feynman, R. B. Leighton, and M. Sands, Reference 5, Section 21-9.
13. L. Solymar, *Superconductive Tunneling and Applications*, John Wiley, New York (1972), Chapters 8–10.
14. L. I. Schiff, Reference 7, pages 101–105.
15. B. D. Josephson, "Superconductive Tunneling," in *Superconductivity in Science and Technology*, M. H. Cohen (editor), University of Chicago Press, Chicago (1968), page 20.
16. J. Clarke, private communication.
17. See, for example, M. Tinkham, Reference 3, page 194.
18. J. Clarke, "Josephson Junction Detectors," *Science*, **184**, 1235–1242 (1974).
19. J. Clarke, "Electronics with Superconducting Junctions," *Physics Today*, **24**, 30–37 (August 1971).
20. M. Tinkham, Reference 3, pages 204–205.
21. J. Clarke, Reference 19, page 31.
22. L. Solymar, Reference 13, page 153; M. Tinkham, Reference 3, pages 195–196.
23. M. Tinkham, Reference 3, pages 194–195.
24. J. Clarke, Reference 6, Section II.2.
25. C. Kittel, Reference 1, pages 388–390.
26. L. Solymar, Reference 13, Chapter 11.
27. L. Solymar, Reference 13, Chapters 8, 15.
28. P. L. Richards, "The Josephson Junction as a Detector of Microwave and Far Infrared Radiation," in *Semiconductors and Semimetals*, R. K. Willardson and A. C. Beer (editors), Academic Press, New York (1977), Volume 12, Chapter 6.
29. J. Clarke, Reference 6, II.3, pages 1078–1080.
30. J. E. Mercereau, "Superconductivity," in *Topics in Solid State and Quantum Electronics*, W. D. Hershberger (editor), John Wiley, New York (1972), Chapter 5, pages 235–236.
31. L. Solymar, Reference 13, pages 157–164.
32. L. Solymar, Reference 13, Chapter 17, pages 271–275.

33. See, for example, F. B. Hildebrand, *Advanced Calculus for Applications*, Prentice-Hall, New York (1962), pages 226–231.
34. P. L. Richards, Reference 29, page 398.
35. L. Solymar, Reference 13, pages 159–160.
36. J. Clarke, Reference 6, Figure 5.
37. See, For example, E. Butkov, *Mathematical Physics*, Addison-Wesley, Reading, Massachusetts (1968), Figure 9.2, page 364; E. Jahnke and F. Emde, *Tables of Functions*, Dover, New York (1943), pages 156–157.
38. F. B. Hildebrand, Reference 33, page 144, equation (70).
39. A. H. Silver and J. E. Zimmerman, "Josephson Weak Link Devices," in *Applied Superconductivity*, V. L. Newhouse (editor), Academic Press, New York (1975), Volume 1, Chapter 1, pages 89–96.
40. See, for example, C. Kittel, Reference 1, pages 380–382.
41. See, for example, C. Kittel, *Introduction of Solid State Physics*, Fourth Edition, John Wiley, New York (1971), Advanced Topic I, pages 727–729.
42. J. D. Jackson, *Classical Electrodynamics*, Second Edition, John Wiley, New York (1975), page 815, equation (A.8).
43. C. Kittel, Reference 1, page 393.
44. J. E. Mercereau, "Macroscopic Quantum Phenomena," in *Superconductivity*, R. D. Parks (editor), Marcel Dekker, New York (1969), Volume 1, Chapter 8.
45. R. C. Jaklevic, J. Lambe, J. E. Mercereau, and A. H. Silver, "Macroscopic Quantum Interference in Superconductors," *Physical Review*, **140**, A1628–A1637 (1965).
46. J. E. Mercereau, Reference 44, page 406; R. C. Jaklevic *et al.*, Reference 45, page A1630.
47. L. Solymar, Reference 13, page 201, equation (13.6).
48. R. C. Jaklevic *et al.*, Reference 45, page A1631.
49. J. Clarke, "Low Frequency Applications of Superconducting Quantum Interference Devices," *Proceedings IEEE*, **61**, 8–19 (1973), Section III.
50. J. Clarke, Reference 49, Section IV.
51. L. Solymar, Reference 13, Chapter 18.
52. (a) J. Clarke, "The Application of Josephson Junctions to Computer Storage and Logic Elements and to Magnetic Measurements," in *Magnetism and Magnetic Materials—1975*, J. J. Becker, G. H. Lander, and J. J. Rhyne (editors), Conference Proceedings No. 29, American Institute of Physics, New York (1975), pages 20–21; (b) W. Anacker, "Computing at 4K," *IEEE Spectrum*, **16**, 26 (May 1979).
53. J. Clarke, "Superconducting Quantum Interference Devices for Low Frequency Measurements," in *Superconductor Applications: Squids and Machines (1977)*, B. B. Schwartz and S. Foner (editors), Plenum Publishing, New York (1978), Chapter 3, pages 67–124.
54. L. Solymar, Reference 13, Chapters 14, 16, 19.
55. A. H. Silver and J. E. Zimmerman, Reference 39, pages 67–89; 96–106.
56. *Future Trends in Superconductive Electronics (Charlottesville, 1978)*, B. S. Deaver, Jr., C. M. Falco, J. H. Harris, and S. A. Wolf (editors), Conference Proceedings No. 44, American Institute of Physics, New York (1978).
57. C. Kittel, Reference 1, pages 378–379.
58. N. W. Ashcroft and N. D. Mermin, Reference 2, pages 739–746.
59. M. Tinkham, Reference 3, Chapter 2.
60. See, for example, C. Kittel, Reference 1, page 363, Figure 6.

61. T. H. Geballe and M. R. Beasley, "Superconducting Materials for Energy-Related Applications," in *Materials Science in Energy Technology*, G. G. Libowitz and M. S. Wittingham (editors), Academic Press, New York (1979), page 522.

62. A. L. Robinson, "Superconductivity: Surpassing the Hydrogen Barrier," *Science*, **183**, 293–296 (January 25, 1974).

63. N. W. Ashcroft and N. D. Mermin, Reference 2, page 743; M. Tinkham, Reference 3, pages 17–21.

64. C. Kittel, *Quantum Theory of Solids*, John Wiley, New York (1963), page 162.

65. R. Meservey and B. B. Schwartz, "Equilibrium Properties: Comparison of Experimental Results with Predictions of BCS Theory," in *Superconductivity*, R. D. Parks (editor), Marcel Dekker, New York (1969), Volume 1, pages 117–191.

66. C. Kittel, Reference 1, page 169, equation (32), with an effective mass m^* in place of the free electron mass m.

67. Calculated by the author.

68. G. Gladstone, M. A. Jensen, and J. R. Schrieffer, "Superconductivity in the Transition Metals: Theory and Experiment," in *Superconductivity*, R. D. Parks (editor), Marcel Dekker, New York (1969), pages 665–816; Table VI, page 734.

69. C. Kittel, Reference 1, pages 164–167.

70. R. M. White and T. H. Geballe, *Long Range Order in Solids*, Supplement 15 to *Solid State Physics*, H. Ehrenreich, F. Seitz, and D. Turnbull (editors), Academic Press, New York (1979), page 221.

71. G. Gladstone *et al.*, Reference 68, pages 682–685. See also R. M. White and T. H. Geballe, Reference 70, page 112, equation (3.46).

72. R. M. White and T. H. Geballe, Reference 70, front end papers.

73. G. Gladstone *et al.*, Reference 68, Figure 30, page 736.

74. B. T. Matthias, "Superconductivity in the Periodic System," in *Progress in Low Temperature Physics*, C. J. Gorter (editor), North-Holland, Amsterdam (1957), Volume II, pages 138–150; Figure 2, page 140.

75. B. T. Matthias, "The Empirical Approach to Superconductivity," in *Applied Solid State Physics*, W. Low and M. Schieber (editors), Plenum Press, New York (1970), pages 179–188; Figure 2, page 184.

76. D. Dew-Hughes, "Practical Superconducting Materials," in *Superconducting Machines and Devices*, S. Foner and B. B. Schwartz (editors), Plenum Press, New York (1974), Chapter 2, pages 91–92.

77. R. M. White and T. H. Geballe, Reference 70, pages 220–246.

78. R. M. White and T. H. Geballe, Reference 70, page 221.

79. T. H. Geballe and M. R. Beasley, "Superconducting Materials for Energy-Related Applications," in *Materials Science in Energy Technology*, G. G. Libowitz and M. S. Wittingham (editors), Academic Press, New York (1979), Chapter 10; pages 520–533.

80. R. M. White and T. H. Geballe, Reference 70, pages 104–113.

81. R. M. White and T. H. Geballe, Reference 70, pages 92–103.

82. D. J. Scalapino, "The Electron–Phonon Interaction and Strong Coupling Superconductors," in *Superconductivity*, R. D. Parks (editor), Marcel Dekker, New York (1969), Volume 1, pages 449–560; equation (135).

83. R. M. White and T. H. Geballe, Reference 70, page 228.

84. R. M. White and T. H. Geballe, Reference 70, pages 112, 226, and 228.

85. R. M. White and T. H. Geballe, Reference 70, pages 93–98.

86. T. H. Geballe and M. R. Beasley, Reference 79, pages 534–537.

87. R. M. White and T. H. Geballe, Reference 70, pages 220–246.
88. P. B. Allen and R. C. Dynes, "Transition Temperatures of Strong-Coupled Super-conductors Reanalyzed," *Physical Review B*, **12**, 905–922 (1975).
89. R. M. White and T. H. Geballe, Reference 70, pages 118–121.
90. See, for example, K. M. Ho, M. L. Cohen, and W. E. Pickett, "Maximum Super-conducting Transition Temperature in A15 Compounds?," *Physical Review Letters*, **41**, 815–818 (1978).

Suggested Reading

C. KITTEL, *Introduction to Solid State Physics*, Fifth Edition, John Wiley, New York (1976). The basic background reference for this chapter is Kittel's Chapter 12, which includes an introduction to the Josephson effects and to superconducting quantum interference.

L. SOLYMAR, *Superconductive Tunneling and Applications*, John Wiley, New York (1972). A discussion, at the intermediate level, of both the basic ideas and many kinds of applications of the Josephson effects.

R. P. FEYNMAN, R. B. LEIGHTON, and M. SANDS, *The Feynman Lectures on Physics*, Addison-Wesley, Reading, Massachusetts (1965), Volume III, Chapter 21. A treatment of the macroscopic quantum mechanical description of superconductivity on which our discussion is based. Highly recommended.

B. S. DEAVER, C. M. FALCO, J. H. HARRIS, and S. A. WOLF, editors, *Future Trends in Superconductive Electronics (Charlottsville, 1978)*, Conference Proceedings No. 44, American Institute of Physics, New York (1978). This collection of papers, both research and review, is suggested for a view of the field as of early 1978.

R. M. WHITE and T. H. GEBALLE, *Long Range Order in Solids*, Supplement 15 to *Solid State Physics*, H. Ehrenreich, F. Seitz, and D. Turnbull (editors), Academic Press, New York (1979). This recent monograph, at the advanced level, offers an excellent discussion of the physics of superconducting materials for the reader already familiar with the BCS theory.

9

Physics and Applications of the Nonlinear Optical Properties of Solids

Introduction

The aim of this chapter is a discussion of some of the physics and applications of the nonlinear optical properties of solids. It begins with a review of electromagnetic wave propagation in solids and a derivation of the familiar linear relation between dielectric polarization and electric field. A more realistic anharmonic oscillator model is next introduced and this is shown to yield a polarization that is a nonlinear (i.e., a quadratic) function of the electric field. A physical picture of the nonlinear polarization and some solid state physics factors affecting the magnitude of the nonlinear susceptibility are also given. The central topic of the chapter is the propagation and interaction of three electromagnetic waves in a nonlinear medium. This results in a set of equations describing the spatial variation of the electric fields of these waves as they move through the crystal. Finally, these equations are used to discuss several applications of nonlinear solids. These are optical second harmonic generation, frequency mixing and up-conversion, and parametric amplification of optical signals.

Review of Electromagnetic Wave Propagation in Solids

We begin by writing Maxwell's equations

$$\text{curl } \mathbf{H} = \frac{4\pi}{c}\,\mathbf{J} + \frac{1}{c}\,\frac{\partial \mathbf{D}}{\partial t} \tag{9.1}$$

$$\text{curl } \mathbf{E} = -\frac{1}{c}\frac{\partial \mathbf{B}}{\partial t} \tag{9.2}$$

$$\text{div } \mathbf{D} = 4\pi\varrho \tag{9.3}$$

$$\text{div } \mathbf{B} = 0 \tag{9.4}$$

using Gaussian units and in macroscopic form. The various quantities in equations (9.1)–(9.4) are macroscopic values[1] obtained by an averaging process which smooths out the microscopic details due to the atomic nature of the matter in the solid being considered. The quantity $\mathbf{E}$ is the electric field, $\mathbf{H}$ is the magnetic intensity, $\mathbf{D}$ is the electric displacement, $\mathbf{B}$ is the magnetic induction, ϱ is the free electric charge density, and $\mathbf{J}$ is the free electric current density. We introduce also the constitutive relations

$$\mathbf{D} = \mathbf{E} + 4\pi\mathbf{P} \tag{9.5}$$

$$\mathbf{H} = \mathbf{B} + 4\pi\mathbf{M} \tag{9.6}$$

where $\mathbf{P}$ is the electric polarization (the electric dipole moment per unit volume) and $\mathbf{M}$ is the magnetization (the magnetic dipole moment per unit volume). Equations (9.5) and (9.6) may be substituted into Maxwell's equations (9.1)–(9.4) to give

$$\text{curl } \mathbf{H} = \frac{1}{c}\frac{\partial \mathbf{E}}{\partial t} + \frac{4\pi}{c}\mathbf{J} + \frac{4\pi}{c}\frac{\partial \mathbf{P}}{\partial t} \tag{9.7}$$

$$\text{curl } \mathbf{E} = -\frac{1}{c}\frac{\partial \mathbf{H}}{\partial t} + \frac{4\pi}{c}\frac{\partial \mathbf{M}}{\partial t} \tag{9.8}$$

$$\text{div}(\mathbf{E} + 4\pi\mathbf{P}) = 4\pi\varrho \tag{9.9}$$

$$\text{div}(\mathbf{H} - 4\pi\mathbf{M}) = 0 \tag{9.10}$$

Equations (9.7)–(9.10) connect the electromagnetic wave specified by the fields $\mathbf{E}$ and $\mathbf{H}$ with the properties $\mathbf{J}$, ϱ, $\mathbf{P}$, and $\mathbf{M}$ of the solid under consideration.

We restrict ourselves to solids that are nonmagnetic, so $\mathbf{M} = 0$, and electrically neutral, so $\varrho = 0$, and equations (9.7)–(9.10) become

$$\text{curl } \mathbf{H} = \frac{1}{c}\frac{\partial \mathbf{E}}{\partial t} + \frac{4\pi}{c}\mathbf{J} + \frac{4\pi}{c}\frac{\partial \mathbf{P}}{\partial t} \tag{9.11}$$

$$\text{curl } \mathbf{E} = -\frac{1}{c}\frac{\partial \mathbf{H}}{\partial t} \tag{9.12}$$

$$\text{div } \mathbf{E} = -4\pi \text{ div } \mathbf{P} \tag{9.13}$$

$$\text{div } \mathbf{H} = 0 \tag{9.14}$$

Equations (9.11)–(9.14) govern the propagation of an electromagnetic wave in an electrically neutral nonmagnetic solid. We may obtain the wave equations for **H** and **E** by taking the curls of equations (9.11) and (9.12). For the electric field, we obtain

$$\text{curl curl } \mathbf{E} = -\frac{1}{c} \frac{\partial}{\partial t} (\text{curl } \mathbf{H}) \tag{9.15}$$

leading to

$$\text{curl curl } \mathbf{E} = -\frac{1}{c} \frac{\partial}{\partial t} \left(\frac{1}{c} \frac{\partial \mathbf{E}}{\partial t} + \frac{4\pi}{c} \mathbf{J} + \frac{4\pi}{c} \frac{\partial \mathbf{P}}{\partial t} \right) \tag{9.16}$$

on substituting equation (9.11) into equation (9.15). Using a standard vector identity for curl curl **E** in equation (9.16) and rearranging terms gives

$$\text{grad div } \mathbf{E} - \nabla^2 \mathbf{E} + \frac{1}{c^2} \frac{\partial^2 \mathbf{E}}{\partial t^2} = -\frac{4\pi}{c^2} \left(\frac{\partial \mathbf{J}}{\partial t} + \frac{\partial^2 \mathbf{P}}{\partial t^2} \right) \tag{9.17}$$

as the wave equation governing the electric field **E** in the solid.

We see that the wave equation (9.17) contains two source terms on its right-hand side. The term in the current density **J** stems from free conduction charges in the solid and the term in the polarization **P** has its origin in the bound polarization charges. The solutions of the wave equation (9.17) for the medium of interest will describe the propagation of electromagnetic waves in that medium. In metals, the source term in **J** is the more important, and the resulting solutions of the wave equation describe the familiar results of the optics of metals. In a nonconducting solid the term in the electric polarization **P** will be more important and leads to a description of dispersion, absorption, etc., in dielectric solids.

For a nonconducting dielectric material, the current density **J** will be zero, so the wave equation (9.17) becomes

$$\text{grad div } \mathbf{E} - \nabla^2 \mathbf{E} + \frac{1}{c^2} \frac{\partial^2 \mathbf{E}}{\partial t^2} = -\frac{4\pi}{c^2} \frac{\partial^2 \mathbf{P}}{\partial t^2} \tag{9.18}$$

The inhomogeneous wave equation (9.18) describes the electric field of an electromagnetic wave propagating in a nonmagnetic dielectric solid. The source term $(\partial^2 \mathbf{P}/\partial t^2)$ involves the polarization, so, in order to solve (9.18), we must set up a model of a dielectric solid from which we can calculate **P** and its time derivatives.

Electric Polarization in an Isotropic Linear Dielectric Solid

Since we are interested in the propagation of electromagnetic waves, we are concerned with the polarization at optical frequencies, and hence with the electronic polarizability.[2] Because of the large mass of atomic nuclei, the ionic and dipolar polarizability cannot respond to the high-frequency electric field of the light wave. We will therefore consider only the polarization of the electron clouds of the atoms in the solid.

To consider the interaction of the incident electromagnetic wave with the electrons, we construct a model of the solid in which the atoms have the following features. The model is one dimensional and the bound electrons in the atoms behave as damped harmonic oscillators in which the damping force $-m\gamma\dot{x}$ is proportional to the electron velocity $\dot{x}$. The coefficient γ is a constant and m is the electron mass. The restoring force on the electron is $-m\omega_0^2 x$ and is thus a *harmonic* Hooke's law force; the quantity ω_0 is a constant characteristic frequency. The force due to the electric field $E(t)$ is $-eE(t)$, where

$$E(t) = E_0 \exp(-j\omega t) \tag{9.19}$$

is the time-dependent electric field of the light wave of circular frequency ω; $-e$ is the charge on the electron and E_0 is the amplitude of the electric field. For this model, the one-dimensional equation of motion for the displacement $x(t)$ of the electron from its equilibrium position is

$$m\ddot{x} = -m\gamma\dot{x} - m\omega_0^2 x - eE \tag{9.20}$$

The solution of equation (9.20) is

$$x(t) = \frac{-e/m}{\omega_0^2 - \omega^2 - j\gamma\omega} E_0 \exp(-j\omega t) \tag{9.21}$$

In obtaining equations (9.20) and (9.21), the force on the electron due to the magnetic field of the light wave has been omitted as negligible compared to the force due to the electric field.

The electric polarization P is the electric dipole moment per unit volume. Since the electronic dipole moment induced by the electric field of the light wave is $-ex$, we have

$$P = -Nex \tag{9.22}$$

where N is the number of electrons per unit volume. Combining equations

(9.21) and (9.22) gives

$$P = \frac{Ne^2/m}{\omega_0^2 - \omega^2 - j\gamma\omega} E_0 e^{-j\omega t} = \frac{Ne^2/m}{\omega_0^2 - \omega^2 - j\gamma\omega} E \qquad (9.23)$$

showing that, for this model of a solid, the electronic polarization P is a *linear* function of the electric field E of the light wave. Equation (9.23) shows also that the polarization has the same time dependence $\exp(-j\omega t)$ as the incident electric field, and hence has the same frequency ω as the light wave. In equations (9.19)–(9.23), we have tacitly assumed that the dielectric is isotropic, and so have written P and E as scalars in those equations.

Strictly speaking, equation (9.23) is not correct because we have neglected the electric field produced by the polarization of the solid. We should rewrite equation (9.23) to read

$$P = \frac{Ne^2/m}{\omega_0^2 - \omega^2 - j\gamma\omega} E_{\text{loc}} \qquad (9.24)$$

where E_{loc} is the local electric field at the atom whose electronic polarization we are considering. The local electric field E_{loc} is made up[3] of the sum of the applied electric field E of the light wave plus the total field at the atom due to the electric dipole moments (i.e., the polarization) of all the other atoms. We will consider the especially simple case of an isotropic solid in a form for which the depolarization field is zero,[3] and for which then the only contributions to E_{loc} are the Lorentz field $4\pi P/3$ and the applied electric field E of the light wave. In this case, we have

$$E_{\text{loc}} = E + 4\pi P/3 \qquad (9.25)$$

so equation (9.24) becomes

$$P = \frac{Ne^2/m}{\omega_0^2 - \omega^2 - j\gamma\omega} \left(E + \frac{4\pi}{3} P \right) \qquad (9.26)$$

which, on solving for P, yields

$$P = \frac{Ne^2/m}{(\omega_0^2 - 4\pi Ne^2/3m) - \omega^2 - j\gamma\omega} E \qquad (9.27)$$

for the polarization P as a function of the electric field E of the light wave in the solid.

Equation (9.27) gives the linear relation between the polarization and electric field for our model of an isotropic *linear* dielectric solid. It is of the form

$$P = \chi(\omega)E \tag{9.28}$$

where

$$\chi(\omega) \equiv \frac{Ne^2/m}{(\omega_0^2 - 4\pi Ne^2/3m) - \omega^2 - j\gamma\omega} \tag{9.29}$$

is the frequency-dependent linear electric susceptibility. Note that, for our model of a solid in which the restoring force on the electron is linear, the susceptibility given by equation (9.29) is independent of the applied electric field. Further, since our model is isotropic, equations (9.28) and (9.29) describe a scalar electric susceptibility. On rewriting equation (9.28) in vector notation, we have

$$\mathbf{P} = \chi(\omega)\mathbf{E} \tag{9.30}$$

where $\chi(\omega)$ is a scalar. For an anisotropic solid,[4] equation (9.30) would become

$$\mathbf{P} = \boldsymbol{\chi}(\omega) \cdot \mathbf{E} \tag{9.31}$$

where $\boldsymbol{\chi}(\omega)$ is the second-rank electric susceptibility tensor, and $\mathbf{P}$ and $\mathbf{E}$ are vectors.

For our isotropic dielectric, then, equation (9.30) allows us to calculate the source term $\partial^2\mathbf{P}/\partial t^2$ in the inhomogeneous wave equation (9.18). We obtain

$$-\frac{4\pi}{c^2}\frac{\partial^2\mathbf{P}}{\partial t^2} = -\frac{4\pi}{c^2}\chi(\omega)\frac{\partial^2\mathbf{E}}{\partial t^2} \tag{9.32}$$

which, when substituted in equation (9.18), yields

$$\text{grad div } \mathbf{E} - \nabla^2\mathbf{E} + \frac{1}{c^2}\frac{\partial^2\mathbf{E}}{\partial t^2} = -\frac{4\pi}{c^2}\chi(\omega)\frac{\partial^2\mathbf{E}}{\partial t^2} \tag{9.33}$$

as the wave equation. Equation (9.33) may be simplified by using the Maxwell equation (9.13)

$$\text{div } \mathbf{E} = -4\pi \text{ div } \mathbf{P} \tag{9.13}$$

which on substituting the relation (9.30) between the polarization and the electric field, becomes

$$\text{div } \mathbf{E} = -4\pi\chi(\omega) \text{ div } \mathbf{E} \tag{9.34}$$

Equation (9.34) may be rewritten as

$$[1 + 4\pi\chi(\omega)] \text{ div } \mathbf{E} = 0 \tag{9.35}$$

an equation whose solution is

$$\text{div } \mathbf{E} = 0 \tag{9.36}$$

Equation (9.36) stating the vanishing of the divergence of the electric field is a consequence of the fact that our solid is a *linear* isotropic dielectric for which the susceptibility is independent of the applied electric field. This is equivalent to characterizing the solid by a uniform dielectric constant independent of the electric field, a condition[5] that leads to the vanishing of div **E**. When equation (9.36) is substituted into the wave equation (9.33), we obtain

$$\nabla^2 \mathbf{E} = \frac{1}{c^2} [1 + 4\pi\chi(\omega)] \frac{\partial^2 \mathbf{E}}{\partial t^2} \tag{9.37}$$

as the wave equation for the electric field of a light wave propagating in a linear isotropic dielectric solid whose electric susceptibility is $\chi(\omega)$. It is worth emphasizing that the wave equation (9.37) is a special case of the more general wave equation (9.18) and that equation (9.37) has the form that it does because our model of a dielectric has the specific features described above, in particular a *linear* relation between the polarization and the applied electric field.

Physically, we may regard the propagation of the light wave governed by equation (9.37) in the linear dielectric in the following way. The electric field of the incident light drives the electrons in the solid. These electrons act as driven, damped harmonic oscillators whose displacement $x(t)$ is given by equation (9.21). This oscillation of electric charge at the frequency ω of the driving electric field constitutes the polarization at ω expressed by equation (9.23) and produces radiation at frequency ω. In this way, the electrons in the solid reradiate the incident light wave, which thus propagates through the solid with frequency ω.

We may also describe the process in terms of the source term in $\partial^2 P/\partial t^2$ in the wave equation (9.18). The linear relation (9.23) between P and E means that the polarization has the same time dependence, and thus the same frequency, as the incident electric field. The source term $\partial^2 P/\partial t^2$ will also be at the frequency ω of the incident light wave, meaning that the light propagates through the solid with an unchanged frequency ω. This fact is a direct result of the linear relation (9.23) between the polarization and the electric field. Later, we will examine the results stemming from a nonlinear relation between P and E, and we go on to consider a picture of a solid in which such a nonlinear relation is the case.

Nonlinear Polarization and Nonlinear Susceptibility

In the preceding section, we examined a particular model of a dielectric solid, namely, a classical model in which the restoring force on the electrons was linear in the coordinate x. One of the properties of this linear model was expressed in equation (9.30),

$$\mathbf{P} = \chi(\omega)\mathbf{E} \tag{9.30}$$

stating that the polarization is a linear function of the electric field. In the limit of weak electric fields, the linear relation between $\mathbf{P}$ and $\mathbf{E}$ is a good approximation for real dielectric solids. However, in general[6] it is true that

$$\mathbf{P} = f(\mathbf{E}) \tag{9.38}$$

where $f(\mathbf{E})$ is a *nonlinear* function of the electric field $\mathbf{E}$. Often the polarization can be written as a power series in the electric field,

$$\mathbf{P} = \boldsymbol{\chi}^{(1)} \cdot \mathbf{E} + \boldsymbol{\chi}^{(2)} : \mathbf{EE} + \cdots \tag{9.39}$$

where $\boldsymbol{\chi}^{(1)}$ is the first-order (linear) electric susceptibility tensor used in equation (9.31) and $\boldsymbol{\chi}^{(2)}$ is the second-order nonlinear susceptibility tensor, etc. The tensor $\boldsymbol{\chi}^{(2)}$ is of third rank and has 27 components. We will not consider nonlinear effects higher[6-8] than the second order. Finally, we rewrite equation (9.39) in the approximate and scalar form

$$P = \chi^{(1)}E + \chi^{(2)}E^2 \tag{9.40}$$

in order to focus attention on the effects of the second-order nonlinearity. We are therefore neglecting the anisotropy of real crystals and confining ourselves to a discussion of second-order nonlinear effects in an hypothetical isotropic dielectric solid. (Strictly speaking, an isotropic optical medium cannot have a second-order nonlinearity because, as discussed later, it has a center of inversion symmetry. However, in the interest of pedagogy and simplicity, we will ignore this fact and use an isotropic solid as our illustrative model.)

Before considering a classical model of a dielectric leading to a relation of the type (9.40) between the polarization and the electric field, it is appropriate to point out *why* we are interested in the nonlinearity expressed by the term $\chi^{(2)}E^2$. Suppose an electromagnetic wave of the form

$$E = E_0 \sin \omega t \tag{9.41}$$

with circular frequency ω is incident on a solid whose polarization–electric-field relation contains a term in E^2, as in equation (9.40). The polarization in the solid will then be of the form

$$P = \chi^{(1)}E_0 \sin \omega t + \chi^{(2)}E_0^2 \sin^2 \omega t \qquad (9.42)$$

which may be rewritten as

$$P = \chi^{(1)}E_0 \sin \omega t - \tfrac{1}{2}\chi^{(2)}E_0^2 \cos(2\omega t) + \tfrac{1}{2}\chi^{(2)}E_0^2 \qquad (9.43)$$

using the trigonometric identity $\cos 2\theta = 1 - 2\sin^2\theta$. From equation (9.43), we can see that, in this case, the source term $\partial^2 P/\partial t^2$ in the wave equation (9.18) will be

$$\frac{\partial^2 P}{\partial t^2} = -\omega^2\chi^{(1)}E_0 \sin \omega t + 2\omega^2\chi^{(2)}E_0^2 \cos(2\omega t) \qquad (9.44)$$

The key result in equations (9.43) and (9.44) is that the factor E^2 in equation (9.40) leads to the presence of the second harmonic frequency 2ω in the polarization P, and thus in the source term $\partial^2 P/\partial t^2$ in the wave equation. This result is equivalent to the existence of a term oscillating at 2ω in the displacement $x(t)$ of the electrons in our model of a solid. We would expect this oscillating electron to radiate at frequency 2ω. We conclude that a light wave of frequency ω, incident on a nonlinear crystal whose polarization is described by equation (9.40), will generate radiation at the second harmonic 2ω as well as at the fundamental frequency ω.

We may also describe this process by saying that the nonlinear term E^2 in the polarization "mixes" the incident electric field, at frequency ω, with itself, leading to a component of the polarization at 2ω as well as at ω. In the same way, a nonlinear optical medium can "mix" incident waves of frequencies ω_1 and ω_2, generating radiation at the sum and difference frequencies $\omega_1 \pm \omega_2$. It is essentially the mixing of electromagnetic waves by the term E^2 in the polarization of a nonlinear medium that is the physical basis of the applications we will consider.

Anharmonic Oscillator Model of a Nonlinear Solid

In order to discuss the physical origin of the nonlinear susceptibility, we examine a model[8–11] of a solid which is more realistic than the linear model considered earlier. In the linear model, the restoring force on the electron was the Hooke's law force $-m\omega_0^2 x$ characteristic of a harmonic oscillator. We now consider a restoring force containing terms in powers of x higher than the first and which has the form

$$-m(\lambda x^2 + \delta x^3 + \cdots) \tag{9.45}$$

so we will be considering the anharmonic oscillator.[12] We will limit ourselves to the first term in the series in equation (9.45). The anharmonic term in the restoring force on the electron will thus have the form

$$-m\lambda x^2 \tag{9.46}$$

where λ is a positive constant and x is the displacement of the electron from its equilibrium position. The equation of motion, which was given by equation (9.20) for the linear or harmonic restoring force, becomes, on adding the term (9.46),

$$m\ddot{x} = -m\gamma\dot{x} - m\omega_0^2 x - m\lambda x^2 - eE \tag{9.47}$$

where the quantities γ, ω_0, m, e, and E were defined in connection with equation (9.20). Equation (9.47) is the equation of motion of a damped anharmonic oscillator driven by the electric field E of the light wave. We choose the form

$$E = E_0 \cos \omega t \tag{9.48}$$

for the electric field, so the equation of motion (9.47) becomes

$$\ddot{x} + \gamma\dot{x} + \omega_0^2 \dot{x} + \lambda x^2 = (-eE_0/m) \cos \omega t \tag{9.49}$$

Equation (9.49) is a nonlinear differential equation because of the presence of the term in x^2.

We will not solve equation (9.49) directly.[13–15] Instead, we will use the solution[16,17] to the simpler nonlinear differential equation

$$\ddot{x} + \omega_0^2 x - \lambda x^2 = 0 \tag{9.50}$$

which describes the undriven, undamped nonlinear oscillator, to suggest the key features of the solution of the equation of motion (9.49). We will assume that the constant λ is a small quantity so the solution to equation (9.50) will be close to that for a simple (linear) harmonic oscillator. The solution of equation (9.50) may be shown to be[17]

$$x(t) = A \cos \omega_0 t - (\lambda A^2/6\omega_0^2) \cos(2\omega_0 t) + (\lambda A^2/2\omega_0^2) \tag{9.51}$$

where A is a constant. Equation (9.51) gives an approximate solution to the differential equation (9.50). The solution (9.51), which neglects terms in λ^2, etc., is therefore correct to first order in λ.

The key physical idea in equation (9.51) is that the presence of the nonlinear term in x^2 in the equation of motion introduces a term at the second harmonic frequency[†] $2\omega_0$ into the expression for $x(t)$. We therefore expect that the first-order solution $x(t)$ to the equation of motion (9.49) for the driven, damped, anharmonic oscillator, will contain terms at the frequency ω of the driving electric field *and* at the second harmonic frequency 2ω. Further, by analogy with the driven, damped *harmonic* oscillator,[18] we expect that the solution to equation (9.49) will contain a resonant denominator in powers of $\omega_0 - \omega$, and will include the damping constant γ.

To obtain a solution of the equation of motion (9.49), we will, using the ideas outlined above and following Yariv,[9,10] assume a solution

$$x(t) = \tfrac{1}{2}(x_1 e^{j\omega t} + x_1{}^* e^{-j\omega t} + x_2 e^{j(2\omega)t} + x_2{}^* e^{-j(2\omega)t}) \qquad (9.52)$$

which contains oscillations at both the driving frequency ω and the second harmonic frequency 2ω. In equation (9.52), x_1 and x_2 are the amplitudes of the electron motion at the frequencies ω and 2ω, respectively, and the asterisk indicates a complex conjugate. In equation (9.52), we have ignored[‡] the nonoscillatory term contained in equation (9.51). We rewrite equation (9.49) in the form

$$\ddot{x} + \gamma\dot{x} + \omega_0{}^2 x + \lambda x^2 = (-eE_0/m)(e^{j\omega t} + e^{-j\omega t}) \qquad (9.53)$$

by using the complex form for $\cos \omega t$. We do not use the usual complex form $E_0 \exp(j\omega t)$ for the driving electric field because[10] of the presence of the term in x^2 in the differential equation.

If the assumed solution (9.52) is substituted into the differential equation (9.53), one obtains[9,10]

$$\begin{aligned}
[&-\tfrac{1}{2}\omega^2(x_1 e^{j\omega t} + x_1{}^* e^{-j\omega t}) - 2\omega^2(x_2 e^{2j\omega t} + x_2{}^* e^{-2j\omega t}) \\
&+ \tfrac{1}{2}j\omega\gamma[x_1 e^{j\omega t} - x_1{}^* e^{-j\omega t}) + j\omega\gamma(x_2 e^{2j\omega t} - x_2{}^* e^{-2j\omega t}) \\
&+ \tfrac{1}{2}\omega_0{}^2(x_1 e^{j\omega t} + x_1{}^* e^{-j\omega t} + x_2 e^{2j\omega t} + x_2{}^* e^{-2j\omega t}) \\
&+ \tfrac{1}{4}\lambda(x_1{}^2 e^{2j\omega t} + x_2{}^2 e^{4j\omega t} + 2x_1 x_1{}^* + 2x_1 x_2 e^{3j\omega t} \\
&+ 2x_1{}^* x_2 e^{j\omega t} + 2x_2 x_2{}^* + x_1{}^{*2} e^{-2j\omega t} \\
&+ x_2{}^{*2} e^{-4j\omega t} + 2x_1{}^* x_2{}^* e^{-3j\omega t} + 2x_1 x_2{}^* e^{-j\omega t})] \\
&= (-eE_0/2m)(e^{j\omega t} + e^{-j\omega t})
\end{aligned} \qquad (9.54)$$

[†] If terms in higher powers of λ had been retained in the solution (9.51), then terms at higher harmonic frequencies would have been present.

[‡] The nonoscillatory or DC term represents a nonzero average displacement of the electron and leads to the effect called optical rectification. See, for example, the book by Baldwin.[19]

Equation (9.54) must hold in order that the assumed equation (9.52) be a solution of the equation of motion (9.53). For equation (9.54) to be true, the coefficients of the term in each power of $\exp(j\omega t)$ must be the same on both sides of the equation. Considering first the term in $\exp(j\omega t)$, the coefficient of $\exp(j\omega t)$ on the left-hand side of equation (9.54) must equal the coefficient on the right-hand side, so we have

$$x_1(-\omega^2 + j\gamma\omega + \omega_0^2) + x_1^* x_2 \lambda = -eE_0/m \qquad (9.55)$$

We now make the assumption that the magnitude of the term $x_1^* x_2 \lambda$ on the left-hand side of equation (9.55) may be neglected compared to the magnitude of the term $x_1(\omega_0^2 - \omega^2 + j\gamma\omega)$. This approximation is expressed[9] as

$$| \lambda x_2 | \ll [(\omega_0^2 - \omega^2)^2 + \gamma^2\omega^2]^{1/2} \qquad (9.56)$$

This assumption is physically reasonable[†] because we are, in equation (9.47), tacitly assuming that the anharmonic contribution $m\lambda x^2$ to the restoring force is small compared to the harmonic term $m\omega_0^2 x$. We would then expect that the magnitude x_2 of the electron displacement at the frequency 2ω will be relatively small, so the product λx_2 will also be small. With the assumption (9.56), equation (9.55) becomes

$$x_1 = \frac{-eE_0/m}{\omega_0^2 - \omega^2 + j\gamma\omega} \qquad (9.57)$$

which is the same result as equation (9.21) obtained for the linear model discussed earlier. This is, of course, a consequence of neglecting the anharmonic term in (9.55) when obtaining equation (9.57).

We may now find the expression for the amplitude x_2 of the electron motion at the second harmonic frequency 2ω. Equating the coefficients of the terms in $\exp[j(2\omega)t]$ on both sides of equation (9.54) gives

$$x_2(-2\omega^2 + j\gamma\omega + \tfrac{1}{2}\omega_0^2) + \tfrac{1}{4}\lambda x_1^2 = 0 \qquad (9.58)$$

[†] We may make an estimate of the validity of the approximation (9.56) as follows. Considering visible frequencies, ω will be of the order of $2\pi \times 10^{15}$ sec^{-1}, so the right-hand side of equation (9.56) will of the order of 10^{31} sec^{-2}. The term in $\gamma^2\omega^2$ is negligible because[20] γ is of the order of 10^8 sec^{-1}. Then, using Garrett's value[21] of $\lambda \approx 10^{39}$ cm^{-1} sec^{-2} (where Garrett's v is our λ), we may estimate $| \lambda x_2 |$. We take the amplitude x_2 as a small fraction, say 1%, of the amplitude x_1, which in turn is estimated to be of the order of 10^{-8} cm, so we take $x_2 \approx 10^{-10}$ cm. This gives a value of $| \lambda x_2 | \approx 10^{29}$ sec^{-2}, which is small compared to the estimated magnitude of 10^{31} sec^{-2} of the right-hand side of equation (9.56). We are therefore justified in using the approximation.

On substituting equation (9.57) for the amplitude x_1 into equation (9.58), we obtain

$$x_2(\omega_0{}^2 - 4\omega^2 + 2j\gamma\omega) + \frac{1}{2}\lambda\frac{(e^2/m^2)E_0{}^2}{(\omega_0{}^2 - \omega^2 + j\gamma\omega)^2} = 0 \qquad (9.59)$$

which may be rewritten as

$$x_2 = \frac{-\lambda(e^2/m^2)E_0{}^2}{2(\omega_0{}^2 - \omega^2 + j\gamma\omega)^2(\omega_0{}^2 - 4\omega^2 + 2j\gamma\omega)} \qquad (9.60)$$

Equation (9.60) gives the amplitude x_2 of the electron motion at the second harmonic frequency 2ω.

Having found the electron displacements x_1 and x_2 at the frequencies ω and 2ω, respectively, we may now find the total polarization P as

$$P = N(-e)[x(t)] = \tfrac{1}{2}N(-e)(x_1 e^{j\omega t} + x_2 e^{j(2\omega)t} + \text{c.c.}) \qquad (9.61)$$

where we have used equation (9.52) for $x(t)$, and have written "c.c." for the complex conjugate terms. Substituting equations (9.57) and (9.60) for x_1 and x_2 into equation (9.61) gives

$$P = \frac{Ne^2 E_0 e^{j\omega t}}{2m(\omega_0{}^2 - \omega^2 + j\gamma\omega)} + \frac{Ne^3\lambda(E_0 e^{j\omega t})^2}{4m^2(\omega_0{}^2 - \omega^2 + j\gamma\omega)^2(\omega_0{}^2 - 4\omega^2 + 2j\gamma\omega)} \qquad (9.62)$$

where we have written $[E_0\exp(j\omega t)]^2$ in place of $E_0{}^2\exp(2j\omega t)$, and have omitted the complex conjugates for simplicity.

Equation (9.62) gives the total time- and frequency-dependent polarization P as a function of the electric field $E_0\exp(j\omega t)$ of the light wave incident on the dielectric solid. Recalling from equation (9.19) that

$$E = E_0\exp(j\omega t) \qquad (9.19)$$

we can write equation (9.62) in the form

$$P = P^{(\omega)} + P^{(2\omega)} = \chi_L{}^{(\omega)}E + d^{(2\omega)}E^2 \qquad (9.63)$$

where

$$\chi_L{}^{(\omega)} \equiv \frac{Ne^2/m}{\omega_0{}^2 - \omega^2 + j\gamma\omega} \qquad (9.64)$$

and

$$d^{(2\omega)} \equiv \frac{\lambda(Ne^3/m^2)}{2(\omega_0{}^2 - \omega^2 + j\gamma\omega)^2(\omega_0{}^2 - 4\omega^2 + 2j\gamma\omega)} \qquad (9.65)$$

and the presence of the complex conjugates in equation (9.62) provides a factor of 2 in the expressions for $\chi_L^{(\omega)}$ and $d^{(2\omega)}$. The quantities $P^{(\omega)}$ and $P^{(2\omega)}$ in equation (9.63) are, respectively, the components of the polarization at the frequencies ω and 2ω. Since $\chi_L^{(\omega)}$ is the ratio of the polarization $P^{(\omega)}$ to the first power of the electric field E, $\chi_L^{(\omega)}$ is the linear electric susceptibility. The expression (9.64) is the same as that given in equation (9.29) if the local field correction is omitted in the latter equation.

The polarization $P^{(2\omega)}$ at the second harmonic frequency 2ω is often called the nonlinear polarization because it is proportional to E^2. The quantity $d^{(2\omega)}$ in equation (9.65) is the ratio of the nonlinear polarization to the square of the electric field and is called the nonlinear electric susceptibility or the nonlinear optical coefficient. We note from equation (9.65) that the nonlinear susceptibility $d^{(2\omega)}$ arises from the presence of the anharmonic force term $m\lambda x^2$ because the expression for $d^{(2\omega)}$ contains the anharmonic force constant λ. We have therefore obtained from the anharmonic oscillator model the nonlinear dependence of the polarization on the electric field expressed by equation (9.63), which is of the same form as equation (9.39) when only second-order nonlinearity is considered. Note also that our assumed solution (9.52) of the nonlinear equation of motion (9.53) is correct only to first order in the anharmonic force constant λ because no terms in $\exp(3j\omega t)$, etc. were used. A perturbation solution correct to higher order in λ would have produced a solution $x(t)$ containing harmonic frequencies higher[17] than 2ω and would have led to an expression for the polarization containing terms in E^3, etc.

Summary of the Physical Picture of Nonlinear Polarization

We may summarize our physical picture of nonlinear polarization as follows. We considered a classical[22] model of a solid containing an anharmonic term $m\lambda x^2$ in the restoring force acting on an electron. The presence of this nonlinear term in the equation of motion led to the electron displacement $x(t)$ having Fourier components at the frequency ω of the incident light wave and at the second harmonic 2ω as well. This, in turn, led to a time-dependent electric polarization with components[23] $P^{(\omega)}$ and $P^{(2\omega)}$ at the frequencies ω and 2ω, respectively. The component $P^{(\omega)}$ of the induced polarization is directly proportional to the incident electric field E and the linear electric susceptibility $\chi_L^{(\omega)}$ is just that found for the harmonic oscillator model of the solid. The component $P^{(2\omega)}$ is proportional to E^2, and the nonlinear electric susceptibility $d^{(2\omega)}$ depends directly on the anharmonic force constant λ.

We now have a simple classical model leading to the second-order nonlinear polarization and nonlinear susceptibility expressed in equation (9.63). With that expression giving the polarization as a function of the electric field, we can next calculate the source term $\partial^2 P/\partial t^2$ in the wave equation (9.18) governing electromagnetic wave propagation in a solid with a nonlinear electric susceptibility. Before doing that, however, we discuss the nonlinear susceptibility $d^{(2\omega)}$ itself.

Tensor Nature of the Nonlinear Susceptibility

In the preceding sections, the results were derived on the basis of a *scalar* model of a dielectric which ignored the anisotropy of real crystals. This viewpoint, while useful for discussing the basic physical ideas involved, is oversimplified. In a vector treatment, we can see from equation (9.39) that equation (9.63) would become

$$\mathbf{P} = \mathbf{P}^{(\omega)} + \mathbf{P}^{(2\omega)} = \boldsymbol{\chi}_L^{(\omega)} \cdot \mathbf{E} + \mathbf{d}^{(2\omega)} : \mathbf{EE} \tag{9.66}$$

where $\boldsymbol{\chi}_L^{(\omega)}$ is the second-rank first-order (linear) susceptibility tensor and $\mathbf{d}^{(2\omega)}$ is the third-rank second-order nonlinear susceptibility tensor. The tensor[24] $\mathbf{d}^{(2\omega)}$ has 27 components $d_{ijk}^{(2\omega)}$, so the components $P_i^{(2\omega)}$ of the nonlinear polarization vector would be expressed[25] as

$$P_i^{(2\omega)} = \sum_j \sum_k d_{ijk}^{(2\omega)} E_j E_k \tag{9.67}$$

where the indices i, j, and k run over the three dimensions x, y, and z. (A contracted notation is often used in which d_{ijk} is replaced by d_{il}, where $i = 1, 2, 3$ represents x, y, z, and $l = 1, 2, 3, 4, 5, 6$ represents xx, yy, zz, yz, xz, xy.) In some crystals,[26] the symmetry of the lattice reduces the number of independent nonzero values of $d_{ijk}^{(2\omega)}$, and, in certain cases (to be discussed later) symmetry causes the tensor to vanish.

For simplicity, however, in our treatment of the applications of the nonlinear susceptibility, we will ignore its tensor nature and use the scalar equations developed so far.

Solid State Physics Factors Affecting the Nonlinear Susceptibility

We now discuss some of the properties of solids that affect the magnitude[27] of the second-order nonlinear susceptibility $d^{(2\omega)}$.

We note from equation (9.65) that the nonlinear susceptibility $d^{(2\omega)}$ is directly proportional to the anharmonic force constant λ. Consider the effect on the magnitude of λ of the crystal in question having a center of symmetry. The anharmonic restoring force $-m\lambda x^2$ corresponds to a term

$$U(x) = m\lambda x^3/3 \qquad (9.68)$$

in the potential energy U of an electron in the crystal. If a center of symmetry is present, the potential energy must be invariant if the position vector $\mathbf{r}$ of a point is replaced by $-\mathbf{r}$. In our one-dimensional discussion, this corresponds to replacing x by $-x$, so a center of symmetry requires that

$$U(x) = U(-x) \qquad (9.69)$$

for the potential energy to be invariant under the transformation. For the potential energy given by equation (9.68), the requirement (9.69) becomes

$$m\lambda x^3 = -m\lambda x^3 \qquad (9.70)$$

an equation which must be true for all values of x. Equation (9.70) requires that the anharmonic force constant λ be equal to zero. The physical consequence of inversion invariance is thus that a crystal with a center of symmetry cannot have a term proportional to x^2 in the restoring force on the electron. If $\lambda = 0$, then equation (9.65) tells us that $d^{(2\omega)} = 0$ in a crystal with a center of symmetry, and such a crystal has no second-order nonlinear terms in its optical polarization. Crystals lacking a center of symmetry may have nonzero values of the components $d_{ijk}^{(2\omega)}$ of the second-order nonlinear susceptibility tensor $d^{(2\omega)}$. An example of a centrosymmetric crystal is NaCl; examples of noncentrosymmetric crystals that have nonzero values[28,29] of some of the $d_{ijk}^{(2\omega)}$ are GaAs, KH_2PO_4 (potassium dihydrogen phosphate, known as KDP), $LiNbO_3$, and SiO_2.

To discuss factors affecting the magnitude of the nonlinear susceptibility $d^{(2\omega)}$, we recall from equation (9.65) that

$$d^{(2\omega)} \equiv \frac{\lambda(Ne^3/m^2)}{2(\omega_0^2 - \omega^2 + j\gamma\omega)^2(\omega_0^2 - 4\omega^2 + 2j\gamma\omega)} \qquad (9.65)$$

and that the linear susceptibility $\chi_L^{(\omega)}$ at frequency ω is

$$\chi_L^{(\omega)} \equiv \frac{Ne^2/m}{\omega_0^2 - \omega^2 + j\gamma\omega} \qquad (9.64)$$

From equation (9.64), we can see that $\chi_L^{(2\omega)}$, the *linear* susceptibility at frequency 2ω, is given by

$$\chi_L^{(2\omega)} \equiv \frac{Ne^2/m}{\omega_0^2 - 4\omega^2 + 2j\gamma\omega} \tag{9.71}$$

On substituting expressions (9.64) and (9.71) into equation (9.65), one obtains

$$d^{(2\omega)} = (m\lambda/2N^2e^3)(\chi_L^{(\omega)})^2(\chi_L^{(2\omega)}) \tag{9.72}$$

Equation (9.72) is an expression for the nonlinear susceptibility $d^{(2\omega)}$ as a function of the linear susceptibilities $\chi_L^{(\omega)}$ and $\chi_L^{(2\omega)}$ at frequencies ω and 2ω, and of the solid state material parameters N and λ. Defining a quantity $\delta^{(2\omega)}$ by the relation

$$\delta^{(2\omega)} \equiv m\lambda/2N^2e^3 \tag{9.73}$$

equation (9.72) can be rewritten in the form

$$\delta^{(2\omega)} = \frac{d^{(2\omega)}}{(\chi_L^{(\omega)})^2(\chi_L^{(2\omega)})} \tag{9.74}$$

It has been found experimentally[21,30] that $\delta^{(2\omega)}$ (or more precisely, its three dimensional analog[30] δ_{ijk}) is remarkably constant for solids whose values of the nonlinear susceptibilities vary by four orders of magnitude. This result is often called Miller's phenomenological rule concerning nonlinear susceptibilities.

The fact that $\delta^{(2\omega)}$ is roughly constant for different materials suggests that the ratio λ/N^2 of the anharmonic force constant to the square of the electron density is approximately constant[11] for different solids. We next rewrite equation (9.74) as

$$d^{(2\omega)} = \delta^{(2\omega)}(\chi_L^{(\omega)})^2(\chi_L^{(2\omega)}) \tag{9.75}$$

Since $\delta^{(2\omega)}$ is approximately constant for different solids, equation (9.75) suggests that the large differences observed in the nonlinear susceptibility $d^{(2\omega)}$ for different materials are due to differing values of the linear susceptibilities $\chi_L^{(\omega)}$ and $\chi_L^{(2\omega)}$. We may carry this idea a little further by assuming that the frequency dependence of $\chi_L^{(\omega)}$ is small between the frequencies ω and 2ω, justifying the approximation of $\chi_L^{(2\omega)}$ by $\chi_L^{(\omega)}$. With this assumption, equation (9.75) becomes

$$d^{(2\omega)} = \delta^{(2\omega)}(\chi_L^{(\omega)})^3 \tag{9.76}$$

stating that the nonlinear susceptibility $d^{(2\omega)}$ is, to a good approximation, directly proportional[31] to the cube of the linear susceptibility $\chi_L^{(\omega)}$ for a given crystal. We would therefore expect large values of the nonlinear susceptibility in solids with large values of the *linear* electric susceptibility at optical frequencies. The linear susceptibility is related to the dielectric constant $\varepsilon(\omega)$ at the frequency ω by the relation

$$\varepsilon(\omega) = 1 + 4\pi\chi_L^{(\omega)} \tag{9.77}$$

and the dielectric constant is related to the refractive index $n(\omega)$ at frequency ω by

$$\varepsilon(\omega) = (n(\omega))^2 \tag{9.78}$$

for frequencies not near a resonant frequency ω_0 and if the damping, and hence the absorption, is small. It is therefore expected that crystals with large values of the refractive index (or, alternatively, of the optical dielectric constant) will have large values of the linear susceptibility, leading, through equation (9.76), to large values of the nonlinear susceptibility.

Table 9.1 illustrates these ideas by showing values of the nonlinear susceptibility[32] d_{xyz} (units are discussed below) for 10.6-μm radiation for the zinc blende cyrstals GaP, GaAs, and GaSb. Also shown are the index of refraction[33] n at a wavelength of 10 μm and the calculated optical dielectric constant $\varepsilon = n^2$. From Table 9.1, we see that the magnitude of the nonlinear susceptibility increases as the optical dielectric constant increases from GaP to GaAs to GaSb. Another example is found in tellurium, which exhibits the very high nonlinear susceptibility[34] of 1600×10^{-9} esu. This figure correlates well with its high values[35] of 6.2 and 4.85 for the refractive index for 8-μm infrared radiation polarized, respectively, parallel and perpendicular to the c axis of the tellurium crystal. It may be noted that this value also correlates well with the high polarizability[36] of the tellurium

Table 9.1. Nonlinear Susceptibility d_{xyz} of GaP, GaAs, and GaSb

Crystal	d_{xyz} (esu)	n	$\varepsilon = n^2$
GaSb	650×10^{-9}	3.84	14.8
GaAs	215×10^{-9}	3.20	10.2
GaP	99×10^{-9}	2.90	8.4

ion, since materials with high electronic polarizabilities should have high values of the linear susceptibility.

To summarize, the solid state factors favoring relatively large values of the second-order nonlinear susceptibility include a lack of (inversion) symmetry and a high *linear* susceptibility, as reflected by a large value of the optical dielectric constant and refractive index. More detailed models of crystals may be used to calculate and explain values of the nonlinear susceptibility. For a discussion of these models, the reader is referred to the literature.[37]

Magnitude of the Nonlinear Susceptibility

We can now make an estimate of the magnitude of the nonlinear susceptibility $d^{(2\omega)}$ using equation (9.72), employing Gaussian units[38] because they are commonly used in the literature. From equation (9.72), we have

$$d^{(2\omega)} = (m\lambda/2N^2e^3)(\chi_L^{(\omega)})^2(\chi_L^{(2\omega)}) \qquad (9.72)$$

so we need an estimate of the linear susceptibilities $\chi_L^{(\omega)}$ and $\chi_L^{(2\omega)}$. From equation (9.77) for the dielectric constant ε,

$$\chi_L^{(\omega)} = (1/4\pi)(\varepsilon(\omega) - 1) \qquad (9.79)$$

Using GaAs as an example, $\varepsilon = 10.2$ from Table 9.1, so $\chi_L^{(\omega)} = 0.732$, and we approximate the value of $\chi_L^{(2\omega)}$ by $\chi_L^{(\omega)}$ on assuming little dispersion. We use Garrett's value[21] of $\lambda \approx 10^{39}$ cm^{-1} sec^{-2} for the anharmonic force constant, take the electron density $N \approx 10^{23}$ cm^{-3}, $m = 9.1 \times 10^{-28}$ g and the charge $e = 1.6 \times 10^{-19}$ C $= 4.8 \times 10^{-10}$ esu (or $g^{1/2}$ cm$^{3/2}$ sec^{-1}). Then, from equation (9.72), one obtains

$$d^{(2\omega)} = \frac{(9.1 \times 10^{-28})(10^{39})(0.732)^3}{2(10^{46})(4.8 \times 10^{10})^3} \quad \frac{\text{g cm}^{-1}\,\text{sec}^{-2}}{\text{cm}^{-6}\,\text{g}^{3/2}\,\text{cm}^{9/2}\,\text{sec}^{-3}}$$

leading to the value

$$d^{(2\omega)} = 161 \times 10^{-9}\,\text{g}^{-1/2}\,\text{cm}^{1/2}\,\text{sec} \qquad (9.80)$$

which is also written as $d^{(2\omega)} = 161 \times 10^{-9}$ esu. The magnitude of this value of $d^{(2\omega)}$ is in agreement with that of the experimental value for GaAs quoted in Table 9.1. The experimental values[29] of nonlinear susceptibilities[†] range from about 1×10^{-9} esu for quartz to 1600×10^{-9} esu for tellurium.

It is instructive to obtain an idea of the magnitude of the induced polarization $P^{(2\omega)}$ at the second harmonic frequency 2ω relative to the polarization $P^{(\omega)}$ at the fundamental frequency ω. From equation (9.63), the ratio

$$\frac{P^{(2\omega)}}{P^{(\omega)}} = \frac{d^{(2\omega)}E^2}{\chi_L^{(\omega)}E} = \frac{d^{(2\omega)}}{\chi_L^{(\omega)}} E \qquad (9.81)$$

is found to depend on the magnitude E of the electric field of the light wave, as well as on the ratio $d^{(2\omega)}/\chi_L^{(\omega)}$ of the nonlinear and linear susceptibilities. Using GaAs as our example again, taking $d^{(2\omega)} = 215 \times 10^{-9}$ esu and $\chi_L^{(\omega)} = 0.732$, one finds that

$$P^{(2\omega)}/P^{(\omega)} = (2.94 \times 10^{-7})E \qquad (9.82)$$

where the ratio will be dimensionless if E is expressed in dynes (esu)$^{-1}$ or statvolts cm^{-1} or g$^{1/2}$ cm$^{-1/2}$ sec^{-1}. From equation (9.82), we can see that the relative magnitude of $P^{(2\omega)}$ will be very small unless the electric field E is quite large. For this reason, nonlinear optics began to develop only with the invention of the laser, which produces intense light beams in which the peak value of the electric field may often be greater than, say, 3×10^9 V m^{-1},[39] which is equal to 10^5 statvolts cm^{-1}. For electric fields of this magnitude, we can see from equation (9.82) that $P^{(2\omega)}$ will be of the order of a few percent of the linear polarization $P^{(\omega)}$.

Wave Equation for the Nonlinear Crystal

We now return to considering the general wave equation (9.18)

$$\text{grad div } \mathbf{E} - \nabla^2\mathbf{E} + \frac{1}{c^2}\frac{\partial^2\mathbf{E}}{\partial t^2} = -\frac{4\pi}{c^2}\frac{\partial^2\mathbf{P}}{\partial t^2} \qquad (9.18)$$

describing the electric field $\mathbf{E}$ of a light wave in a crystal whose time-dependent polarization is $\mathbf{P}(t)$. We consider specifically the nonlinear crystal model developed above, for which the (scalar) polarization is given

---

[†] It is interesting to note here that the *third-order* nonlinear susceptibility is of the order[39] of 10^{-12} esu, several orders of magnitude smaller than the values of the second-order susceptibility quoted above.

by equation (9.63) as

$$P = P^{(\omega)} + P^{(2\omega)} = \chi_L{}^{(\omega)}E + d^{(2\omega)}E^2 \tag{9.63}$$

and where the time dependence of P is contained in the expression $E_0 \exp(j\omega t)$ given by equation (9.19) for the electric field E.

Equation (9.63) for $P(t)$ is now used to calculate the driving or source term $(\partial^2 P/\partial t^2)$ in the scalar version of the wave equation (9.18). One obtains

$$\frac{\partial^2 P}{\partial t^2} = \frac{\partial^2}{\partial t^2}(\chi_L{}^{(\omega)}E + d^{(2\omega)}E^2) = \chi_L{}^{(\omega)}\frac{\partial^2 E}{\partial t^2} + d^{(2\omega)}\frac{\partial^2}{\partial t^2}(E^2) \tag{9.83}$$

Substitution of equation (9.83) into equation (9.18) yields

$$\text{grad}(\text{div } E) - \nabla^2 E + \frac{1}{c^2}(1 + 4\pi\chi_L{}^{(\omega)})\frac{\partial^2 E}{\partial t^2} + \frac{4\pi}{c^2}d^{(2\omega)}\frac{\partial^2}{\partial t^2}(E^2) = 0 \tag{9.84}$$

Earlier, in equations (9.34)–(9.36), it was shown that div $E = 0$ for an isotropic solid characterized by the linear relation $P = \chi(\omega)E$ between the polarization and electric field vectors. It is now necessary to consider the term in div E for a nonlinear crystal for which

$$P = \chi_L{}^{(\omega)}E + d^{(2\omega)}E^2 \tag{9.63}$$

The Maxwell equation (9.13) requires that

$$\text{div } E = -4\pi \text{ div } P \tag{9.13}$$

In a one-dimensional problem this becomes[40]

$$\frac{dE}{dx} = -4\pi\frac{dP}{dx} \tag{9.85}$$

again using scalar notation for simplicity. From equation (9.63),

$$\frac{dP}{dx} = \chi_L{}^{(\omega)}\frac{dE}{dx} + 2d^{(2\omega)}E\frac{dE}{dx} \tag{9.86}$$

which, on substitution into equation (9.85), yields the result

$$\frac{dE}{dx}(1 + 4\pi\chi_L{}^{(\omega)} + 8\pi d^{(2\omega)}E) = 0 \tag{9.87}$$

For equation (9.87) to be true, one of the two factors must be zero for all values of the electric field E. Since the term in square brackets will not vanish for an arbitrary value of E, it is concluded[40] that the term dE/dx must equal zero. Our result is thus, that the one-dimensional version (9.85) of Maxwell's equation (9.13) requires that dE/dx equal zero. Extending this to three dimensions, it is concluded that

$$\text{div } \mathbf{E} = 0 \tag{9.88}$$

for a solid with a nonlinear relation of the type (9.63) between the polarization and the electric field.

Substitution of the relation (9.88) into (9.84) gives

$$\nabla^2 E - \frac{1}{c^2}(1 + 4\pi\chi_L{}^{(\omega)})\frac{\partial^2 E}{\partial t^2} - \frac{4\pi}{c^2}d^{(2\omega)}\frac{\partial^2}{\partial t^2}(E^2) = 0 \tag{9.89}$$

as the wave equation governing electromagnetic wave propagation in the nonlinear crystal. It should be kept in mind that the scalar equation (9.89) is oversimplified in that we have tacitly assumed isotropy of the solid considered in using a scalar rather than a vector wave equation. Finally, substituting the dielectric constant ε defined by equation (9.77) into (9.89) gives

$$\nabla^2 E - \frac{1}{c^2}\varepsilon(\omega)\frac{\partial^2 E}{\partial t^2} - \frac{4\pi}{c^2}d^{(2\omega)}\frac{\partial^2}{\partial t^2}(E^2) = 0 \tag{9.90}$$

Next, equation (9.90) is used to consider the propagation and interaction, via the nonlinear term in the polarization, of electromagnetic waves in a nonlinear crystal.

Wave Propagation and Interaction in a Nonlinear Crystal

To discuss the physics of a number of nonlinear optical devices, it is necessary to consider the propagation and interaction of several electromagnetic waves in a nonlinear crystal. The propagation of these waves is governed by the wave equation (9.90). We consider the waves

$$E_1(\omega_1, z) = \tfrac{1}{2}\{E_{10}(z)\exp[j(\omega_1 t - k_1 z)] + \text{c.c.}\} \tag{9.91}$$

$$E_2(\omega_2, z) = \tfrac{1}{2}\{E_{20}(z)\exp[j(\omega_2 t - k_2 z)] + \text{c.c.}\} \tag{9.92}$$

$$E_3(\omega_3, z) = \tfrac{1}{2}\{E_{30}(z)\exp[j(\omega_3 t - k_3 z)] + \text{c.c.}\} \tag{9.93}$$

with frequencies $\omega_1, \omega_2, \omega_3$. In equations (9.91)–(9.93), k_1, k_2, and k_3

are the wave numbers, c.c. means complex conjugate, and we indicate an explicit z dependence of the amplitudes $E_{10}(z)$, $E_{20}(z)$, and $E_{30}(z)$ of these plane waves moving in the z direction. This is done because, as we will see, the nonlinear interaction between the various waves can change the amplitudes. The aim is to find expressions giving the z dependence of the amplitudes E_{10}, E_{20}, and E_{30}.

We will not proceed in the most general manner[41] but instead will ignore the anisotropy of real solids and consider the three waves above, all of whose wave vectors are in the same direction. Following Yariv,[42] the electric field E at any point z is given by

$$E = E_1(\omega_1, z) + E_2(\omega_2, z) + E_3(\omega_3, z) \tag{9.94}$$

We want to consider the wave equation (9.90), which we rewrite as

$$\nabla^2 E - \frac{1}{c^2}\, \varepsilon(\omega)\, \frac{\partial^2 E}{\partial t^2} = \frac{4\pi}{c^2}\, d^{(2\omega)}\, \frac{\partial^2}{\partial t^2}\, (E^2) \tag{9.95}$$

where E is given by equation (9.94), E^2 is given by

$$E^2 = EE^* \tag{9.96}$$

and the asterisk indicates the complex conjugate. The next step is to substitute expressions (9.94) and (9.96) for E and E^2 into the wave equation (9.95); but, since this leads to an extremely long result, we shall merely describe what happens to the various terms in (9.95).

First, since (9.94) states that E is not a function of x or y, the term in $\nabla^2 E$ becomes

$$\nabla^2 E = \frac{\partial^2}{\partial z^2}\, E_1(\omega_1, z) + \frac{\partial^2}{\partial z^2}\, E_2(\omega_2, z) + \frac{\partial^2}{\partial z^2}\, E_3(\omega_3, z) \tag{9.97}$$

Second, the term in $\partial^2 E/\partial t^2$ is given by

$$\frac{\partial^2 E}{\partial t^2} = \frac{\partial^2}{\partial t^2}\, E_1(\omega_1, z) + \frac{\partial^2}{\partial t^2}\, E_2(\omega_2, z) + \frac{\partial^2}{\partial t^2}\, E_3(\omega_3, z) \tag{9.98}$$

The result of equations (9.97) and (9.98) is that the left-hand side of the wave equation (9.95) is a sum of terms, each of which is a function of one of the frequencies ω_1 or ω_2 or ω_3.

If we consider the right-hand side

$$\frac{4\pi}{c^2}\, d^{(2\omega)}\, \frac{\partial^2}{\partial t^2}\, (E^2) \tag{9.99}$$

of the wave equation, we can see that E^2 will contain all of the sums and differences of the three frequencies ω_1, ω_2, and ω_3. Thus, E^2 will contain terms oscillating at the frequencies zero, $2\omega_1$, etc., $\omega_1 + \omega_2$, etc., and $\omega_1 - \omega_2$, etc. The conclusion is that the driving term (9.99) on the right-hand side of the wave equation (9.95) will *not* in general contain terms oscillating at the frequencies ω_1, ω_2, and ω_3 of the left-hand side of the equation. In general then, there will not be traveling wave solutions of (9.95) because the driving term (9.99) is not synchronous (i.e., does not contain the same frequencies) with the left-hand side of the wave equation.

However, if the driving term *is* synchronous with the right-hand side of the wave equation, then there will be traveling wave solutions of (9.95). We may consider one such situation by choosing the case[43] for which

$$\omega_1 + \omega_2 = \omega_3 \tag{9.100}$$

A result of equation (9.100) is that the driving term (9.99), which includes a factor

$$-\tfrac{1}{2}(\omega_1 + \omega_2)^2 E_{10} E_{20} \exp[j(\omega_1 + \omega_2)t - j(k_1 + k_2)z] \tag{9.101}$$

(and its complex conjugate) in the sum frequency $\omega_1 + \omega_2$, will, because of the condition expressed by (9.100), contain a term

$$-\tfrac{1}{2}\omega_3^2 E_{10} E_{20} \exp[j\omega_3 t - j(k_1 + k_2)z] \tag{9.102}$$

Since the requirement (9.100) has made the expression (9.101) oscillate at the frequency ω_3, as in (9.102), this term can "drive" the terms at ω_3 on the left-hand side of the wave equation. In the same way, the terms oscillating at $\omega_3 - \omega_2$ and $\omega_3 - \omega_1$ in the source term (9.99) can drive oscillations at frequencies ω_1 and ω_2, respectively.

We may consider the requirement (9.100), rewritten

$$\hbar\omega_1 + \hbar\omega_2 = \hbar\omega_3 \tag{9.103}$$

as expressing the conservation of energy in a process in which photons of energies $\hbar\omega_1$ and $\hbar\omega_2$ are destroyed and a photon of energy $\hbar\omega_3$ is created. Equation (9.103) can also be thought of as describing the inverse process, in which a photon $\hbar\omega_3$ is destroyed with the creation of photons of energies $\hbar\omega_1$ and $\hbar\omega_2$. Equations (9.100) and (9.103) thus describe processes in which energy flows from electromagnetic fields of frequency ω_3 to fields at frequencies ω_1 and ω_2, or vice versa. It is this interchange of energy between electromagnetic fields of different frequencies, due to the nonlinear susceptibility of the solid, that is the physical basis of the various applications we will discuss.

We return to consideration of the wave equation subject to the condition that $\omega_1 + \omega_2 = \omega_3$. There will be terms $\partial^2 E_3/\partial z^2$ and $\partial^2 E_3/\partial t^2$, oscillating at frequency ω_3, on the left-hand side of the wave equation. The wave equation for E_3 at frequency ω_3 is

$$\frac{\partial^2 E_3}{\partial z^2} - \frac{1}{c^2}\, \varepsilon(\omega_3)\, \frac{\partial^2 E_3}{\partial t^2} = \frac{4\pi}{c^2}\, d^{(2\omega)}\left(-\frac{1}{2}\, \omega_3^2 E_{10}E_{20}e^{j[\omega_3 t-(k_1+k_2)z]} + \text{c.c.}\right)$$

$$(9.104)$$

where $\varepsilon(\omega_3)$ is the dielectric constant at frequency ω_3, E_3 is a function of ω_3 given by (9.93), and c.c. means complex conjugate.

We now consider the various terms in equation (9.104) in order to obtain an expression for the variation in space (i.e., with z coordinate) of the electric field $E_3(\omega_3, z)$. From equation (9.93), we have

$$\frac{\partial^2 E_3}{\partial z^2} = \frac{1}{2}\left[(-k_3^2)E_{30} - 2jk_3\frac{dE_{30}}{dz} + \frac{d^2 E_{30}}{dz^2}\right]e^{j(\omega_3 t-k_3 z)} + \text{c.c.} \quad (9.105)$$

and that

$$\frac{\partial^2 E_3}{\partial t^2} = \frac{1}{2}\left(-\omega_3^2 E_{30}e^{j(\omega_3 t-k_3 z)} + \text{c.c.}\right) \quad (9.106)$$

Results (9.105) and (9.106) are to be substituted into the form (9.104) of the wave equation describing the electromagnetic field oscillating at frequency ω_3.

We next make the assumption[44] that the relative change in the amplitude E_{30} per wavelength is small because the nonlinear susceptibility is small compared to the linear susceptibility. Since dE_{30}/dz is the change in E_{30} per unit length, $\lambda_3(dE_{30}/dz)$ is the change in E_{30} per wavelength λ_3, and the assumption states that this is small compared to the amplitude E_{30} itself. This condition[45] is expressed as

$$E_{30} \gg \lambda_3\left|\frac{dE_{30}}{dz}\right| \quad (9.107)$$

which can be rewritten as

$$k_3\left|\frac{dE_{30}}{dz}\right| \gg \left|\frac{d^2 E_{30}}{dz^2}\right| \quad (9.108)$$

The assumption contained in equation (9.108) allows us to neglect the term in $d^2 E_{30}/dz^2$ in equation (9.105), which then becomes

$$\frac{\partial^2 E_3}{\partial z^2} = \frac{1}{2}\left(-k_3^2 E_{30} - 2jk_3 \frac{dE_{30}}{dz}\right)e^{j(\omega_3 t - k_3 z)} + \text{c.c.} \qquad (9.109)$$

On substituting the expressions (9.109) and (9.106) for $\partial^2 E_3/\partial z^2$ and $\partial^2 E_{30}/\partial t^2$ into the wave equation (9.104), one obtains, on cancelling the factors of $\frac{1}{2}$,

$$\left(-k_3^2 E_{30} - 2jk_3 \frac{dE_{30}}{dz}\right)e^{j(\omega_3 t - k_3 z)} + \left(-k_3^2 E_{30}^* + 2jk_3 \frac{dE_{30}^*}{dz}\right)e^{-j(\omega_3 t - k_3 z)}$$

$$+ \frac{\varepsilon(\omega_3)}{c^2}\omega_3^2(E_{30}e^{j(\omega_3 t - k_3 z)} + E_{30}^* e^{-j(\omega_3 t - k_3 z)})$$

$$= -\frac{4\pi}{c^2}d^{(2\omega)}\omega_3^2(E_{10}E_{20}e^{j[\omega_3 t - (k_1 + k_2)z]} + E_{10}^* E_{20}^* e^{-j[\omega_3 t - (k_1 + k_2)z]}) \quad (9.110)$$

which can be rewritten as

$$\left[\left(-k_3^2 E_{30} - 2jk_3 \frac{dE_{30}}{dz} + \frac{\omega_3^2 \varepsilon(\omega_3)}{c^2}E_{30}\right)e^{j(\omega_3 t - k_3 z)}\right.$$

$$\left. + \left(-k_3 E_{30}^* + 2jk_3 \frac{dE_{30}^*}{dz} + \frac{\omega_3^2 \varepsilon(\omega_3)}{c^2}E_{30}^*\right)e^{-j(\omega_3 t - k_3 z)}\right]$$

$$= \frac{-4\pi}{c^2}d^{(2\omega)}\omega_3^2(E_{10}E_{20}e^{j[\omega_3 t - (k_1 + k_2)z]} + E_{10}^* E_{20}^* e^{-j[\omega_3 t - (k_1 + k_2)z]}) \quad (9.111)$$

We note next from the wave equation (9.95) that the phase velocity v of a wave is given by

$$v^2 = c^2/\varepsilon \qquad (9.112)$$

Since it is also true that

$$v = \omega/k \qquad (9.113)$$

we have the result that

$$k^2 = \varepsilon\omega^2/c^2 \qquad (9.114)$$

Applying this result to the wave E_3 we are considering,

$$k_3^2 = \varepsilon(\omega_3)\omega_3^2/c^2 \qquad (9.115)$$

which when substituted into equation (9.111), gives

$$\left(-2jk_3 \frac{dE_{30}}{dz}\right)\exp[j(\omega_3 t - k_3 z)] + \left(2jk_3 \frac{dE_{30}^*}{dz}\right)\exp[-j(\omega_3 t - k_3 z)]$$

$$= -\frac{4\pi}{c^2}d^{(2\omega)}\omega_3^2(E_{10}E_{20}e^{j[\omega_3 t - (k_1 + k_2)z]} + E_{10}^* E_{20}^* e^{-j[\omega_3 t - (k_1 + k_2)z]}) \quad (9.116)$$

Equating coefficients of $\exp(j\omega_3 t)$ on both sides of equation (9.116) leads to

$$e^{-jk_3 z} \frac{dE_{30}}{dz} = \frac{4\pi}{2jc^2} \frac{\omega_3^2}{k_3} d^{(2\omega)} E_{10} E_{20} \exp[-j(k_1 + k_2)z] \quad (9.117)$$

which can be written as

$$\frac{dE_{30}}{dz} = -j \frac{2\pi}{c^2} \frac{\omega_3^2}{k_3} d^{(2\omega)} E_{10} E_{20} \exp[-j(k_1 + k_2 - k_3)z] \quad (9.118)$$

Equating the coefficients of $\exp(-j\omega_3 t)$ in equation (9.116) leads to the complex conjugate of equation (9.118). In view of equation (9.115), one finds that

$$\omega_3^2 / c^2 k_3 = \omega_3 / c\varepsilon_3^{1/2} \quad (9.119)$$

where the notation $\varepsilon_3 \equiv \varepsilon(\omega_3)$ has been introduced.

Using result (9.119), the expression (9.118) can be rewritten as

$$\frac{dE_{30}}{dz} = -j \frac{2\pi\omega_3}{c\varepsilon_3^{1/2}} d^{(2\omega)} E_{10} E_{20} \exp[-j(k_1 + k_2 - k_3)z] \quad (9.120)$$

an equation that describes the variation in space of the amplitude $E_{30}(z)$ of the electric field $E_3(\omega_3, z)$ of frequency ω_3. Using the units of $d^{(2\omega)}$ discussed in an earlier section, a dimensional analysis of equation (9.120) shows that the right-hand side has the same units (i.e., electric field per unit length) as does the left-hand side.

In the same manner as that in which equation (9.120) was obtained, expressions for the spatial variation of the amplitudes $E_{10}(z)$ and $E_{20}(z)$ may be found, giving the full set of equations below. These are

$$\frac{dE_{10}^*}{dz} = j \frac{2\pi\omega_1}{c\varepsilon_1^{1/2}} d^{(2\omega)} E_{20} E_{30}^* \exp[-j(\Delta k)z] \quad (9.121)$$

$$\frac{dE_{20}^*}{dz} = j \frac{2\pi\omega_2}{c\varepsilon_2^{1/2}} d^{(2\omega)} E_{10} E_{30}^* \exp[-j(\Delta k)z] \quad (9.122)$$

$$\frac{dE_{30}}{dz} = -j \frac{2\pi\omega_3}{c\varepsilon_3^{1/2}} d^{(2\omega)} E_{10} E_{20} \exp[-j(\Delta k)z] \quad (9.123)$$

where $\varepsilon_1 \equiv \varepsilon(\omega_1)$, etc. are the dielectric constants of the nonlinear crystal at the frequencies ω_1, ω_2, and ω_3, and where the notation

$$\Delta k \equiv k_1 + k_2 - k_3 \quad (9.124)$$

has been introduced.

Equatións (9.121)–(9.123) describe the spatial variation of the amplitudes $E_{10}(z)$, $E_{20}(z)$, and $E_{30}(z)$ of the three interacting waves $E_1(\omega_1, z)$, $E_2(\omega_2, z)$, and $E_3(\omega_3, z)$ in a dielectric,[46] subject to the condition (9.100) that $\omega_1 + \omega_2 = \omega_3$. The interaction is via the nonlinear term $d^{(2\omega)}E^2$ in the polarization given by equation (9.63); this is reflected in the presence of the nonlinear susceptibility $d^{(2\omega)}$ in each of these equations. The fact that the three waves are interacting is seen in the result that equations (9.121)–(9.123) are coupled differential equations because the spatial variation of any one amplitude is a function of the other amplitudes. Finally, as we shall discuss in detail later, the spatial variation of each amplitude is a function of the quantity Δk defined by equation (9.124).

Equations (9.21)–(9.123) are the basic equations governing the interaction of three electromagnetic waves (of frequencies ω_1, ω_2, ω_3) in the nonlinear solid, subject to the requirement $\omega_1 + \omega_2 = \omega_3$. We now apply these equations to the discussion of several processes of device interest in nonlinear solids.

Optical Second Harmonic Generation

The first application of the nonlinear optical properties of solids that we shall consider is the generation of the second harmonic frequency 2ω of a light wave of frequency ω. The physical process considered is one in which light of frequency ω is incident on a crystal with a nonlinear susceptibility $d^{(2\omega)}$. The nonlinearity produces a component of the electric polarization at frequency 2ω, resulting in the generation of radiation at the second harmonic frequency 2ω. One can also speak of the incident radiation of frequency ω being mixed with itself to produce radiation at the sum frequency $\omega + \omega = 2\omega$. Experimentally, a laser is necessary as the source of incident radiation in order that the incident electric field be large. The original experiment[47] involved a ruby laser beam, with a wavelength of 6943 Å, incident on a quartz crystal. The radiation emerging from the crystal was found to contain radiation at one-half the incident wavelength, i.e., at the second harmonic frequency.

We will use equations (9.121)–(9.123) to discuss the process of optical second harmonic generation. Consider a nonlinear crystal of length L, upon whose face at $z = 0$ is incident an input wave

$$E_1(\omega, z) = E_{10} \exp[j(\omega t - k_1 z)] \qquad (9.125)$$

of frequency ω. Equation (9.125) is obtained from equation (9.91) by setting $\omega_1 = \omega$. This input wave is then mixed with itself in the nonlinear crystal to produce the output wave at the second harmonic frequency 2ω. Using the condition

$$\omega_3 = \omega_1 + \omega_2 \qquad (9.100)$$

and the fact that $\omega_2 = \omega$ since identical frequencies are being mixed, we obtain $\omega_3 = 2\omega$, leading to the choice of

$$E_3(2\omega, z) = E_{30} \exp[j(2\omega t - k_3 z)] \qquad (9.126)$$

as the output wave. Equation (9.126) is obtained from (9.93). The complex conjugate term has been ignored for simplicity, and we keep in mind that the amplitudes E_{10} and E_{30} of the input and output waves are functions of the space coordinate z.

With $\omega_3 = 2\omega$, equation (9.123) becomes

$$\frac{dE_{30}}{dz} = \frac{-4\pi j\omega \, d^{(2\omega)}}{c\varepsilon_3^{1/2}} E_{10}^2 \exp[-j(2k_1 - k_3)z] \qquad (9.127)$$

since $E_{20} = E_{10}$ because two identical waves are mixed in this process. In addition, because $\omega_1 = \omega_2$, the wave numbers k_1 and k_2 are equal, so, in this case,

$$\Delta k \equiv k_1 + k_2 - k_3 = 2k_1 - k_3 \qquad (9.128)$$

a result that has also been included in equation (9.127). The nonlinear susceptibility $d^{(2\omega)}$ is given by (9.65) or (9.72). Equation (9.127) describes the spatial variation of the amplitude E_{30} of the second harmonic wave in the crystal. Our aim is to integrate this equation in order to obtain the amplitude $E_{30}(z)$ as a function of space coordinate z in the crystal.

In general, as seen from equation (9.91), the amplitude E_{10} of the input wave at frequency ω will be a function of z. In order to integrate (9.127) easily, we consider the simple case in which the fractional decrease in the energy of the input wave E_1, due to conversion to the second harmonic, is small. This case[48] is a physically reasonable one to discuss because the amplitude E_{10} of the incident wave (from, say, a laser) is large, and the magnitude of the nonlinear interaction is small. With this assumption, we may treat the amplitude E_{10} in equation (9.127) as a constant. If we integrate that equation over the length of the crystal from $z = 0$ to $z = L$, we obtain

$$E_{30}(L) - E_{30}(0) = \frac{-4\pi j\omega \, d^{(2\omega)} E_{10}^2}{c\varepsilon_3^{1/2}} \int_0^L \exp[-j(\Delta k)z] \, dz \qquad (9.129)$$

In (9.129), $E_{30}(z)$ is the magnitude of the amplitude E_{30} of the second harmonic wave at the point z of the crystal, and $\Delta k = 2k_1 - k_3$ from equation (9.128).

We next choose the boundary condition that there be no input wave at the second harmonic frequency at $z = 0$, so we have

$$E_{30}(0) = 0 \tag{9.130}$$

The amplitude $E_{30}(L)$ of the second harmonic wave at the output $z = L$ of the crystal is found by integrating (9.129), giving

$$E_{30}(L) = \frac{-4\pi j\omega\, d^{(2\omega)}E_{10}^2}{c\varepsilon_3^{1/2}} \frac{e^{-j(\Delta k)L} - 1}{-j(\Delta k)} \tag{9.131}$$

Multiplying (9.131) by its complex conjugate and multiplying and dividing by L^2 gives

$$E_{30}(L)E_{30}^*(L) = \frac{16\pi^2\omega^2[d^{(2\omega)}]^2E_{10}^4L^2}{c^2\varepsilon_3} \frac{(e^{-j(\Delta k)L} - 1)(e^{j(\Delta k)L} - 1)}{(\Delta k)^2L^2} \tag{9.132}$$

which, using the trigonometric identity $1 - \cos x = 2\sin^2(x/2)$, becomes

$$E_{30}(L)E_{30}^*(L) = \frac{16\pi^2\omega^2[d^{(2\omega)}]^2E_{10}^4L^2}{c^2\varepsilon_3} \left(\frac{\sin[\frac{1}{2}(\Delta k)L]}{\frac{1}{2}(\Delta k)L} \right)^2 \tag{9.133}$$

We now relate the square $E_{30}E_{30}^*$ of the electric field amplitude of the second harmonic wave to the intensity. Recalling[49] that the intensity† I of a plane electromagnetic wave is the time-averaged energy per unit time per unit area transmitted by the wave, we have the result that the intensity $I_{30}(L)$ of the second harmonic wave at $z = L$ is given by[49]

$$I_{30}(L) = (n_3c/8\pi)E_{30}(L)E_{30}^*(L) \tag{9.134}$$

In equation (9.134), n_3 is the index of refraction of the crystal at the second harmonic frequency $\omega_3 = 2\omega$. Substituting (9.134) into (9.133), and using the fact that $n_3^2 = \varepsilon_3$, we find

$$I_{30}(L) = \frac{2\pi\omega^2[d^{(2\omega)}]^2E_{10}^4L^2}{c\varepsilon_3^{1/2}} \left(\frac{\sin[\frac{1}{2}(\Delta k)L]}{\frac{1}{2}(\Delta k)L} \right)^2 \tag{9.135}$$

This equation gives the intensity $I_{30}(L)$ of the wave of frequency 2ω at the output end $z = L$ of the crystal.

---

† Strictly speaking, the term irradiance[49] should be used for the (average) energy per unit time per unit area transmitted by the wave. However, intensity is a more common usage.

Equation (9.135) shows that the intensity $I_{30}(L)$ is proportional to the factor

$$\left(\frac{\sin\left[\frac{1}{2}(\Delta k)L\right]}{\frac{1}{2}(\Delta k)L}\right)^2 \tag{9.136}$$

indicating, by comparison with results[50] of optics, that an interference effect is taking place. We consider first the variation of $I_{30}(L)$ as a function of the length L of the crystal for a given *nonzero* value of Δk. From equation (9.135), we have

$$I_{30}(L) = \frac{2\pi\omega^2[d^{(2\omega)}]^2 E_{10}^4 L^2}{c\varepsilon_3^{1/2}}\left(\frac{2}{\Delta k}\right)^2 \sin^2\left[\frac{1}{2}(\Delta k)L\right] \tag{9.137}$$

showing that $I_{30}(L)$ is proportional to $\sin^2[\frac{1}{2}(\Delta k)L]$, a function that has zeros at $(\Delta k)L = 0, 2\pi, 4\pi, \ldots$ and has maxima at $(\Delta k)L = \pi, 3\pi, 5\pi, \ldots$ etc. Figure 9.1 shows $I_{30}(L)$ from equation (9.137) plotted (in arbitrary units) as a function of L [which is expressed in units of $\pi/(\Delta k)$], for an arbitrary nonzero value of $\Delta k = 2k_1 - k_3$. From this figure, it can be seen that the intensity of the second harmonic wave varies periodically with the length L of the crystal, reaching a maximum at $L = \pi/\Delta k$, $3\pi/\Delta k, \ldots$ etc. (We continue to keep in mind that we are considering the case for which $\Delta k \neq 0$; the significance of the case for which $\Delta k = 0$

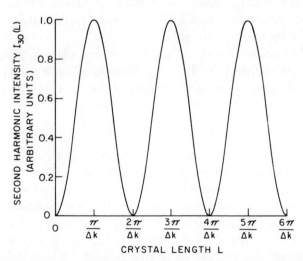

Figure 9.1. Second harmonic intensity $I_{30}(L)$ from equation (9.137) plotted (in arbitrary units) as a function of crystal length L [in units of $\pi/(\Delta k)$] for an arbitrary nonzero value of $\Delta k = 2k_1 - k_3$.

will be discussed later.) The length

$$L_c \equiv \pi/\Delta k \tag{9.138}$$

at which a maximum in the second harmonic intensity occurs is called the coherence length. The periodicity shown in Figure 9.1 is due to the term of the form (9.136) in the expression (9.135) for the second harmonic intensity $I_{3\omega}(L)$, and this periodicity is due to the presence of an interference effect. It can be shown[51,52] that, for $\Delta k \neq 0$, interference takes place between the second harmonic wave generated at one point $z = z_1$ of the crystal and the second harmonic wave generated at a second point $z = z_2$. The phase differences between these waves are such that the following takes place. Starting from $z = 0$ (the input face of the crystal), the input wave at frequency ω will generate the second harmonic at 2ω over the first coherence length $0 \leq z \leq L_c$ of the crystal. The second harmonic intensity $I_{3\omega}(L)$ reaches a maximum for a crystal length L equal to L_c. However, if L is greater than the coherence length, the second harmonic wave will beat with the input wave in the second coherence length $L_c \leq z \leq 2L_c$. The result of this interference is the generation of the difference frequency $2\omega - \omega = \omega$. Energy is thus transferred from the second harmonic to the fundamental and the intensity of the second harmonic wave decreases, reaching zero at a crystal length $L = 2L_c$. In this way, for $\Delta k \neq 0$, energy is interchanged between the fundamental and second harmonic waves. Since $I_{3\omega}(L)$ has its maximum value for a crystal length L equal to the coherence length L_c, the coherence length is the maximum length of the crystal useful in producing the second harmonic.

We may discuss the magnitude of the coherence length, using equation (9.138), by calculating $\Delta k = 2k_1 - k_3$. Since $k_1 = \omega n_1/c$ and $k_3 = 2\omega n_3/c$, where n_1 and n_3 are the refractive indices of the crystal at the frequencies ω and 2ω, respectively, we have

$$\Delta k = (2\omega/c)(n_1 - n_3) \tag{9.139}$$

for second harmonic generation. Because of dispersion, the refractive indices n_1 and n_3 will not usually be equal. In general, then, Δk will not be zero and the process of second harmonic generation will not be efficient because of the interference effects which result in energy transfer back and forth between the fundamental and second harmonic waves. This inefficiency is reflected in the typical magnitude of the coherence length L_c, which, from (9.138) and (9.139), can be written as

$$L_c = \frac{\pi c}{2\omega(n_1 - n_3)} = \frac{\lambda_0}{4(n_1 - n_3)} \tag{9.140}$$

where λ_0 is the wavelength in free space of the fundamental wave of frequency ω. From equation (9.140), we can see that the coherence length will be of the order of $100 \, \lambda_0$ on using an estimate of 0.01 for the quantity $n_1 - n_3$. This leads to a magnitude of perhaps 0.01 cm for the coherence length if λ_0 is a visible wavelength. We conclude that L_c is generally small unless the difference Δk in wave number can be very small, or preferably, equal to zero.

Next, we consider the intensity $I_{30}(L)$ when $\Delta k = 0$. From equation (9.135), we see that the second harmonic intensity $I_{30}(L)$ is proportional to

$$\left(\frac{\sin[\frac{1}{2}(\Delta k)L]}{[\frac{1}{2}(\Delta k)L]} \right)^2 \tag{9.141}$$

Figure 9.2 shows $I_{30}(L)$ from equation (9.135) plotted (in arbitrary units) as a function of Δk, which is expressed in units of π/L where L is the fixed value of the crystal length. We recall[50] that the function (9.141) has its maximum value of unity for $\frac{1}{2}(\Delta k)L = 0$, and has the minimum value of zero for $\frac{1}{2}(\Delta k)L = \pm\pi, \pm2\pi, \ldots$ etc. This means that the intensity $I_{30}(L)$ has its maximum for $\Delta k = 0$ and has zeros at $\Delta k = 2\pi/L, 4\pi/L$, etc., as shown in the figure. It is clear that, for a given crystal length L, the second harmonic intensity $I_{30}(L)$ is largest when $\Delta k = 0$, and falls rapidly with increasing nonzero values of Δk. It is therefore desirable to arrange things such that $\Delta k = 0$, a procedure called phase matching or index matching, and which is discussed briefly later. Using equation (9.138), the same conclusions may be restated in terms of the coherence length. The second

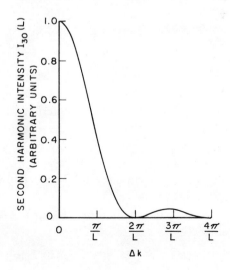

Figure 9.2. Second harmonic intensity $I_{30}(L)$ from equation (9.135) plotted (in arbitrary units) as a function of $\Delta k = 2k_1 - k_3$ for a fixed value of the crystal length L. The quantity Δk is plotted in units of π/L.

harmonic intensity is a maximum when the coherence length is infinite (for $\Delta k = 0$), and falls sharply as the coherence length decreases (for $\Delta k \neq 0$).

Having examined the effect of Δk (or, alternatively, of the coherence length) on the second harmonic intensity $I_{30}(L)$, we may now examine the influence of other factors on the intensity. From the definition of intensity introduced in equation (9.134), the intensity $I_{10}(0)$ of the input fundamental wave of frequency ω at $z = 0$ is given by

$$I_{10}(0) = (n_1 c/8\pi)E_{10}(0)E_{10}^*(0) \tag{9.142}$$

where n_1 is the refractive index at frequency ω. Recalling our assumption that the amplitude E_{10} is approximately constant, equation (9.142) becomes

$$I_{10}(0) = (n_1 c/8\pi)E_{10}^2 \equiv I_{10} \tag{9.143}$$

Using equation (9.143), the expression (9.135) for $I_{30}(L)$ may be written as

$$I_{30}(L) = \frac{128\pi^3\omega^2(d^{(2\omega)})^2 L^2}{c^3 \varepsilon_1 \varepsilon_3^{1/2}} I_{10}^2 \left(\frac{\sin x}{x}\right)^2 \tag{9.144}$$

where $x \equiv (\Delta k)L/2$ and $\varepsilon_1 = n_1^2$ is the dielectric constant at frequency ω. From equation (9.144), we see that the second harmonic intensity is proportional to the *square* of the input intensity at the fundamental frequency. Further, if we assume that the situation in which $\Delta k = 0$ has been achieved, then $[(\sin x)/x]^2 = 1$, and equation (9.144) gives us

$$[I_{30}(L)]_{\text{max}} = \frac{128\pi^3\omega^2(d^{(2\omega)})^2 L^2 I_{10}^2}{c^3 \varepsilon_1 \varepsilon_3^{1/2}} \tag{9.145}$$

as the expression for the *maximum* second harmonic intensity. For the case for which $\Delta k = 0$ and interference effects do not decrease the output, equation (9.145) shows that the second harmonic intensity $I_{30}(L)$ is proportional to the square of the crystal length L. Finally, we see from the expressions above that, in any case, $I_{30}(L)$ is proportional to the square of the nonlinear susceptibility $d^{(2\omega)}$ and thus will depend on the crystal used in the experiment.

We may calculate the efficiency e of second harmonic generation, defined by

$$e \equiv I_{30}(L)/I_{10}(0) = I_{30}(L)/I_{10} \tag{9.146}$$

From equation (9.144), we find that

$$e = \frac{128\pi^3\omega^2(d^{(2\omega)})^2L^2I_{10}}{c^3\varepsilon_1\varepsilon_3^{1/2}} \left(\frac{\sin x}{x}\right)^2 \tag{9.147}$$

showing that the *efficiency* of second harmonic generation is proportional to the intensity I_{10} of the input wave at the fundamental frequency. If the condition $\Delta k = 0$ is achieved, then the factor $[(\sin x)/x]^2 = 1$ and the maximum efficiency is given by

$$e_{\max} = \frac{128\pi^3\omega^2(d^{(2\omega)})^2L^2I_{10}}{c^3\varepsilon_1\varepsilon_3^{1/2}} \tag{9.148}$$

From equations (9.147) and (9.148), the efficiency e is seen to depend on the input intensity I_{10}, so one way to increase e is to increase I_{10}. A method[53] of achieving this is to place the nonlinear crystal inside a laser resonator, which then supplies the input wave. In such a situation, the input intensity can be very large, thereby increasing the efficiency e.

It might appear, from equations (9.147) and (9.148), that the conversion efficiency e could become larger than unity. This difficulty is only apparent, however, and stems from the assumption that the input wave is of constant amplitude, used in deriving these results. Actually, calculation[48] of the conversion efficiency for the case of a depleted input wave shows that the efficiency approaches unity asymptotically as the input intensity becomes very large. The assumption of constant input amplitude is a good approximation[48] for conversion efficiencies e less than about 0.1.

To summarize, we can see from the results developed in this section that efficient second harmonic generation is favored by the following factors. First, use of a crystal with a high value of the nonlinear susceptibility. Second, a high value of the intensity of the input wave at the fundamental frequency. Third, achievement as closely as possible of the condition $\Delta k = 0$, in order that the second harmonic intensity $I_{30}(L)$ and the efficiency e be maximized.

The aim of this section has been the use of the coupled-amplitude equations (9.121)–(9.123) to discuss, in an introductory and approximate way, some aspects of the physics of harmonic generation. For more complete discussions, including more realistic treatments, the reader is referred to the books by Yariv[9,10] and the review article by Akhmanov et al.[54] The latter paper includes a discussion of nonlinear materials for second harmonic generation applications.

Phase Matching (Index Matching) in Second Harmonic Generation

In our discussion of second harmonic generation, we considered the effect on the second harmonic intensity $I_{30}(L)$ of the quantity Δk defined by equation (9.124) and given by

$$\Delta k = 2k_1 - k_3 \tag{9.128}$$

for the case of second harmonic generation. It was shown that $I_{30}(L)$ and the efficiency e were both maximized for $\Delta k = 0$. The aim of this section is a brief discussion of the meaning of the condition $\Delta k = 0$ in the process of second harmonic generation.

If $\Delta k = 0$, then equation (9.128) states that

$$2k_1 = k_3 \tag{9.149}$$

where k_1 is the wave vector of the fundamental of frequency ω and k_3 is that of the second harmonic of frequency 2ω. Since the phase velocity v is given by

$$v = \omega/k \tag{9.150}$$

for a wave of frequency ω, the condition (9.149) is equivalent to

$$v(\omega) = v(2\omega) \tag{9.151}$$

where $v(\omega)$ and $v(2\omega)$ are the phase velocities of the waves of frequencies ω and 2ω, respectively. The condition (9.149) that $\Delta k = 0$ in second harmonic generation therefore is equivalent to requiring that the phase velocities $v(\omega)$ and $v(2\omega)$ of the fundamental and second harmonic be equal. Since the phase velocity $v = c/n$, where n is the index of refraction, equation (9.151) is equivalent to

$$n(\omega) = n(2\omega) \tag{9.152}$$

where $n(\omega)$ and $n(2\omega)$ are the refractive indices of the nonlinear crystal at the frequencies ω and 2ω, respectively. The condition that $\Delta k = 0$ is thus equivalent to requiring that the refractive indices at the fundamental and second harmonic frequencies be equal. For this reason, the process of making $\Delta k = 0$ is called index matching. Since, as described in the previous section, having $\Delta k \neq 0$ results in a phase difference and interference between the fundamental and the second harmonic, the attainment of $\Delta k = 0$ is also referred to as phase matching.

The condition for phase matching in the second harmonic generation process is found, from (9.152), to be the equality of the refractive index for the fundamental and second harmonic waves. Since the nonlinear crystal will in general exhibit dispersion, $n(\omega)$ and $n(2\omega)$ will not generally be equal. For this reason, phase matching will not be achieved unless special steps are taken to satisfy the condition contained in equation (9.152). A variety of methods have been used experimentally to obtain phase matching. A discussion of these techniques, while containing much interesting physics,[56-58] is outside the scope of this introduction, and the interested reader is referred to the literature.[52,54,55]

From the point of view of applications, phase matching is important in order to maximize the intensity and efficiency of second harmonic generation by making $\Delta k = 0$ so the factor $[(\sin x)/x]^2$ in (9.144) and (9.147) is close to unity. The same phase-matching requirement exists for high efficiency in other nonlinear mixing processes, including parametric amplification and frequency up-conversion.

Frequency Mixing and Up-Conversion

We consider in this section the mixing of two different frequencies ω_1 and ω_2 to give a third frequency ω_3, again subject to the condition (9.100) that

$$\omega_3 = \omega_1 + \omega_2 \tag{9.100}$$

This process converts a lower frequency ω_1 to a higher frequency ω_3 and is often called frequency up-conversion.[59,60] It is the mixing process of which second harmonic generation is the special case for which $\omega_1 = \omega_2$ and $\omega_3 = 2\omega_1$. We will refer to ω_1 as the input frequency which is mixed with the pump frequency ω_2 to produce the output frequency $\omega_3 = \omega_1 + \omega_2$. An example is the mixing of infrared photons (ω_1) with a strong laser source (ω_2) to produce visible photons (ω_3) for which sensitive detectors are available.

We return to the coupled-amplitude equations (9.121)–(9.123) to discuss this process and assume that a negligible fraction of the pump frequency (ω_2) photons is converted. The pump signal amplitude E_{20} may thus be considered as approximately constant, so we have

$$dE_{20}/dz = 0 \tag{9.153}$$

It is assumed also that phase matching has been achieved,[61] so we have

$$\Delta k \equiv k_1 + k_2 - k_3 = 0 \tag{9.154}$$

from the definition (9.124) of Δk. Following Yariv,[59] the coupled-amplitude equations (9.121) and (9.123) then become

$$\frac{dE_{10}^*}{dz} = j \frac{2\pi\omega_1}{c\varepsilon_1^{1/2}} d^{(2\omega)} E_{20} E_{30}^* \tag{9.155}$$

$$\frac{dE_{30}}{dz} = -j \frac{2\pi\omega_3}{c\varepsilon_3^{1/2}} d^{(2\omega)} E_{10} E_{20} \tag{9.156}$$

where $E_{10}(z)$ and $E_{30}(z)$ are, respectively, the amplitudes of the input and output waves. Defining a quantity β_1 by the relation

$$\beta_1 \equiv 2\pi\omega_1 d^{(2\omega)} E_{20}/c\varepsilon_1^{1/2} \tag{9.157}$$

where E_{20} is the constant electric field of the pump signal, and β_3 by the analogous relation

$$\beta_3 \equiv 2\pi\omega_3 d^{(2\omega)} E_{20}/c\varepsilon_3^{1/2} \tag{9.158}$$

the differential equations (9.155) and (9.156) become

$$\frac{dE_{10}}{dz} = -j\beta_1 E_{30} \tag{9.159}$$

$$\frac{dE_{30}}{dz} = -j\beta_3 E_{10} \tag{9.160}$$

In (9.159), the complex conjugate of (9.155) was taken, and both E_{10} and E_{30} are functions of the space coordinate z along the nonlinear crystal.

Equations (9.159) and (9.160) are a pair of coupled first-order differential equations for the electric fields $E_{10}(z)$ and $E_{30}(z)$ of the input and output waves. The solutions are

$$E_{10}(z) = A \cos \lambda z + B \sin \lambda z \tag{9.161}$$

$$E_{30}(z) = C \cos \lambda z + D \sin \lambda z \tag{9.162}$$

where A, B, C, and D are constants to be determined by the boundary conditions on the problem, and the quantity λ is defined by

$$\lambda^2 \equiv \beta_1\beta_3 = 4\pi^2\omega_1\omega_3[d^{(2\omega)}]^2 E_{20}^2/c\varepsilon_1^{1/2}\varepsilon_3^{1/2} \tag{9.163}$$

Using the symbols $E_{10}(0)$ and $E_{30}(0)$ for the values of $E_{10}(z)$ and $E_{30}(z)$ at the input face $z = 0$ of the crystal, satisfaction of either of the original differential equations (9.159) or (9.160) determines A, B, C, and D, leading to

$$E_{10}(z) = E_{10}(0) \cos \lambda z - j(\beta_1/\beta_3)^{1/2} E_{30}(0) \sin \lambda z \qquad (9.164)$$

$$E_{30}(z) = E_{30}(0) \cos \lambda z - j(\beta_3/\beta_1)^{1/2} E_{10}(0) \sin \lambda z \qquad (9.165)$$

We choose next the boundary condition that there shall be no photons of frequency ω_3 at the input face $z = 0$ of the crystal, so

$$E_{30}(0) = 0 \qquad (9.166)$$

and the solutions (9.164) and (9.165) then become

$$E_{10}(z) = E_{10}(0) \cos \lambda z \qquad (9.167)$$

$$E_{30}(z) = -j(\beta_3/\beta_1)^{1/2} E_{10}(0) \sin \lambda z \qquad (9.168)$$

These equations give the input electric field $E_{10}(z)$ and the output electric field $E_{30}(z)$ as functions of distance z in the nonlinear crystal. From equation (9.163), these solutions are functions of the frequencies ω_1 and ω_3, the nonlinear susceptibility $d^{(2\omega)}$, the pump electric field E_{20}, and the dielectric constants ε_1 and ε_3, through the dependence on β_1, β_3, and λ in equations (9.167) and (9.168).

We may calculate the intensities of the input (ω_1) and output (ω_3) waves as functions of z, obtaining

$$I_{10}(z) = (n_1 c/8\pi) E_{10}(z) E_{10}^*(z) = (n_1 c/8\pi)[E_{10}(0)]^2 \cos^2 \lambda z \qquad (9.169)$$

$$I_{30}(z) = (n_3 c/8\pi) E_{30}(z) E_{30}^*(z) = (n_3 c/8\pi)(\beta_3/\beta_1)[E_{10}(0)]^2 \sin^2 \lambda z \qquad (9.170)$$

where $n_1 = \varepsilon_1^{1/2}$ and $\varepsilon_3 = n_3^{1/2}$ are the refractive indices of the nonlinear crystal at the frequencies ω_1 and ω_3. From equations (9.157) and (9.158), one finds that

$$\beta_3/\beta_1 = \omega_3 \varepsilon_1^{1/2}/\omega_1 \varepsilon_3^{1/2} = \omega_3 n_1/\omega_1 n_3 \qquad (9.171)$$

Further, the input intensity $I_{10}(0)$ at $z = 0$ is given by

$$I_{10}(0) = (n_1 c/8\pi)[E_{10}(0)]^2 \qquad (9.172)$$

Substituting (9.171) and (9.172) into (9.169) and (9.170) gives

$$I_{10}(z) = I_{10}(0) \cos^2 \lambda z \qquad (9.173)$$

$$I_{30}(z) = (\hbar\omega_3/\hbar\omega_1) I_{10}(0) \sin^2 \lambda z \qquad (9.174)$$

for the input intensity $I_{10}(z)$ and the output intensity $I_{30}(z)$ as functions of distance z in the crystal. In obtaining (9.174), numerator and denominator were multiplied by $\hbar$ in order to convert the frequencies into photon energies.

We can discuss the magnitude of the quantity λ defined by equation (9.163) by rewriting that expression as

$$\lambda^2 = 32\pi^3 \omega_1 \omega_3 (d^{(2\omega)})^2 I_{20}/c^3 n_1 n_2 n_3 \tag{9.175}$$

on using the relation $I_{20} = (cn_2/8\pi)E_{20}^2$ for the intensity I_{20} of the pump wave of frequency ω_2, and the fact that $n_1{}^2 = \varepsilon_1$ and $n_3{}^2 = \varepsilon_3$. A dimensional analysis of equation (9.175) shows that λ has units of cm^{-1}, as it should, because λz in equation (9.173) and (9.174) must be dimensionless.

We may calculate λ for a typical example[62] of up-conversion in which the 10.6 µm wavelength output of a CO_2 laser is to be mixed with the 1.06 µm output of a neodymium doped YAG laser to produce a near-infrared signal which can be readily detected with a photomultiplier tube. In this example, then, the input (CO_2 laser) frequency $\omega_1 = 1.78 \times 10^{14}$ sec^{-1}, the pump (YAG laser) frequency $\omega_2 = 1.78 \times 10^{15}$ sec^{-1}, and the output frequency $\omega_3 = \omega_2 + \omega_1 = 1.96 \times 10^{15}$ sec^{-1}, corresponding to a wavelength of 0.964 µm in the near infrared. The nonlinear crystal used is Ag_3AsS_3, known as "proustite," which is quite transparent[63] between 0.6 and 13 µm and has a value[64] of $d^{(2\omega)}$ of approximately 2.5×10^{-22} (MKS) $\cong 7 \times 10^{-8}$ esu. Taking the refractive index values[63] $n_1 \cong n_2 \cong n_3 \cong 2.6$, and using a value of 10^{11} ergs sec^{-1} cm^{-2} for the pump intensity I_{20} of the YAG laser, we calculate a value of $\lambda = 1.8 \times 10^{-2}$ cm^{-1} from equation (9.175).

Since, from equation (9.173),

$$I_{10}(1/\lambda) = I_{10}(0)[\cos 1]^2 = 0.292 I_{10}(0) \tag{9.176}$$

we can see the physical meaning of λ. The quantity $1/\lambda$ is the distance in the crystal over which the input intensity $I_{10}(z)$ at ω_1 decreases to 0.292 of its initial value $I_{10}(0)$. For this example, the distance $1/\lambda = 56$ cm. Using the value of $\hbar\omega_3/\hbar\omega_1 = 11.0$ for this example, the output intensity at frequency ω_3 is given by

$$I_{30}(z) = (11.0)\{\sin^2[(1.80 \times 10^{-2})z]\}I_{10}(0) \tag{9.177}$$

a function that is plotted in Figure 9.3. This figure shows $I_{30}(z)$ in units of the initial input intensity $I_{10}(0)$, as a function of crystal length z in centimeters. For a crystal length of 1 cm, the curve shows that I_{30} is equal to $0.00356 I_{10}(0)$; for a crystal length of 5 cm, $I_{30} = 0.0889 I_{10}(0)$.

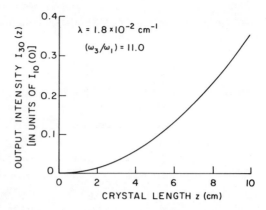

Figure 9.3. Output intensity $I_{30}(z)$, at frequency ω_3, from equation (9.177) [in units of the initial input intensity $I_{10}(0)$] plotted as a function of crystal length z (in cm) for the case of $\lambda = 1.8 \times 10^{-2}\,\text{cm}^{-1}$ and $\omega_3/\omega_1 = 11.0$ used in the up-conversion example in the text.

Further, the conversion efficiency e may be defined as

$$e \equiv I_{30}(z)/I_{10}(0) = (\omega_3/\omega_1)\sin^2 \lambda z \qquad (9.178)$$

on using equation (9.174). For the example, given above, $e = 0.0036$ for a length of 1 cm, and $e = 0.089$ for a length of 5 cm, showing that useful percentages of the input intensity $I_{10}(0)$ may be up-converted in this situation. Since the maximum value of $\sin^2 \lambda z$ is unity, the largest value of the conversion efficiency is $e_{max} = \omega_3/\omega_1$, which we may write as

$$e_{max} = \hbar\omega_3/\hbar\omega_1 \qquad (9.179)$$

This equation shows that the maximum value of e is the ratio of the energies of the output and input photons and hence is larger than unity. This value of $e_{max} = \hbar\omega_3/\hbar\omega_1$ corresponds to the situation in which *all* of the input photons of energy $\hbar\omega_1$ are converted to output photons of energy $\hbar\omega_3$. The energy necessary to do this comes, of course, from the pump photons of energy $\hbar\omega_2$, which are destroyed whenever a photon of energy $\hbar\omega_3$ is created and a photon of energy $\hbar\omega_1$ is destroyed.

Further, if λz is small compared to unity, then we may approximate

$$\sin \lambda z \simeq \lambda z \qquad (9.180)$$

and equations (9.174) and (9.178) become

$$I_{30}(z) \simeq (\hbar\omega_3/\hbar\omega_1)\lambda^2 I_{10}(0)z^2 \qquad (9.181)$$

$$e \simeq (\omega_3/\omega_1)\lambda^2 z^2 \qquad (9.182)$$

for the output intensity at frequency ω_3 and for the conversion efficiency. For the example plotted in Figure 9.3, λz is less than 0.18 and equation (9.181) for $I_{30}(z)$ is a good approximation to equation (9.177) for the values of z shown in the figure.

Finally, it should be emphasized that results obtained in this section for the output intensity and for the conversion efficiency refer to the largest possible values of these quantities. This is because we assumed phase matching ($\Delta k = 0$) in setting up the equations (9.155) and (9.156) which were solved to describe the up-conversion process. If $\Delta k \neq 0$, then the values of the output intensity and the conversion efficiency will be reduced[65] below the largest possible values.

For a discussion of experimental details of up-conversion processes and various applications, the reader is referred to the references, particularly those by Zernike and Midwinter[59] and by Warner.[60]

Parametric Amplification

In this section, we consider a signal wave of frequency ω_1 and an intense pump wave of frequency ω_3 both incident on a nonlinear crystal. Assuming that ω_3 is greater than ω_1, parametric amplification is the process by which energy is transferred from the pump wave to the signal wave, thereby amplifying the latter. At the same time, an idler wave of frequency

$$\omega_2 = \omega_3 - \omega_1 \tag{9.183}$$

is produced. In terms of photons, a pump photon $\hbar\omega_3$ is destroyed with the simultaneous creation of a signal photon $\hbar\omega_1$ and an idler photon $\hbar\omega_2$.

Since equation (9.183) is equivalent to the condition that $\omega_1 + \omega_2 = \omega_3$, we may again use the coupled amplitude equations (9.121)–(9.123) to discuss parametric amplification.[66,67] We will assume that the pump wave is sufficiently intense that it is undepleted, so the amplitude $E_{30}(z)$ is a constant, and

$$dE_{30}/dz = 0 \tag{9.184}$$

For simplicity, we assume also that phase matching[68] has been achieved, so

$$\Delta k = k_3 - k_1 - k_2 = 0 \tag{9.185}$$

With the conditions (9.184) and (9.185), the coupled amplitude equations

(9.121) and (9.122) become

$$\frac{dE_{10}^*}{dz} = j \frac{2\pi\omega_1}{c\varepsilon_1^{1/2}} d^{(2\omega)} E_{20} E_{30}^* \tag{9.186}$$

$$\frac{dE_{20}^*}{dz} = j \frac{2\pi\omega_2}{c\varepsilon_2^{1/2}} d^{(2\omega)} E_{10} E_{30}^* \tag{9.187}$$

These equations describe the variation in space of the amplitudes $E_{10}(z)$ and $E_{20}(z)$ of, respectively, the electric fields of the signal and idler waves. As were equations (9.155) and (9.156) for the up-conversion process, they are a pair of coupled differential equations.

Introducing the symbols α_1 and α_2 by

$$\alpha_1 = 2\pi\omega_1 d^{(2\omega)} E_{30}^* / c\varepsilon_1^{1/2} \tag{9.188}$$

$$\alpha_2 = 2\pi\omega_2 d^{(2\omega)} E_{30}^* / c\varepsilon_2^{1/2} \tag{9.189}$$

we may rewrite equations (9.186) and (9.187) as

$$\frac{dE_{10}^*}{dz} = j\alpha_1 E_{20} \tag{9.190}$$

$$\frac{dE_{20}^*}{dz} = j\alpha_2 E_{10} \tag{9.191}$$

It is useful to note the differences (signs and complex conjugates) between these equations and the analogous ones [(9.159) and (9.160)] for the up-conversion process. These differences lead, of course, to very different solutions.

The coupled differential equations (9.190) and (9.191) have the solution

$$E_{10}(z) = Ae^{\delta z} + Be^{-\delta z} \tag{9.192}$$

$$E_{20}(z) = Ce^{\delta z} + De^{-\delta z} \tag{9.193}$$

where A, B, C, and D are constants, and the quantity δ is defined by

$$\delta^2 \equiv \alpha_1\alpha_2 = 4\pi^2\omega_1\omega_2(d^{(2\omega)})^2 E_{30}^2 / c^2 \varepsilon_1^{1/2} \varepsilon_2^{1/2} \tag{9.194}$$

Next, we prescribe the boundary condition that there is no idler wave input to the crystal at the input face $z = 0$. This means that

$$E_{20}(0) = 0 \tag{9.195}$$

leading to the result that $D = -C$, so equation (9.193) for $E_{20}(z)$ becomes

$$E_{20}(z) = C(e^{\delta z} - e^{-\delta z}) = 2C \sinh(\delta z) \qquad (9.196)$$

Substituting equation (9.196) for $E_{20}(z)$ into the differential equation (9.191) for (dE_{20}^*/dz) leads to the results

$$A = B \qquad (9.197)$$

$$C = (ja_2 A/\delta) = j(\alpha_2/\alpha_1)^{1/2} A \qquad (9.198)$$

which, when substituted into equations (9.192) and (9.196), give

$$E_{10}(z) = 2A \cosh(\delta z) \qquad (9.199)$$

$$E_{20}(z) = 2jA(\alpha_2/\alpha_1)^{1/2} \sinh(\delta z) \qquad (9.200)$$

From (9.199), we see that the value $E_{10}(0)$ of $E_{10}(z)$ at $z = 0$ is equal to the constant $2A$, and we finally obtain the solutions

$$E_{10}(z) = E_{10}(0) \cosh(\delta z) \qquad (9.201)$$

$$E_{20}(z) = j(\alpha_2/\alpha_1)^{1/2} E_{10}(0) \sinh(\delta z) \qquad (9.202)$$

These equations describe the spatial variation of the electric fields $E_{10}(z)$ and $E_{20}(z)$ of the signal and idler waves for the case in which there is no idler (ω_2) wave input at $z = 0$.

We next use the definition of the intensity I as

$$I(z) = (cn/8\pi)EE^* \qquad (9.203)$$

where n is the refractive index, to obtain the intensities $I_{10}(z)$ and $I_{20}(z)$ of the signal and idler waves as functions of position z in the crystal. We find

$$I_{10}(z) = (cn_1/8\pi)E_{10}E_{10}^* = (cn_1/8\pi)[E_{10}(0)]^2 \cosh^2(\delta z) \qquad (9.204)$$

$$I_{20}(z) = (cn_2/8\pi)E_{20}E_{20}^* = (cn_2/8\pi)(\alpha_2/\alpha_1)[E_{10}(0)]^2 \sinh^2(\delta z) \qquad (9.205)$$

where $n_1 = \varepsilon_1^{1/2}$ and $n_2 = \varepsilon_2^{1/2}$ are the refractive indices of the crystal at the signal and idler frequencies ω_1 and ω_2. The signal intensity at the input $z = 0$ is, from (9.204),

$$I_{10}(0) = (cn_1/8\pi)[E_{10}(0)]^2 \qquad (9.206)$$

so the expressions for the intensities become

$$I_{10}(z) = I_{10}(0) \cosh^2(\delta z) \qquad (9.207)$$

$$I_{20}(z) = (n_2\alpha_2/n_1\alpha_1)I_{10}(0) \sinh^2(\delta z) \qquad (9.208)$$

From the behavior of the functions $\cosh(\delta z)$ and $\sinh(\delta z)$ as z increases from zero, we see that the intensities $I_{10}(z)$ and $I_{20}(z)$ increase as the waves move along the crystal (i.e., as z increases). Since $\cosh^2(\delta z)$ is greater than unity for z greater than zero, the signal intensity increases from its initial value $I_{10}(0)$ as the signal wave moves through the crystal. In this manner, the signal wave is amplified, the energy to do so coming, of course, from the pump wave. Similarly, since $\sinh^2(\delta z)$ is greater than zero for positive values of z, the idler wave intensity $I_{20}(z)$ increases from its input value of zero as the wave moves through the crystal. From equation (9.207), the signal wave undergoes, in a distance z, an amplification by a factor a, where

$$a \equiv I_{10}(z)/I_{10}(0) = \cosh^2(\delta z) \qquad (9.209)$$

Let us make some numerical estimates concerning parametric amplification in a representative case. From equation (9.194),

$$\delta^2 = 4\pi^2\omega_1\omega_2(d^{(2\omega)})^2 E_{30}^2/c^2 n_1 n_2 \qquad (9.210)$$

which, on substituting the expression

$$I_{30} = (cn_3/8\pi)E_{30}^2 \qquad (9.211)$$

for the constant intensity of the pump wave, becomes

$$\delta^2 = 32\pi^3\omega_1\omega_2(d^{(2\omega)})^2 I_{30}/c^3 n_1 n_2 n_3 \qquad (9.212)$$

We calculate the value of δ for an amplification experiment[69] in which a signal wavelength of 0.633 μm was amplified using a pump wavelength of 0.348 μm. The nonlinear crystal was ammonium dihydrogen phosphate (ADP), with a nonlinear susceptibility value[70] $d^{(2\omega)} = 1.50 \times 10^{-9}$ esu, and a refractive index[71] in the visible of approximately 1.5. The pump power was 2×10^{13} ergs sec^{-1} cm^{-2}. Using equation (9.212), we obtain $\delta = 0.06$ cm^{-1}. From equation (9.209), we can see that the physical significance of δ is that $1/\delta$ is the distance z over which the signal intensity $I_{10}(z)$ is amplified by a factor of $(\cosh^2 1) = 2.38$ over its input value $I_{10}(0)$. For the experiment described here, the ADP crystal was 8 cm long, so the factor $\delta z = 0.48$, and the amplification factor a is equal to $\cosh^2 0.48 = 1.25$. This is only a modest degree of amplification. From (9.209), we see that the amplification factor a depends on δ, which itself depends on, among other things, the pump intensity I_{30}. For this reason,[71] optical parametric amplification does not usually lead to large values of gain unless the pump intensity is extremely high. Because of this, one of the main reasons for interest in parametric amplification is its connection with optical parametric oscillation. In this process, the nonlinear crystal is placed in an appropriate optical resonant cavity, generating oscillations at the signal and idler frequencies. For a discussion of the parametric oscillator, the reader is referred to the references.[72-74]

Summary

How does the physics of optically nonlinear solids result in useful applications?

The key physical point is that the nonlinear term in the relation between polarization P and electric field E mixes waves of different frequencies in the crystal. The term in P that is proportional to E^2 gives rise to a component of the polarization at frequency $\omega_1 \pm \omega_2$ when the frequencies ω_1 and ω_2 are present. Since the wave equation is "driven" by a term $\partial^2 P/\partial t^2$, the result is an oscillating electric field at frequency (for example), $\omega_1 + \omega_2$, as well as at ω_1 and ω_2. This mixing process is exploited to generate second harmonic $(\omega_1 = \omega_2 = \omega; \ \omega_1 + \omega_2 = 2\omega)$ and sum $(\omega_1 + \omega_2 = \omega_3)$ frequencies. For high conversion efficiencies it is necessary to use large (i.e., laser) input electric fields since the nonlinear susceptibilities are small.

Problems

9.1. *Nonlinear Polarization in Tellurium.* (a) Estimate the value of the electric field for which the magnitude of the second harmonic polarization $P^{(2\omega)}$ in Te is equal to 0.1% of the linear polarization $P^{(\omega)}$. Use the values $d^{(2\omega)} = 1600 \times 10^{-9}$ esu and 6.2 for the refractive index n. (b) Calculate the value of $P^{(2\omega)}$ for this value of the applied electric field. Compare this value with some value of the polarization observed in a solid.

9.2. *Dimensional Analysis.* Using the units for electric field and nonlinear susceptibility introduced in the text, show that the right-hand side of equation (9.120) has units appropriate for dE_{30}/dz, i.e., has units of electric field per unit length.

9.3. *Derivation of Coupled Amplitude Equations.* Using the derivation of equation (9.120) for dE_{30}/dz as a guide, derive equation (9.121) for dE_{10}^*/dz, or the equivalent complex conjugate equation for dE_{10}/dz. It will be necessary to calculate the driving term $d^2(E^2)/dt^2$, where E is given by (9.94), in the wave equation (9.95). This tedious task is required because it is necessary to find the driving terms at frequency ω_1 on the right-hand side of the equation analogous to (9.104) which describes the electric field E_1.

9.4. *Efficiency of Second Harmonic Generation in Quartz.* Calculate the maximum efficiency of second harmonic generation in quartz, using as an input the $\lambda = 6940$ Å line from a ruby laser for which the electric field amplitude is 10^5 V cm^{-1}. (Find the input intensity in ergs sec^{-1} cm^{-2} by converting this electric field to statvolts cm^{-1}). Take the index of refraction of quartz as 1.5, the crystal length as 1 cm, and the nonlinear susceptibility as 10^{-9} esu.

9.5. *Photon Fluxes in Up-Conversion.* For the up-conversion of photons of frequency ω_1 to frequency ω_3, show that, at any point z of the crystal, the flux of ω_1 photons plus the flux of ω_3 photons is equal to the flux of ω_1 photons at the point $z = 0$.

9.6. *Manley–Rowe Relations.* Consider the coupled amplitude equations (9.121)–(9.123) and multiply each derivative by the complex conjugate of the electric field, thus obtaining the quantities $E_{10}(dE_{10}^*/dz)$, etc. Relate the three quantities so obtained to the intensities $I_{10} = (cn_1/8\pi)E_{10}E_{10}^*$, etc., and thereby to the fluxes N_1, N_2, N_3 of photons of frequency ω_1, ω_2, ω_3. Show that

$$\frac{dN_1}{dz} = \frac{dN_2}{dz} = -\frac{dN_3}{dz}$$

These equations are one form of the Manley–Rowe relations. (These useful results are discussed, for example, in the book by Zernike and Midwinter.[52])

References and Comments

1. J. M. Stone, *Radiation and Optics*, McGraw-Hill, New York (1963), Chapter 15; E. M. Purcell, *Electricity and Magnetism*, McGraw-Hill, New York (1965), Chapters 9 and 10.
2. C. Kittel, *Introduction to Solid State Physics*, Fifth Edition, John Wiley, New York (1976), page 410.
3. C. Kittel, Reference 2, pages 400–408.
4. J. F. Nye, *Physical Properties of Crystals*, Oxford University Press (1957), Chapter 4.
5. A. Sommerfeld, *Electrodynamics*, Academic, New York (1952), page 33.
6. Y. R. Shen, "Recent Advances in Nonlinear Optics," *Reviews of Modern Physics*, **48**, 1–32 (1976).
7. J. Ducuing, in *Quantum Optics*, R. J. Glauber (editor), Proceedings of the International School of Physics "Enrico Fermi," Course XLII, Academic Press, New York (1969), page 448.
8. N. Bloembergen, *Nonlinear Optics*, W. A. Benjamin, New York (1965).
9. A. Yariv, *Quantum Electronics*, Second Edition, John Wiley, New York (1975), pages 413–418.
10. A. Yariv, in *Topics in Solid State and Quantum Electronics*, W. D. Hershberger (editor), John Wiley, New York (1972), pages 280–290.
11. C. G. B. Garrett, "Nonlinear Optics, Anharmonic Oscillators, and Pyroelectricity," *IEEE Journal of Quantum Electronics*, **QE-4**, 70–84 (1968).

12. See, for example, C. Kittel, W. D. Knight, M. A. Ruderman, A. C. Helmholz, and B. J. Moyer, *Mechanics* (Berkeley Physics Course, Volume 1), Second Edition, McGraw-Hill, New York (1973), pages 224–226.

13. J. J. Stoker, *Nonlinear Vibrations*, Interscience Publishers, New York (1950), Chapter 4.

14. L. A. Pipes and L. R. Harvill, *Applied Mathematics for Engineers and Physicists*, Third Edition, McGraw-Hill, New York (1970), Chapter 15.

15. G. C. Baldwin, *An Introduction to Nonlinear Optics*, Plenum Press, New York (1969), page 141.

16. F. W. Constant, *Theoretical Physics*, Addison-Wesley, Reading, Massachusetts (1954), Section 6-6, pages 93–97.

17. J. B. Marion, *Classical Dynamics of Particles and Systems*, Second Edition, Academic Press, New York (1970), Section 5.5, pages 165–167.

18. See, for example, C. Kittel, W. D. Knight, M. A. Ruderman, A. C. Helmholz and B. J. Moyer, Reference 12, page 226.

19. G. C. Baldwin, Reference 15, pages 98–99.

20. F. Seitz, *Modern Theory of Solids*, McGraw-Hill, New York (1940), page 637.

21. C. G. B. Garrett and F. N. H. Robinson, "Miller's Phenomenological Rule for Computing Nonlinear Susceptibilities," *IEEE Journal of Quantum Electronics*, **QE-2**, 328–329 (1966).

22. A. Yariv, Reference 9, Appendix 4, and N. Bloembergern, Reference 8, Chapter 2, give quantum mechanical discussions.

23. A. Yariv, Reference 10, pages 283–285, gives figures exhibiting the polarization components at the fundamental and second harmonic frequencies.

24. J. F. Nye, Reference 4, Chapter 7, pages 110–115.

25. A. Yariv, Reference 10, page 287.

26. A. Yariv, Reference 9, pages 410–411, gives a table of the form of the nonlinear susceptibility tensor for various crystal classes.

27. J. Ducuing, Reference 7, Sections 3.2.1–3.2.3, pages 435–439.

28. A. Yariv, Reference 9, page 416, gives a table of values of $d_{ijk}^{(2\omega)}$ for several crystals using the contracted d_{il} notation mentioned earlier. Yariv explains this notation on page 409. J. F. Nye, Reference 4, page 113, also discusses it, as do T. S. Moss, G. J. Burrell, and B. Ellis, *Semiconductor Opto-Electronics*, John Wiley, New York (1973), page 249.

29. B. F. Levine, "Bond-Charge Calculation of Nonlinear Optical Susceptibilities for Various Crystal Structures," *Physical Review B*, **7**, 2600–2626 (1973), gives values of nonlinear susceptibilities (with references) for a large number of crystals. See also the paper by B. F. Levine and C. G. Bethea, *Applied Physics Letters*, **20**, 272–274 (1972). The book by Yariv, Reference 9, page 416, gives a table of values of $d^{(2\omega)}$ in units of $\frac{1}{9} \times 10^{-22}$ MKS. (To convert $d^{(2\omega)}$ in these units to $d^{(2\omega)}$ in esu, divide $d^{(2\omega)}$ (MKS) by 3.68×10^{-15}, as described by F. Zernike and J. E. Midwinter, *Applied Nonlinear Optics*, John Wiley, New York (1973), pages 52–53.) Finally, mention should be made of the article by S. K. Kurtz, "Measurement of Nonlinear Optical Susceptibilities," in *Quantum Electronics: A Treatise*, H. Rabin and C. L. Tang (editors), Academic Press, New York (1975), Volume 1 (Nonlinear Optics, Part A), Section 3, pages 209–281. This article describes methods of experimental measurement and gives values of $d^{(2\omega)}$ for a number of crystals important in nonlinear applications.

30. R. C. Miller, "Optical Second Harmonic Generation in Piezoelectric Crystals," *Applied Physics Letters*, **5**, 17–19 (1964).

31. J. A. Giordmaine, "Nonlinear Optics," *Physics Today*, **21**, 39–44 (January 1969).

32. B. F. Levine, Reference 29, page 2608, Table III.

33. B. O. Seraphin and H. E. Bennett, in *Semiconductors and Semimetals*, R. K. Willardson and A. C. Beer (editors), Volume 3, Academic Press, New York (1967), Chapter 12, pages 526, 520, and 511. The value of n quoted for GaAs is the average of the two values given in this reference.

34. B. F. Levine, Reference 29, page 2621, Table XVI.

35. R. A. Smith, *Semiconductors*, Cambridge University Press (1959), page 386.

36. C. Kittel, Reference 2, pages 409–412.

37. Y. R. Shen, Reference 6, Section II, pages 2–4.

38. See, for example, C. Kittel, W. D. Knight, M. A. Ruderman, A. C. Helmholz, and B. J. Moyer, Reference 12, pages 67–70. A useful table of conversion factors is given by G. Joos, *Theoretical Physics*, Second Edition, Hafner Publishing Co., New York (1950), page 836.

39. N. Bloembergen, "The Stimulated Raman Effect," *American Journal of Physics*, **35**, 989–1023 (1967), page 996.

40. The author thanks B. Black for pointing out this approach.

41. See, for example, J. Ducuing, Reference 7, Sections 2.3 and 3.3 pages 433–435 and 439–448. This paper also gives references to the original work, including that of J. A. Armstrong, N. Bloembergen, J. Ducuing, and P. S. Pershan, "Interactions between Light Waves in a Nonlinear Dielectric," *Physical Review*, **127**, 1918–1939 (1962). This paper is reproduced in the book by N. Bloembergen, Reference 8.

42. A. Yariv, Reference 10, pages 290–292.

43. See N. Bloembergen, Reference 39, page 995, for a treatment of the interaction of four waves (with frequencies $\omega_1, \omega_2, \omega_3, \omega_4$) subject to the condition that $\omega_1 + \omega_4 = \omega_2 + \omega_3$.

44. J. A. Armstrong, N. Bloembergen, J. Ducuing, and P. S. Pershan, "Interactions between Light Waves in a Nonlinear Dielectric," *Physical Review*, **127**, 1918–1939 (1962), page 1928.

45. J. Ducuing, Reference 7, page 434, equation (31).

46. See, for example, A. Yariv, Reference 9, page 421, for a treatment of the case of a solid with a nonzero value of the electrical conductivity.

47. P. A. Franken, A. E. Hill, C. W. Peters, and G. Weinreich, "Generation of Optical Harmonics," *Physical Review Letters*, **7**, 118–119 (1961).

48. See, for example, A. Yariv, Reference 9, Section 16.6, for a discussion of second harmonic generation for the case of a depleted input wave.

49. See, for example, G. R. Fowles, *Introduction to Modern Optics*, Second Edition, Holt, Rinehart and Winston, New York (1975), page 25.

50. See, for example, G. R. Fowles, Reference 49, Section 5.4.

51. See, for example, G. C. Baldwin, Reference 15, Section 4.3; J. Ducuing, Reference 7, page 441.

52. F. Zernike and J. E. Midwinter, *Applied Nonlinear Optics*, John Wiley, New York (1973), Chapter 3.

53. See A. Yariv, Reference 10, pages 298–303, for a discussion.

54. S. A. Akhmanov, A. I. Kovrygin, and A. P. Sukhorukov, "Optical Harmonic Generation and Optical Frequency Multipliers," in *Quantum Electronics: A Treatise*, H.

Rabin and C. L. Tang (editors), Academic Press, New York (1975), Volume 1 (Nonlinear Optics, Part B), Section 8, pages 475–586.

55. See, for example, A. Yariv, Reference 9, pages 424–428; A. Yariv, Reference 10, pages 294–298, G. C. Baldwin, Reference 15, pages 88–97.

56. See, for example, M. Born and E. Wolf, *Principles of Optics*, Fourth Edition, Pergamon Press, New York (1970), page 671.

57. See, for example, G. R. Fowles, Reference 49, Section 6.7.

58. See, for example, A. Yariv, Reference 9, Section 5.4.

59. See, for example, F. Zernike and J. E. Midwinter, Reference 52, Chapter 6; A. Yariv, Reference 9, pages 454-456; A. Yariv, Reference 10, pages 321–325.

60. J. Warner, "Difference Frequency Generation and Up-Conversion," in *Quantum Electronics: A Treatise*, H. Rabin and C. L. Tang (editors), Academic Press, New York (1975), Volume 1 (Nonlinear Optics, Part B), Section 10, pages 703–737.

61. F. Zernike and J. E. Midwinter, Reference 52, Section 6.4; J. Warner, Reference 60, pages 707–712. These references also discuss the situation when Δk given by (9.154) is not zero.

62. A. Yariv, Reference 10, page 325.

63. F. Zernike and J. E. Midwinter, Reference 52, page 99.

64. A. Yariv, Reference 9, Table 16.2, page 416.

65. J. Warner, Reference 60, page 705.

66. See, for example, A. Yariv, Reference 9, Section 17.1; A. Yariv, Reference 10, pages 304–313; F. Zernike and J. E. Midwinter, Reference 52, Sections 7.1–7.3.

67. J. A. Giordmaine in *Quantum Optics*, R. J. Glauber (editor), Proceedings of the International School of Physics "Enrico Fermi," Course XLII, Academic Press, New York (1969), pages 499–502.

68. See especially A. Yariv, Reference 10, pages 311–313, and F. Zernike and J. E. Midwinter, Reference 52, Section 7.3, for discussions of phase mismatching in parametric amplification.

69. J. A. Giordmaine, Reference 67, page 502, Table II.

70. A. Yariv, Reference 9, Table 16.2, page 416.

71. A. Yariv, Reference 10, page 311.

72. See, for example, A. Yariv, Reference 10, Sections 7.9–7.11.

73. R. L. Byer, "Optical Parametric Oscillators," in *Quantum Electronics: A Treatise*, H. Rabin and C. L. Tang (editors), Academic Press, New York (1975), Volume 1 (Nonlinear Optics, Part B), Section 9, pages 587–702.

74. J. A. Giordmaine, Reference 67, pages 502–509.

Suggested Reading

G. R. Fowles, *Introduction to Modern Optics*, Second Edition, Holt, Rinehart and Winston, New York (1975). Chapter 6 of this senior-level optics text provides an introduction to the optics of solids; our treatment of polarization and wave propagation in a linear dielectric is similar to Fowles' sections 6.2–6.4.

A. Yariv, in *Topics in Solid State and Quantum Electronics*, W. D. Hershberger (editor), John Wiley, New York (1972), Chapter 7. This collection of articles contains a chapter by Yariv on optical second harmonic generation which discusses, among other things, nonlinear polarization in solids. Our discussion parallels the treatment of Yariv.

A. YARIV, *Quantum Electronics*, Second Edition, John Wiley, New York (1975). This advanced textbook covers many topics in quantum electronics including, in Chapters 16 and 17, a detailed discussion of nonlinear optics, second harmonic generation, and parametric amplification more extensive than that in Yariv's article quoted above.

N. BLOEMBERGEN, *Non-Linear Optics*, W. A. Benjamin, New York (1965). This lecture note and reprint volume discusses its subject, including the quantum theory of non-linear susceptibilities, at the advanced level. It includes reprints of several important original papers.

G. C. BALDWIN, *Non-Linear Optics*, Plenum Press, New York (1969). This short introductory book has brief but useful discussions of many of the topics we have covered.

J. J. STOKER, *Nonlinear Vibrations*, Interscience Publishers, New York (1950). This short book is a good introduction to nonlinear oscillations and is clear and readable.

R. J. GLAUBER (editor), *Quantum Optics* [Proceedings of the International School of Physics "Enrico Fermi," Course LXII] Academic Press, New York (1969). This collection of tutorial lectures at the advanced level includes several on various aspects of nonlinear optics. Especially pertinent is that by J. Ducuing on nonlinear optical processes.

N. BLOEMBERGEN, "The Stimulated Raman Effect," *Am. J. Phys.* **35**, 989–1023 (1967). A tutorial and review article discussing many topics in the physics of nonlinear optics.

H. RABIN and C. L. TANG (editors), *Quantum Electronics: A Treatise*, Academic Press, New York (1975), Volume 1 on *Nonlinear Optics*, Parts A and B. For those wishing to delve deeply into nonlinear optics at the advanced level, these books offer articles on, among other topics, the measurement of nonlinear optical susceptibilities, optical harmonic generation, optical parametric oscillators, and frequency up-conversion.

F. ZERNIKE and J. E. MIDWINTER, *Applied Nonlinear Optics*, John Wiley, New York (1973)· This short book is rather terse in style, but covers a wide variety of topics in the physics and applications of nonlinear optics.

Appendix:
References on Some Other Topics

There were a number of topics that I would have liked to include, but did not for lack of space and time. These topics concern magnetism, and the following is a list of references I have found useful and suggest to the interested reader.

The first topic is that of the magnitude of the critical magnetic field in superconductors. As an introduction, I recommend "High Field Superconductors," by O. Fischer, in *Bulletin of the European Physical Society*, **7**, 1–4 (1976). The article by Geballe and Beasley (Reference 79 of Chapter 8) discusses this topic in detail and at the advanced level.

The second topic is the applications of ferromagnetism. For background, I suggest Kittel's *Introduction to Solid State Physics*, Fifth Edition, John Wiley, New York (1976), Chapter 15. A good introductory reference on the applications of ferro- and ferrimagnetism is W. R. Beam's *Electronics of Solids*, McGraw-Hill, New York (1965), Chapters 8 and 9. An older book which discusses many applications is *Solid State Magnetic and Dielectric Devices*, edited by H. W. Katz, John Wiley, New York (1959). Microwave ferrite devices are treated at the advanced level by B. Lax and K. L. Button in *Microwave Ferrites and Ferrimagnetics*, McGraw-Hill, New York (1962). A topic of current interest is covered in *Magnetic Bubbles*, by A. H. Bobeck and E. Della Torre, North Holland Publishing, Amsterdam (1975). Magnetic computer memories of several types are discussed in *Physics of Computer Memory Devices*, by S. Middelhoek, P. K. George, and P. Dekker, Academic Press, New York (1976). Finally, the article by P. J. Wojtowicz, "Magnetic Materials," in *Topics in Solid State and Quantum Electronics*, W. D. Hershberger (editor), John Wiley, New York (1972), provides an overview of its field.

Author Index

Adler, D., 172, 173
Akhmanov, S. A., 291, 305
Allan, R., 172
Allen, P. B., 255
Altman, L., 138
Armstrong, J. A., 305
Ashcroft, N. W., 23, 24,
  77, 78, 245, 252, 253,
  254
Aspnes, D. E., 171

Baldwin, G. C., 267, 304,
  305, 306, 307
Bardeen, J., 171
Bate, R. T., 172
Beam, W. R., 24, 77,
  212, 309
Beasley, M. R., 254, 309
Bell, R. L., 171
Bennett, H. E., 305
Bertram, W. J., 138
Bethea, C. G., 304
Bixby, B., 137
Black, B., 251, 305
Blakemore, J. S., 23
Bloembergen, N., 303,
  304, 305, 307
Bobeck, A. H., 309
Bohm, D., ix, 252
Born, M., 306
Boyle, W. S., 124, 137
Bratt, P. R., 210, 211
Brophy, J. J., 107, 108,
  137
Bube, R. H., 211, 212
Burrell, G. J., 210, 211,
  212, 213, 304

Butkov, E., 253
Button, K. L., 309
Byer, R. L., 306

Carver, L., 137
Chang, K. K. N., 107
Chelikowsky, J. R., 5, 23
Clarke, J., 225, 230, 231,
  241, 242, 252, 253
Cohen, M. H., 168, 172
Cohen, M. L., 5, 23, 255
Constant, F. W., 304
Cooper, D., 137
Copeland, J. A., 172
Cowley, A. M., 171
Curran, L., 211

Dalven, R., 211
Dash, W. C., 210
Davis, E. A., 172
Davison, S. G., 170, 171
Dean, P. J., 212
Deaver, B. S., 255
Dekker, A. J., 23, 24, 124
Dekker, P., 137, 138, 309
Della Torre, E., 309
Dew-Hughes, D., 254
duBow, J., 211
Ducuing, J., 303, 304,
  305, 307
Duke, C. B., 212
Dynes, R. C., 255

Economou, E. N., 172
Ehrenreich, H., 172
Ellis, B., 210, 211, 212,
  213, 304

Emde, F., 253
Eyring, H., 212

Falco, C. M., 255
Fan, H. Y., 198, 212
Feucht, D. L., 124
Feynmann, R. P., ix,
  216, 252, 255
Fischer, O., 309
Fisher, D. G., 171
Fowles, G. R., ix, 305,
  306
Frank, N. H., 77
Franken, P. A., 305
Freed, K. F., 172
Fritzsche, H., 168, 172

Garrett, C. G. B., 268,
  275, 303, 304
Geballe, T. H., ix, 254,
  255, 309
George, P. K., 137, 138,
  309
Giordmaine, J. A., 305,
  306
Gladstone, G., 245, 254
Glauber, R. J., 307
Goetzberger, A., 124
Goldstein, Y., 170, 173
Greenaway, D. L., 211
Grove, A. S., 23, 24, 68,
  75, 76, 77, 78, 107,
  108, 124, 137, 170, 171
Grover, N. B., 170, 173
Gutsche, E., 182, 211

Halkias, C. C., 77, 106,
  107, 108, 137, 138

Harbeke, G., 211
Harman, T. C., 184, 188, 211, 213
Harris, J. H., 255
Harvill, L. R., 304
Hayashi, I., 213
Helmholz, A. C., 304
Henisch, H. K., 124, 125, 171, 173
Hershberger, W. D., ix
Hildebrand, F. B., 252
Hill, A. E., 305
Ho, K. M., 255
Hogarth, C. A., 24
Hoges, D. A., 137
Holonyak, N., 212
Hovel, H. J., 22, 24, 211

Ing, S. W., 211

Jackson, J. D., 253
Jahnke, E., 253
Jaklevic, R. C., 253
James, T. H., 212
Javetski, J., 211
Jensen, M. A., 245, 254
Jones, M. E., 172
Joos, G., 305
Josephson, B. D., 252

Kamins, T. I., 77, 107, 124, 125, 137, 138, 171
Katz, H. W., 309
Kimball, G. E., 212
Kino, G. S., 171
Kirkpatrick, E. S., 172
Kittel, C., ix, 1, 6, 23, 24, 76, 77, 106, 107, 170, 172, 210, 211, 212, 245, 252, 253, 255, 303, 304, 305, 309
Knight, S., 172
Knight, W. D., 304, 305
Koechner, W., 212, 213
Kovrygin, A. I., 305
Kressel, H., 212
Kroemer, H., 172, 173
Kruse, P. W., 211
Kurtz, S. K., 304

Ladany, I., 212
Lambe, J., 253
Lapidus, G., 172

Lax, B., 213, 309
Leighton, R. B., ix, 252, 255
Levine, B. F., 304, 305
Levine, J. D., 170, 171
Levinstein, H., 211
Long, D., 6, 23, 24, 210, 211, 212
Lynton, E. A., 252

Many, A., 170, 173
Marion, J. B., 304
Martinelli, R. U., 171
Matthias, B. T., 248, 254
McGill, T. C., 107
McKelvey, J. P., 124, 170
Mead, C. A., 107
Mees, C. E. K., 212
Meindl, J. D., 137
Melngailis, I., 184, 188, 211, 213
Mercereau, J. E., 252, 253
Mermin, N. D., 23, 24, 77, 78, 245, 252, 253, 254
Meservey, R., 254
Middelhoek, S., 137, 138, 309
Midwinter, J. E., 298, 303, 304, 305, 306, 307
Milford, F. J., ix
Miller, R. C., 273, 305
Millman, J., 77, 106, 107, 108, 137, 138
Milnes, A. G., 124, 211
Moll, J. L., 69, 75, 76, 77, 78
Mooradian, A., 198, 212
Moss, T. S., 183, 210, 211, 212, 213, 304
Mott, N. F., 168, 172, 173
Moyer, B. J., 304, 305
Muller, R. S., 77, 107, 124, 125, 137, 138, 171

Neuse, C. J., 212
Newman, R., 210
Nye, J. F., 303, 304

Okean, H. C., 107
Oldham, W. G., 106, 107, 108, 137
Ost, E., 211
Ovshinsky, S., 168

Panish, M. B., 213
Pankove, J. I., 124, 171, 210, 212, 213
Pershan, P. S., 305
Peters, C. W., 305
Phillips, J. C., 24
Pickett, W. E., 255
Pincherle, L., 23
Pipes, L. A., 304
Purcell, E. M., 303

Rabin, H., 307
Reitz, J. R., ix
Richards, P. L., 252, 253
Robinson, A. L., 254
Robinson, F. N. H., 304
Rosenberg, H. M., 23, 210
Ruch, J. G., 171
Ruderman, M. A., 304, 305

Sands, M., ix, 252, 255
Scalapino, D. J., 254
Scharfe, M. E., 211
Scheer, J. J., 170, 171, 173
Schiff, L. I., 107, 212, 252
Schmidlin, F. W., 211
Schmidt, J. L., 211
Schrieffer, J. R., 245, 254
Schultz, M. L., 23
Schwartz, B. B., 254
Schwarz, S. E., 106, 107, 108, 137
Sealer, D. A., 138
Seitz, F., 304
Séquin, C. H., 137, 138
Seraphin, B. O., 305
Shen, Y. R., 303, 305
Shockley, W., 67, 75, 77, 78, 140
Silver, A. H., 253
Simon, R. E., 171
Slater, J. C., 77
Smith, G. E., 124, 137
Smith, R. A., 23, 24, 76, 305
Smith, W. V., 212, 213
Solymar, L., 252, 253, 255
Sommerfeld, A., 303
Sorokin, P. P., 212

Stoker, J. J., 304, 307
Stone, J. M., ix, 210, 303
Streetman, B. G., ix, 24,
    69, 75, 76, 77, 107,
    108, 137, 172, 211, 212,
    213
Sukhorukov, A. P., 305
Sze, S. M., 23, 75, 76, 77,
    78, 107, 108, 124, 125,
    137, 171

Tabak, M. D., 211
Tamm, I., 140
Tang, C. L., 307
Tauc, J., 172, 173

Thornber, K. K., 107
Tinkham, M., 23, 252,
    253, 254
Tompsett, M. F., 137, 138

van der Ziel, A., 77, 78,
    124, 125, 171
Van Laar, J., 170, 171, 173
Voight, J., 211

Walter, J., 212
Wang, S., 23, 24, 77, 78,
    107, 170, 172, 212
Warner, J., 298, 306
Weinreich, G., 305

White, R. M., ix, 254, 255
Wojtowicz, P. J., 309
Wolf, E., 306
Wolf, S. A., 255

Yariv, A., ix, 212, 213,
    267, 279, 291, 303, 304,
    305, 306, 307
Yu, A. Y. C., 124, 137

Zener, C., 73, 75, 81
Zernike, F., 298, 303, 304,
    305, 306, 307
Ziman, J. M., viii
Zimmerman, J. E., 253

Subject Index

Absorption of photons in semiconductors:
see Photon absorption in
semiconductors
Acceptor, 15
  ionization, 15
  ionization energy, 15
  surface, 141
Accumulation layer, 114
Activation energy, amorphous semiconductor, 166, 167, 168
AlAs–GaAs solid solutions, 209
Amorphous semiconductors, 164–169
  activation energy, 166, 167, 168
  conductivity, 166, 168, 169
  density of states, 166–167
  devices, 168–169
  electronic structure, 165–168
  extended states, 165, 167
  impurities and defects, 166–167
  localized states, 165–167
  memory device, 168
  mobility, 167–168
    dependence on energy, 167
  mobility gap, 167
  Mott–CFO model, 168
  switching device, 169
  values of parameters, 168
Amplification, light, 200
Amplification factor, junction field-
  effect transistor (JFET), 105–106
Amplifier
  common base, 90
  common emitter, 89
  current-controlled, 85
  voltage-controlled, 85, 102

Anharmonic oscillator
  anharmonic force constant, 265–266,
    270
  equation of motion, 265–267
  solution of equation of motion, 267–
    269
Applications
  of amorphous semiconductors,
    168–169
  of bipolar junction transistor, 85, 88–90,
    91
  of charge-coupled devices, 136
  of DC squid, 241–242, 251
  of degenerate p–n junction, 101
  of ferromagnetism (references), 309
  of intrinsic photoconductivity,
    188–195
  of Josephson junction, 232, 233,
    241–242
  of junction field-effect transistor, 106
  of luminescence in semiconductors,
    199–200, 210
  of metal–insulator–semiconductor
    (MIS) structure
    as charge-coupled device, 134–136
    as MOSFET, 128–133
  of metal–semiconductor junction
    as contact to semiconductor, 118–119
    as diode, 127–128
  of MOSFET, 133
  of nonlinear optical properties of solids
    frequency mixing, 293–298
    optical second harmonic generation,
      284–293
    parametric amplification, 298–302
    up-conversion, 293–298

Applications (*cont.*)
  of photovoltaic effect, 191
  of *p-n* junction diode, 81
  of superconducting quantum inter-
      ference, 239–242
  of tunnel diode, 101
Applied potential, effect on energy bands,
    48
Approximations
  bipolar junction transistor, 91
  ideal abrupt junction, 36, 63–64, 69,
      71, 75
  ideal metal–insulator–semiconductor
      structure, 119
Avalanche breakdown in *p-n* junction,
    74–75
Avalanche multiplication, 74

Band
  impurity: *see* Impurity band
  parabolic: *see* Parabolic band
Band bending
  calculation, 145–147, 170
  in GaP, 170
  and negative electron affinity, 154–155
  and photoemission from semiconduc-
      tors, 151–153
  at semiconductor surface, 142–143
Band gap: *see* Energy gap
Band structure, 1–5
  amorphous semiconductors, 165–168
  CdS, 181–182
  diamond, 3–4
  GaAs, 158–159
  InSb, 5–6
  insulator, 2
  and intrinsic optical absorption,
      175–179
  lead salt semiconductors, 183–184
  metal, 1–2
  PbTe–SnTe solid solutions, 183–184
  semiconductor, 2–3
  semiconductor surface, 141–145
  silicon, 4–5
  and transferred-electron effect,
      159–160
Band structure diagram
  bipolar junction transistor
    with applied bias, 83
    at equilibrium, 83
  cold cathode, 157

Band structure diagram (*cont.*)
  degenerate *p-n* junction
    at equilibrium, 98
    under forward bias, 99
  heterojunction laser, 208–209
  metal, 2, 110
  metal–insulator–semiconductor
      structure
    equilibrium, 119–120
    under applied potential, 120–122
  metal–semiconductor junction
    equilibrium, 112, 115
    showing effect of Fermi level pinning
        at surface, 148–149
    under applied potential, 116, 118
  negative electron affinity, 154–155
  n-type semiconductor, 18, 110
  photoemission from degenerate semi-
      conductor, 150–151
  photovoltaic effect, 190
  p-n junction
    at equilibrium, 30, 55
    injection laser, 207
    luminescence, 199
    under forward bias, 56
    under reverse bias, 56
  p-type semiconductor, 18–19
  semiconductor surface
    negative electron affinity, 154–155
    n-type with surface acceptors, 143
    p-type with surface donors, 144
    showing pinned Fermi level, 145
Bardeen–Cooper–Schrieffer: *see* BCS
Base of bipolar junction, 82
Base transit time, 86
Base transport factor, 86–88
  factors affecting, 87–88
Base width, 91–92
BCS "one square well" model, 243
BCS theory, 215–216, 242–244
Bipolar junction transistor
  alpha, 90
  amplification, 85–86
  applications, 85, 88–90, 91
  approximations made, 91
  band structure diagram
    with applied bias, 83–84
    at equilibrium, 82–83
  base, 82
  base current, 84–85, 94–95
    calculation, 94–95

Bipolar junction transistor (*cont.*)
  base current (*cont.*)
    in forward active mode, 94–95
    as function of base width, 94
    as function of emitter voltage, 95
  base-to-collector current gain (β), 87–88,
      95–96
    factors affecting, 88
    as function of base width, 95–96
    as function of minority carrier
        diffusion length, 95–96
    as function of minority carrier life-
        time, 96
  base transit time, 86
  base transport factor, 86–88
  base width, 91–92
  beta, 87, 90
  bias applied to junctions, 83, 88, 94
  calculation of currents, 91–95
  circuits for amplification, 88–90
  collector current, 85, 93–95
    calculation, 93–95
    in forward active mode, 94–95
    as function of base width, 94
    as function of emitter voltage, 95
  collector junction, 82–83
  collector voltage, 94, 95
  current gain, 86–88, 90
  currents, 84–85, 93–95
  electric fields, 84
  emitter current, 84–85, 93–95
    calculation, 93–95
    in forward active mode, 94–95
    as function of base width, 94
    as function of emitter voltage, 95
  emitter injection efficiency, 87–88
  emitter junction, 82–83
  emitter-to-collector current gain (α),
      90, 95–96
    as function of base width, 95
    as function of minority carrier
        diffusion length, 95
  emitter voltage, 94, 95
  factors affecting current gain, 87–88
  forward active mode, 88
  geometry, 91
  injected excess carrier density in base, 92
  npn, 106
  particle flows, 83–84
  pnp, 82–85
  summary of physics, 96

Brillouin zone, 3
  special points, 3

Capacitance, *p–n* junction, 75
Carrier density
  excess: *see* Excess carrier density
  increase in photoconductivity, 186
  in intrinsic semiconductor, 12–14
  in *n*-type semiconductor, 16
  in *p–n* junction, 37–39, 59–62
  in *p*-type semiconductor, 17
Carrier lifetime, 21–22, 50, 197; *see also*
      Minority carriers, lifetime
  in direct gap semiconductor, 21
  effect of impurities, 22
  in indirect gap semiconductor, 21
  for recombination, 197
  in silicon, 22
Carriers, photogenerated: *see* Photo-
      generated carriers
$CaWO_4$, 206
CCD: *see* Charge-coupled device
CdS
  band structure, 181–182
  energy threshold for photon absorption,
      181–182
  optical absorption spectrum, 182
Chalcogenide glass, 164
Channel
  diffused in MOSFET, 128
  induced in MOSFET, 131
  in junction field-effect transistor, 102,
      104–105
Charge-coupled device, 134–136
  applications, 136
  effect of applied potentials, 135
  energy wells, 134–135
  imaging with, 136
  introduction of minority carriers, 136
  inversion layer, 134
  metal–insulator–semiconductor
      structure, 134, 135
  storage of minority carriers, 134–135
  transfer of minority carriers, 135–136
Charge density, in *p–n* junction, 39
Charge transfer device: *see* Charge-coupled
      device
Coherence, of laser emission, 205
Cold cathode, 156–157
Collector junction, 82–83
Collector voltage 94, 95

Collision time, 11
Concentration gradient, 34
Conduction band, 3
Conductivity
  amorphous semiconductor, 166, 168,
    169
  extrinsic, 14–17
  impurity, 14–17
  negative: *see* Negative differential
    conductivity
  semiconductors, 10–12
Contact potential, 30
  in terms of Fermi energies, 32
  in terms of impurity densities, 43,
    45–46
Contacts, metal–semiconductor, 118–119,
    147–149
  effect of semiconductor surface, 147–
    149, 170
  ohmic, 118–119, 147–148
  rectifying, 118–119, 147–148
Cooper pairs: *see* Superconductors
Coulomb repulsion, 249, 251
Current
  in bipolar junction transistor, 83–85,
    86–87, 91–95
    base current, 84–85, 86–87, 94–95
    calculation, 91–95
    collector current, 85, 86–87, 93–95
    emitter current, 84–85, 86–87,
      93–95
  in DC squid, 239–241
  in degenerate *p-n* junction, 98–101
    excess, 101
    injection, 101
    tunnel, 98–101
  in Josephson junction, 223–229
    alternating supercurrent, 226–228
    critical, 225, 226, 227, 228
    effect of electromagnetic radiation,
      229
    normal electrons, 226
    pair tunneling, 223–225
  in magnetic field, 223–234
  in metal–semiconductor junction,
    116–117
  in *p-n* junction, 59, 62–72
    calculation, 62–69
    effect of applied potential, 59, 67–
      68
    majority and minority carrier
      components, 69–72

Current (*cont.*)
  in superconductor, 219–220
  in two Josephson junctions on super-
    conducting loop, 238–239
Current density: *see also* Current
  in magnetic field, 233–234
  in semiconductor, 10
  in superconductor, 217, 218
Current–voltage characteristic
  amorphous semiconductor devices, 169
  degenerate *p-n* junction, 100, 101
  diffused channel MOSFET, 129–130
  induced channel MOSFET, 132
  Josephson junction, 228, 229
    effect of electromagnetic radiation,
      229, 231–232
  junction field-effect transistor, 102–
    105
  metal–semiconductor junction, 117–
    118, 127
  photodiode, 189
  p-n junction, 59, 68
  p-n junction reverse breakdown, 73
  tunnel diode: *see* Degenerate *p-n*
    junction

Debye temperature, 243, 244, 250
Degenerate diode: *see* Tunnel diode
Degenerate *p-n* junction, 97–101
  applications, 101
  band structure diagram
    at equilibrium, 98
    under forward bias, 99
  current–voltage characteristic
    under forward bias, 100–101
    under reverse bias, 101, 106
  effect of band filling, 99
  effect of forward bias, 98–100
  effect of reverse bias, 101
  excess current, 101
  factors affecting tunneling, 97–98
  injection current, 101
  injection laser 206–209
  negative conductivity, 99
  speed, 101
  summary, 101
  tunnel current, 98–101
  tunneling, 97–98
  valley voltage, 100
Density of states
  amorphous semiconductor, 166–167
  impurity band, 19–20

Density of states (*cont.*)
  n-type semiconductor, 16
  p-type semiconductor, 16
Depletion layer
  diffused-channel MOSFET, 129–130
  in induced-channel MOSFET, 131–
    132
  in junction field-effect transistor,
    102–104
  in metal–insulator–semiconductor
    structure, 121–123
  in metal–semiconductor junction, 111,
    112
  in p-n junction, 27
Detectivity, 187
Detectors of electromagnetic radiation
  broadband Josephson junction, 232
  extrinsic semiconductor, 184–185
  infrared, 183, 184, 185
  intrinsic semiconductor, 181–185
  Josephson junction, 229–233
  lead salt semiconductor, 183–184
  PbTe–SnTe solid solutions, 187, 191
  photovoltaic, 191
  polycrystalline lead salt, 183
  summary of semiconductor properties,
    194
Dielectric relaxation time, 161, 162, 163
Diffused-channel MOSFET, 128–131
Diffusion coefficient, 34
  related to mobility, 36
Diffusion current, 33, 49, 53, 57–59,
    64–66
Diffusion length, 51, 52, 64–66
  effect on reverse current in p-n junction,
    69
Diffusion of excess carriers, 49–53
Diffusion potential, 30; *see also* Contact
  potential
Diode, degenerate: *see* Tunnel diode
Diode, metal–semiconductor, 127–128
  applications, 127–128
  current–voltage characteristic, 127
  as majority carrier device, 128
  speed, 128
Diode, p-n junction, 79–81
  applications, 81
  reverse saturation current, 79–81
Diode, Schottky, 109, 127–128; *see also*
  Diode, metal–semiconductor
Diode, tunnel, 96, 101; *see also*
  Degenerate p-n junction
Diode, Zener, 81

Direct transition, 176–120; *see also*
  Intrinsic photon absorption
Domains, space charge, 162–164
Donor, 14
  ionization, 14
  ionization energy, 14
  surface, 141
Drift current, 33, 49, 53, 57, 64
Drift velocity
  and current density, 10
  dependence on electric field, 157–158
Effective mass, 7–9
  and band curvature, 7
  defined, 7
  electron
    in conduction band, 8–9
    in silicon, 13
    in valence band, 9
  hole, 6, 9
    in GaAs, 9
    in silicon, 13
Einstein relation, 36, 80
Electrical conductivity: *see* Conductivity
Electric field
  built-in, 28, 40–42
  in metal–insulator–semiconductor
    structure, 120–121
  in metal–semiconductor junction, 111
  in photovoltaic effect, 190
  in p-n junction, 28, 40–42
  near semiconductor surface, 142, 146
Electric polarization
  as function of electric field, 264
  in isotropic linear dielectric, 260–263
    harmonic oscillator model, 260–261
  nonlinear
    as function of electric field, 276
    importance, 264–265
    physical picture, 270–271
    at second harmonic frequency, 269,
      270, 276
    as source term in wave equation,
      265, 277
  in nonlinear solid, 269–270
    anharmonic oscillator model, 265–269
    as source term in wave equation, 259,
      262–263, 271, 277
Electric susceptibility
  linear, 262, 269–270
  local field correction, 261–262
  nonlinear
    effect of inversion symmetry, 272

Electric susceptibility (*cont.*)
  nonlinear (*cont.*)
    expression for, 269–270
    magnitude, 275–276
    Miller's phenomenological rule, 273
    notation for, 271
    related to linear susceptibility,
      273–274
    related to optical dielectric constant,
      274–275
    solid state factors affecting, 271–275
    tensor nature, 271
    units, 275–276
    values, 274, 275, 276
Electromagnetic radiation, detectors:
    see Detectors of electromagnetic
    radiation
Electromagnetic waves in solids
  conservation of energy, 280
  electric field, 262–263, 277–278
  interaction in nonlinear crystal, 278–284
    assumptions, 279, 281
    equations for amplitudes, 283
    frequency mixing, 293–298; *see also*
      Frequency mixing
    optical second harmonic generation,
      284–293; *see also* Second harmonic
      generation
    parametric amplification, 298–302;
      see also Parametric amplification
    summary of physics, 302
    synchronous terms, 280
    up-conversion, 293–298; *see also*
      Frequency mixing
  physical picture of propagation
    linear dielectric, 263
    nonlinear dielectric, 270–271
  propagation, 257–259
  wave equation
    linear dielectric, 263
    nonconducting dielectric, 259
    nonlinear crystal, 276–278
    nonmagnetic solid, 258–259
    source terms, 259, 262–263, 271, 277
Electron affinity
  negative, in semiconductor, 154–157;
    see also Photoemission from semi-
    conductors
  semiconductor, 110, 149
Electron–hole pair, produced by intrinsic
  absorption, 176
Electronic specific heat, 244

Electrophotography, 192–193
Electrostatic potential
  in metal–semiconductor junction, 111
  in *p–n* junction, 29–31, 42–44, 55–56
  near semiconductor surface, 142
Eliashberg equations, 249
Emission
  of photons in semiconductors, 195–198
    direct-gap, 196–197
    InAs, 198
    indirect-gap, 197
    lowest-energy photon, 197
    p–n junction, 199–200
    spectrum of photon energies, 197–198
    via band-to-band recombination,
      196–198
    via impurity state, 198
  stimulated, 200, 205
Emitter injection efficiency, 87–88
  factors affecting, 88
Emitter junction, 82–83
Emitter voltage, 94, 95
Energy bands, effect of applied potential,
  48
Energy barrier
  in metal–semiconductor junction
    effect of pinning of Fermi level at
      surface, 149
    at equilibrium, 111–112, 113–114
    under applied potential, 115–116, 118
  in photovoltaic effect, 190
  p–n junction: *see p–n* junction, energy
    barrier
Energy gap
  direct, 5
  indirect, 5
  in insulator, 2
  in semiconductor, 2–3
  in superconductor, 216
Energy wells, in charge-compled device,
  134–135
Enhancement mode
  diffused channel MOSFET, 129–130
  induced channel MOSFET, 132
Equation of continuity, 49, 160–161, 216
Excess carrier density
  in bipolar transistor base, 91–92
  electrons, 52
  extraction in *p–n* junction, 59–62
  holes, 50, 51
  injection in *p–n* junction, 59–62
  steady state, 50–53

Extraction of carriers, 53
Extrinsic photon absorption, 180–181,
    184–185
  by acceptors, 180, 184–185
  connection with applications, 181
  by deep impurity levels, 181, 184–185
  by donors, 180, 184–186
  energy threshold, 180, 184–185
  generation of free carriers, 181
  germanium, 184–185
  optical absorption coefficient,
      180–181
  silicon, 185

Fermi–Dirac distribution, 12–13
  above 0 K, 21
  at 0 K, 20
  classical limit, 12–13
Fermi energy, 12
Fermi level
  metal, 110
  pinned at semiconductor surface,
      144–145
  position
    in extrinsic semiconductor, 17–21
    as function of impurity content,
        18–19
    in intrinsic semiconductor, 13–14
    in n-type semiconductor, 17–18
    in p-type semiconductor, 18–19
    temperature dependence, 20–21
  at semiconductor surface, 143–145
Ferromagnetism (references), 309
FET, 128, 132; see also Field-effect
    transistor
Field-effect transistor (FET): see Junction
    field-effect transistor, insulated-gate
    field-effect transistor; MOSFET
Flux of particles, 34
Flux quantization: see Quantization of
    magnetic flux
Flux quantum, 235
Forward bias, defined, 54
Fourier–Bessel series, 230
Frequency mixing, 293–298
  approximations made, 293, 298
  effect of phase matching, 298
  efficiency, 297–298
  example, 296
  output intensity, 295–297
  photon picture, 297, 303

GaAs
  band structure, 158–159
  p-n junction cold cathode, 156–157
  solar cell, 191
  solid solution with AlAs, 209
  transferred electron effect, 157–160
Gain, in bipolar junction transistor,
    85–88, 95–96
  base-to-collector (β), 87, 90, 95–96
  emitter-to-collector (α), 90, 95–96
  factors affecting, 87–88
Generation current, 33, 57–59, 67–69,
    79–81
Generation of carriers in p-n junction,
    64–65, 71, 80
Germanium
  energy thresholds for photon absorp-
      tion, 183, 185
  extrinsic photon absorption, 184–185
Gunn effect, see: Transferred electron
    effect

Harmonic oscillator, 260
Heterojunction injection laser, 208, 209
HgTe–CdTe solid solutions, 184
Holes, 6–7
  charge, 6
  effective mass, 6
  energy, 6
  in GaAs, 9
  heavy, 9
  light, 9
  on plot of electron energy, 7
  wave vector, 6

IGFET, 128–133; see also MOSFET
Imaging, charge-coupled device, 136
Impact ionization, 74
Impurities
  ionization energies in Ge and Si, 14–15
  in semiconductors, 14–17
Impurity band, 19–20
Impurity photon absorption, 180–181;
      see also Extrinsic photon absorption
InAs, luminescence, 198
Index matching, 289–290, 292–293, 298
Indirect transition, 177; see also Intrinsic
    photon absorption
Induced-channel MOSFET, 131–132
Inductance, self-, 240–241
Injected carrier density in p-n junction,
    59–62

Injected carrier density in *p-n* junction
  (*cont.*)
  low-level assumption, 63–64
  related to applied potential, 61–62
Injection efficiency, of emitter, 87–88
Injection of carriers, 53, 59
Input impedance
  diffused channel MOSFET, 131
  induced channel MOSFET, 132
  junction field-effect transistor, 105
InSb
  band structure, 5–6
  energy threshold for photon absorption,
    183
Insulated-gate field-effect transistor
  (IGFET), 128–133; *see also*
  MOSFET
Integrated circuit, 133
Interaction of electromagnetic waves: *see*
  Electromagnetic waves in solids
Intrinsic constant, 13
  value for silicon at 300 K, 13
Intrinsic photon absorption, 175–179
  connection with applications, 179
  defined, 12
  direct transition, 176–177, 210
    energy threshold, 177
  electron–hole pairs, 176
  indirect transition, 177–179
    conservation of wave vector,
      177–178
    energy threshold, 178–179, 210
    phonon absorption, 177–179
    phonon emission, 210
  optical absorption coefficient
    direct transition, 176–177
    indirect transition, 178–179
    values, 181
  photon absorption, 175–179; *see also*
    Intrinsic photon absorption
  spectrum
    direct transition, 177
    indirect transition, 179
  threshold energy, 176, 181–184
Inversion layer
  in charge-coupled device, 134
  in induced channel MOSFET, 131–132
  in metal–insulator–semiconductor
    structure, 122–123
Inversion symmetry, 272
Ionization, of impurities in semiconductors, 14–15

JFET: *see* Junction field-effect transistor
Josephson effect, AC, 226–228
  alternating supercurrent, 227, 228
    frequency, 227
  current density, 227, 228
  pair wave functions, 226–227
  phase difference, 227
  summarized, 228
  tunneling by normal electrons, 226
  tunneling by pairs, 226
Josephson effect, DC, 220–225
  critical current density, 225
  current density, 225, 228
    related to phase difference, 225
  pair wave functions, 220–222
  phase difference, 224–225
  summarized, 225
  tunneling by pairs, 221–222
Josephson effects, summarized, 220, 225,
  228
Josephson junction
  applications, 232, 233, 241–242
  broad-band detector, 232
  critical current density, 225
    effect of electromagnetic radiation,
      232
  current density, 223–224, 225
    effect of electromagnetic radiation,
      230–232
  current flow, 223–224
  current–voltage characteristic, 228–229
    effect of electromagnetic radiation,
      231–232
    point-contact junction, 229
  detector of electromagnetic radiation,
    232, 233
  effect of electromagnetic radiation,
    229–233
  phase difference across junction, 224–
    225, 227
    effect of electromagnetic radiation,
      230
    effect of magnetic flux, 236–238
  phase of wave function, 222, 225, 226,
    227
  point-contact, 229
  structure, 220, 229
  tunneling in, 221–222, 226–227, 151
  two on superconducting loop,
    236–239
    current density, 238
    effect of magnetic flux, 237–238

Josephson junction (*cont.*)
  two on superconducting loop (*cont.*)
    oscillations in current density, 238–239
    phase differences, 236–238
    voltage across, 226–228
Junction
  degenerate: *see* Degenerate *p-n* junction
  metal–insulator–semiconductor: *see* Metal–insulator–semiconductor structure
  metal–semiconductor: *see* Metal–semiconductor junction
  p-n: *see p-n* junction
Junction field-effect transistor (JFET), 101–106
  amplification factor, 105–106
  applications, 106
  basic mechanism, 102
  channel, 102
  channel resistance, 104–105
  connections, 103
  current–voltage characteristic, 102–105
  depletion layers, 102–104
  effect of gate voltage, 104–105
  input impedence, 105
  majority carrier device, 105
  pinch-off, 104–105
  p-n junctions, 102–104
  space charge region, 102–104
  structure, 102
  values of parameters, 105–106
  voltage controlled amplifier, 102
Junction transistor: *see* Bipolar junction transistor

Kronig–Penney model, terminated, 140

Laser, 200–209
  coherence of emission, 205
  four level, 203–204
    advantages, 203
    population inversion, 203–204
    relaxation times, 203
    transitions, 203–204
  p-n junction injection, 206–209
    band structure diagrams, 206–207
    confinement of electrons, 209
    heterojunction, 208, 209
    injection process, 206
    optical resonant cavity, 208

Laser (*cont.*)
  p-n junction injection (*cont.*)
    photon energy range, 207
    population inversion, 206
  ruby, 204–205
  solid state, 204–206
    factors affecting, 205–206
  summary, 209
  three level, 200–203
    amplification, 201–202
    population inversion, 202–203
    relaxation times, 202
    ruby, 204–205
    transition probabilities, 201, 202
    transitions, 200–201
Lead salt semiconductors (PbS, PbSe, PbTe)
  band structure, 184–185
  energy gaps, 183
  energy thresholds for photon absorption, 183
Lifetime, carrier: *see* Carrier lifetime
Light-emitting diode (LED), 199–200, 210
Linear electric susceptibility: *see* Electric susceptibility, linear
Localized states, in amorphous semiconductor, 165–167
Luminescence
  applications, 199–200
  in InAs, 198
  in *p-n* junction, 199–200; *see also p-n* junction, luminescence
  in semiconductors, 196–198

Magnetic devices (references), 309
Magnetic flux
  effect on DC squid, 240–241
  quantization, 233–235
  quantum, 235
  through superconducting ring, 235
Magnetic vector potential, 233–235
Magnetometer, 241
Majority carriers
  defined, 53
  in junction field-effect transistor, 105
  in metal–semiconductor junction, 116, 128
  in MOSFET, 130–131
  in *p-n* junction, 59, 69–72
Manley–Rowe relations, 303
Maxwell's equations, 257–259

McMillan equation, 250–251
Memory
  band structure, 1–2
  band structure diagram, 2, 110
  charge-coupled device, 136
  semiconductor, 133
  work function, 110
Metal–insulator–semiconductor (MIS)
      structure, 119–123
  applications
    charge-coupled device, 134–136; see
        also Charge-coupled device
    MOSFET, 128–133; see also MOSFET
  band structure diagram
    at equilibrium, 119–120
    under applied potential, 120–122
  in charge-coupled device, 134, 135
  depletion layer, 121–123
  electric field, 120–121
  ideal, 119
  inversion layer, 122–123
  nonideal, 123
  space charge density, 121–122
  space charge region, 121–123
  surface charge density, 121, 122
  under applied potential, 120–123
Metal–oxide–semiconductor field-effect
      transistor (MOSFET), 128–133;
      see also MOSFET
Metal–semiconductor junction, 109–119
  accumulation layer, 114
  applications
    as contact to semiconductor, 118–
        119
    as diode, 127–128
  band structure diagram
    at equilibrium, 112, 115
    showing effect of Fermi level pinning
        at surface, 148–149
    under applied potential, 116, 118
  contacts
    effect of surface states, 147–149
    ohmic and rectifying, 118–119,
        147–148
  current in, 116–117, 127
  current–voltage characteristic, 117–118,
      127
  depletion layer, 111, 112
  effect of semiconductor surface,
      147–149
  electric field, 111
  electrostatic potential, 111

Metal–semiconductor junction (cont.)
  energy barriers
    effect of semiconductor surface,
        148–149
    at equilibrium, 111–112, 113–114
    under applied potential, 115–116, 118
  formation, 110–115
    diffusion of carriers during, 111–112,
        114–115
    majority carriers, 116
    space charge region, 111, 112, 115
    surface charge density, 111, 114–115
    under applied potential, 115–119
    with n-type semiconductor, 111–115
    with p-type semiconductor, 115, 123
Minority carriers
  in charge-coupled device, 135–136
  defined, 53
  equilibrium density in p–n junction,
      59–60
  injected density in p–n junction, 61–62
  injection in p–n junction, 59–62
  lifetime, 64, 81
    effect on reverse current in p–n
        junction, 69, 81
  in neutral regions of p–n junction,
      64–66
  in p–n junction, 59, 62, 64–66,
      69–72, 75
  storage in p–n junction, 75, 81
MIS device: see Metal–insulator–
      semiconductor structure
Mobility, 10–12
  amorphous semiconductor, 167–168
  electron defined, 10
  as function of effective mass, 12
  hole, defined, 10
  and transferred electron effect, 159
Mobility gap, 167
Momentum
  field, 233
  kinetic, 233
MOSFET, 128–133
  applications, 133
  compared to JFET, 132–133
  diffused-channel, 128–131
    channel resistance, 129, 130
    comparison with JFET, 130–131
    current–voltage characteristic,
        129–130
    depletion mode, 129–130
    effect of applied potential, 128–130

MOSFET (*cont.*)
  diffused-channel (*cont.*)
    enhancement mode, 129–130
    input impedance, 131
    majority carriers, 130–131
    structure, 128
  enhancement mode: *see* Induced-
      channel
  induced-channel, 131–132
    current–voltage characteristic, 132
    depletion layer, 131–132
    input impedance, 132
    inversion layer, 131–132
    structure, 131
  speed, 136
  summary of physics, 132–133
  switch, 133
  as voltage-controlled amplifier, 130,132
Mott–CFO model, 168

Negative differential conductivity
  in bulk semiconductor, 157–158; *see
      also* Transferred electron effect
  and transferred electron effect
    production of space charge instabili-
        ties, 160–162
    related to band structure, 158–159
  in tunnel diode, 96–100
Neutral regions of *p-n* junction, 28
  currents in, 70–72
Noise equivalent power, 187
Nonlinear electric susceptibility: *see*
    Electric susceptibility, nonlinear
Nonlinear optical coefficient, 270; *see also*
    Electric susceptibility, nonlinear

Optical absorption coefficient
  defined, 176
  intrinsic direct transitions, 176–177
  intrinsic indirect transitions, 177–179
  value for extrinsic absorption, 181
  value for intrinsic absorption, 181
Optical absorption spectrum
  direct transitions, 177
  indirect transitions, 179
Optical resonant cavity, 208
Optical second harmonic generation,
    284–293
  approximations made, 285, 291
  coherence length, 288–289
  efficiency, 290–291, 303
  factors affecting, 291

Optical second harmonic generation
    (*cont.*)
  index matching, 289–290, 292–293
  intensity of second harmonic, 286–290
  phase matching, 289–290, 292–293

Parabolic band, 8
Parametric amplification, 298–302
  amplification factor, 301–302
  approximations made, 298
  example, 301
  signal intensity, 300–301
PbS, PbSe, PbTe: *see* Lead salt semi-
    conductors
PbSe–SnSe solid solutions, 184
PbTe–SnTe solid solutions
  energy gap, 183–184
  photoconductivity, 187
  photovoltaic effect, 191
Phase matching, 289–290, 292–293, 298
Phonon frequencies
  effect on lasers, 206
  effect on superconductors, 243–244,
      250–251
Photoconductive gain, 186–187
  dependence on carrier lifetime, 186–187
  factors affecting, 187
  related to carrier mobility, 186–187
  related to carrier transit time, 186–187
Photoconductivity, 185–187
  applications, 188–195
  measures of sensitivity, 187
  in PbTe–SnTe solid solutions, 187–188
  photoconductive gain, 186–187
  photogenerated carrier density, 186
Photodiode, 188–189
  current–voltage characteristic, 189
  energy threshold for photon absorption,
      189
Photoemission from semiconductors,
    149–156
  band structure diagrams
    degenerate semiconductor, 150–151
    negative electron affinity, 154–155
  effect of surface, 151–154
    band bending, 151–153
    band structure diagram, 152–153
    electron affinity at surface, 152
    metal layer, 154–155
    negative electron affinity, 154–157
    on bulk electron affinity, 152
  escape distance *d*, 153

Photoemission from semiconductors (*cont.*)
  from degenerate semiconductor,
    150–151
  dependence on band filling, 151
  infrared, 170
  negative electron affinity, 154–157
    applications, 156
    band structure diagram, 155
    surface preparation, 155–156
  p-type GaP, 153–154
  threshold photon energy, 150–151,
    155
    for degenerate semiconductors,
      150–151
    negative electron affinity, 155
Photogenerated carriers
  extrinsic absorption, 181
  intrinsic absorption, 176
  in photoconductivity, 185, 186
Photography, 194–195
Photomultiplier tube, 156
Photon absorption in semiconductors
  extrinsic, 180–181; *see also* Extrinsic
    photon absorption
  intrinsic, 175–179; *see also* Intrinsic
    photon absorption
Photon emission: *see* Emission, of photons
    in semiconductors
Photovoltaic effect, 189–191
  applications, 191
  band structure diagram, 190
  electric fields in, 190
  energy barrier, 190
  and Fermi levels, 190–191
  optimum value of energy gap, 191
  in PbTe–SnTe solid solutions, 191
  photon detectors, 191
  in selenium, 191
  solar cell, 191
  voltage produced, 190–191
Pinch-off, 104–105
p-n junction
  abrupt, 26, 37, 63
  approximation in ideal abrupt, 36,
    63–64, 69, 71, 75
  avalanche breakdown, 74–75
  band structure diagram
    at equilibrium, 30, 55
    under forward bias, 56
    under reverse bias, 56
  built-in electric field, 28, 40–42, 54–55
    effect of applied potential, 54–55

p-n junction (*cont.*)
  built-in electric field (*cont.*)
    forces on carriers, 28
    from Poisson's equation, 40–42
    maximum magnitude, 42
    related to impurity density, 41
  capacitance, 75
  carrier densities, 37–39, 59–62
    related to applied potential, 60–62
    related to diffusion potential, 38, 59
  cold cathode, 156–157
  concentration gradient, 34
  contact potential, 30–43
    in terms of Fermi energies, 32
    in terms of impurity densities, 43,
      45–46
  current flow
    at equilibrium, 33–35, 58
    under applied potential, 57–59
  current–voltage characteristic, 59, 68, 73
  degenerate, 97–101; *see also* Degenerate
    p-n junction
  depletion layer, 27; *see also* *p-n* junc-
    tion, space charge region
  diffusion current, 33, 49, 53, 57–59,
    64–66
  diffusion potential, 30; *see also* Contact
    potential
  diode, 79–81
  drift current, 33, 49, 53, 57, 64
  effect of applied potential, 53–56
  effect of forward bias, 53–56
  effect of reverse bias, 53–56
  electron currents, 33, 57–59, 62
    effect of forward bias, 58, 62
    effect of reverse bias, 58–59, 62
  electrostatic potential
    at equilibrium, 29–31, 42–44
    under applied potential, 55–56
  electrostatic potential energy
    at equilibrium, 30, 44–45
    under applied potential, 56
  electrostatic potential gradient
    at equilibrium, 29, 43–44
  energy barrier
    at equilibrium, 30–31, 45
    effect of applied potential, 56, 62
    effect of forward bias, 56, 62
    effect of reverse bias, 56, 62
    for holes, at equilibrium, 30–31, 45
  equilibrium conditions, 34, 35, 53
  extracted carrier density, 59–62
    related to applied potential, 61–62

p-n junction (*cont.*)
  extraction of minority carriers, 59–62
  Fermi level at equilibrium, 32
  formation, 26–28
    diffusion of carriers during, 26–27
    production of space charge regions,
      27
    recombination during, 27
    in terms of Fermi level, 31–32
  forward bias of, 54
  generation and recombination in deple-
    tion layer, 64–65, 71
  generation current, 33, 57–59, 67–69
  hole currents, 34, 57–59, 62
    effect of forward bias, 58
    effect of reverse bias, 58
  ideal abrupt, 26, 36–37, 47, 63–64,
    69, 75
  injected carrier density, 59–62
    related to applied potential, 61–62
  injection laser, 206–209
  injection of minority carriers, 59–62
  in junction field-effect transistor, 102–
    104
  luminescence, 199–200
    applications, 199–200
    band structure diagrams, 199
    injection process, 199
    quantum efficiency, 200
  majority carriers, 69–70
  minority carrier density
    at equilibrium, 59–60
    injected or extracted, 61–62
    in neutral regions, 64–66
  minority carrier lifetime, 64, 69,
    81
  minority carrier storage, 75, 81
  neutral regions, 28
    currents in, 70–72
  p^+n, 81–82
  reverse bias
    critical value for breakdown, 74
    defined, 54
  reverse biased
    used as particle detector, 189
    used as photodiode, 188–189
  reverse breakdown, 73–75
    avalanche, 74–75
    critical voltage, 74–75, 81
    tunneling in, 73–74
    Zener mechanism, 73–74
  reverse saturation current, 59, 68–69,
    79–81

p-n junction (*cont.*)
  reverse saturation current (*cont.*)
    dependence on diffusion length, 69,
      80–81
    dependence on illumination, 189
    dependence on minority carrier life-
      time, 69, 80–81
    effect on energy gap, 80–81
    estimate for silicon, 68
    factors affecting magnitude, 69, 80–
      81
    sign convention, 68
  reverse tunnel current, 73–74
  Shockley equation, 67
  in silicon vidicon, 194
  space charge region, 27, 39, 45, 47,
    53–55
    capacitance, 75
    charge density, 39–40
    dependence on contact potential,
      45–47
    effect of applied potential, 53–55
    electrical neutrality, 33, 43
    variation with electrostatic potential,
      47
    variation with impurity density, 47
    width, 32–33, 43, 45–47, 54–55
  summary of basic physics, 72–73
  total current, 59, 62–72
    calculation, 62–69
    effect of applied potential, 59, 67–68
    majority and minority carrier compo-
      nents, 69–72
  transient response, 75
  tunneling in, 73–74
  use of Poisson's equation, 40–42, 63
  Zener breakdown, 73–74, 81
p^+n junction, 81–82, 88
  current flow, 81–82
  impurity content, 81
  injection efficiency, 88
  space charge regions, 82
Poisson's equation
  near semiconductor surface, 145–147
  in *p-n* junction, 40–42, 63
Polarizability, 274
Polarization: *see* Electric polarization
Population inversion
  in laser, 202, 203
  in *p-n* junction laser, 206
Potential
  electrostatic: *see* Electrostatic
    potential

Potential (*cont.*)
  surface, 142, 146–147; *see also*
    Semiconductor surface

Quantization of magnetic flux, 233–235
Quantum interference: *see* Superconduct-
    ing quantum interference

Recombination of electrons and holes,
    195–200
  band-to-band, 196–198
  between band and impurity state, 198
  in direct-gap semiconductor, 196–197
  in InAs, 198
  in indirect-gap semiconductor, 197
  lifetime, 197
  with photon emission, 196–200
  in *p-n* junction, 199–200
Recombination of excess carriers, 49–53
  in *p-n* junction, 64–65, 71
Responsivity, 187
Reverse bias
  critical value for breakdown, 74
  defined, 54
Reverse breakdown in *p-n* junction,
    73–75, 81
  avalanche, 74–75
  critical voltage, 74–75, 81
  tunneling in, 73–74
  Zener mechanism, 73–74
Reverse saturation current in *p-n* junction,
    59, 68–69, 79–81
  effect of energy gap, 80–81
  estimate for silicon, 68
  factors affecting magnitude, 69, 80–81
  sign convention, 68
Reverse tunnel current in *p-n* junction,
    73–74
Ruby (Al$_2$O$_3$:Cr^{+3}), 204, 205

Scattering
  of carriers in semiconductor, 12
Schottky diode, 109, 127, 128; *see also*
    Metal–semiconductor diode
Schottky junction, 109
Second harmonic generation: *see* Optical
    second harmonic generation
Secondary electron emission, 156
Selenium
  amorphous, 192
  crystalline hexagonal, 191

Semiconductor
  compared with metal, 2–3
  compensated, 17
  degenerate, 13, 19–20, 97
    band structure diagram, 20, 97
  electron affinity, 110
  intrinsic, 12
  nondegenerate, 13
  n-type, 16
  p-type, 16
  work function, 110
Semiconductor memory, 133
Semiconductor surface
  band bending, 142–143
    calculation, 145–147
  band structure, 141–145
    effect of surface acceptors, 141–143
    effect of surface donors, 143
    n-type with surface acceptors, 143
    p-type with surface donors, 144
    showing pinned Fermi level, 145
  effect on metal–semiconductor contacts,
      147–149, 170
  effect on metal–semiconductor junction,
      147–149
  effect on photoemission, 151–154
  Fermi level, 143–145
    pinning at surface, 144–145
  negative electron affinity, 154–157
  pinning of Fermi level, 144–145
  space charge region, 141–143, 145–147
    charge density, 145
    electric field, 142, 146
    electrostatic potential gradient, 142
    width, 146
  states, 139–141
    acceptor, 141
    density, 140
    donor, 141
    fast, 140–141
    Shockley, 140
    slow, 140–141
    Tamm, 140
  surface potential, 142, 146–147
    calculation, 146–147
    dependence on impurity density, 147
  in terminated Kronig–Penney model, 140
Shockley equation, 67
Silicon
  band structure, 4–5
  energy thresholds for photon absorp-
      tion, 183, 185

Silicon (*cont.*)
  extrinsic photon absorption, 185
  vidicon, 193–194
Silver halides, 194–195
SnTe, 183
Solar cell, 191
Space charge density
  metal–insulator–semiconductor
    structure, 121–122
  p-n junction (abrupt), 39–40
Space charge instabilities, 160–162
Space charge region
  in junction field-effect transistor,
    102–104
  in metal–insulator–semiconductor
    structure, 121–123
  in metal–semiconductor junction, 111,
    112, 115
  near semiconductor surface, 141–143,
    145–147
  in *p-n* junction: *see p-n* junction, space
    charge region
Squid: *see* Superconducting quantum
  interference device
Storage time, minority carrier, 75
Superconducting quantum interference,
  236–239
Superconducting quantum interference
  device (DC squid), 239–242
  applications, 241–242
  critical current, 239–241
    as function of magnetic flux, 240,
      241
  effect of magnetic flux, 240–241
  effect of self-inductance, 240–241
  magnetometer, 241
  voltmeter, 241–242, 251
Superconductors
  A15 crystal structure, 248, 251
  BCS "one square well" model, 243
  BCS theory, 215–216, 242–243
  condensate of pairs, 216
  Cooper pair, 216
  critical magnetic field (references), 309
  critical temperature: *see* Transition
    temperature
  current density, 217, 218
  current flow, 219–220
  density of states at Fermi energy, 243,
    244–248
    and number of valence electrons,
      246–247

Superconductors (*cont.*)
  electron–electron interaction, 215–216,
    243–246
    and number of valence electrons,
      245–246
  energy gap, 216, 243–244
  Matthias' rules, 248
  maximum transition temperature,
    251
  McMillan equation, 250–251
  particle momentum, 218
  phase of wave function, 218–220
    and particle momentum, 218
  and resistivity, 244–245
  strong coupling, 248–251
  transition temperature, 242–251; *see
    also* Transition temperature,
    superconducting
    strong coupling, 248–251
    in weak coupling limit, 243–248
  wave function of condensate of pairs,
    216–220
    and density of pairs, 216–219
    gradient of phase, 218
    in Josephson junction 220–225,
      226–228
    macroscopic significance, 217–218,
      220
    phase, 218–220, 224–225, 227
  weak-coupling, 243–248
Surface charge density
  metal–insulator–semiconductor
    structure, 121–122
  metal–semiconductor junction, 111,
    114–115
Surface potential: *see* Semiconductor
  surface, surface potential
Surface states: *see* Semiconductor surface,
  states

Transition temperature (superconducting), 242–251
  BCS picture, 243–244, 248–251
    effect of phonon frequencies, 250
    strong coupling, 248–251
    weak coupling limit, 243–248
  and Debye temperature, 243–244
  equations for
    strong coupling, 250, 251
    in weak coupling limit, 243
  factors affecting
    density of states, 246–248

Transition temperature (superconducting) (*cont.*)
  factors affecting (*cont.*)
   electron–electron interaction,
      244–246
   in strong coupling, 250–251
   high values, 248
   maximum, 251
   and number of valence electrons, 248
Transferred electron effect, 157–164
  band structure requirements, 159–160
  critical requirement, 164
  dielectric relaxation time, 161, 163–164
  domain formation, 162–164
  exploitation, 160, 164
  and GaAs band structure, 158–159
  negative differential conductivity,
      157–158, 160–162
  in *n*-type GaAs, 157–160
  production of electrical oscillations,
      158, 163
  related to band structure, 158–160
  related to carrier mobility, 159
  space charge instabilities, 160–162
  transit time, 163
Transient response of *p–n* junction, 75
Transistor
  bipolar junction: *see* Bipolar junction
      transistor
  insulated-gate field-effect; *see* MOSFET
  junction field-effect: *see* Junction field-
      effect transistor
Transit time
  and photoconductive gain, 186–187
  transferred electron effect, 163
  transistor base, 86
Tunnel diode, 96–101; *see also* Degenerate
      p–n junction
  applications, 101
Tunneling in Josephson junction,
      221–222, 226

Tunneling in *p–n* junctions, 97–98
  in degenerate junction, 97–98
  in reverse breakdown, 73–74

Up-conversion: *see* Frequency mixing

Vacuum level energy, 26, 110, 149
Valence band, 3
  diamond, 4
  InSb, 5
Velocity
  electron, 10
  hole, 10
Vidicon
  conventional, 193
  silicon, 193–194
Voltmeter, superconducting, 241–242,
      251

Wave function in superconductor
  gradient of phase, 218
   across Josephson junction, 223–224
  and Josephson effects, 220–225,
      226–228
  macroscopic significance, 217–218, 220
  and particle density, 216–219
  phase, 218–220
  phase difference, 224–225, 227
   around superconducting ring, 234
  in superconducting ring, 234
Work function
  metal, 110
  semiconductor, 110, 149

Xerox process: *see* Electrophotography

Yttrium aluminum garnet (YAG), 204,
      206

Zener breakdown in *p–n* junction, 73–74,
      81